KU-663-287

FUNDAMENTALS OF FOOD CHEMISTRY

ELLIS HORWOOD SERIES IN FOOD SCIENCE AND TECHNOLOGY

This publishing programme provides an organised coverage on food science for professional technologists, research workers, and students in universities and polytechnics.

Series Editors:

Professor I. D. Morton, Head of Food Science, Queen Elizabeth College, University of London;

Dr R. Scott, formerly Reading University

Dr R. Osner, Sheffield City Polytechnic

FUNDAMENTALS OF FOOD CHEMISTRY
W. Heimann, Director, Institute for Food Chemistry, University of Karlsruhe, Germany

HYGIENIC DESIGN AND OPERATION OF FOOD PLANT
Edited by R. Jowitt, National College of Food Technology, University of Reading

HANDBOOK OF DOMESTIC ENGINEERING
D. Kirk and A. Milson, Sheffield City Polytechnic

PRINCIPLES OF DESIGN AND OPERATION OF CATERING EQUIPMENT
A. Milson and D. Kirk, Sheffield City Polytechnic

FUNDAMENTALS OF FOOD CHEMISTRY

Dr.-Ing. WERNER HEIMANN
Professor and Director of the
Institute of Food Chemistry
University of Karlsruhe

Translator
CHLOE MORTON, M.A., F.I.L.

Translation Editors
Professor I. MORTON and Dr. R. SCOTT

ELLIS HORWOOD LIMITED
Publishers · Chichester

AVI PUBLISHING COMPANY
West Port, Connecticut, USA

First published in 1980 by

ELLIS HORWOOD LIMITED
Market Cross House, Cooper Street, Chichester, West Sussex, PO19 1EB, England

The publisher's colophon is reproduced from James Gillison's drawing of the ancient Market Cross, Chichester.

Distributors:

Australia, New Zealand, South-east Asia:
Jacaranda-Wiley Ltd., Jacaranda Press,
JOHN WILEY & SONS INC.,
G.P.O. Box 859, Brisbane, Queensland 40001, Australia.

Europe, Africa:
JOHN WILEY & SONS LIMITED
Baffins Lane, Chichester, West Sussex, England.

The Americas, Japan, The Phillipines:
AVI PUBLISHING COMPANY
250 Post Road East, Westport, Connecticut 06880, U.S.A.

British Library Cataloguing in Publication Data
Heimann, Werner
Fundamentals of food chemistry. –
(Ellis Horwood series in food science and technology).
1. Food – Composition
I. Title
641.1 TX531 79–41651

ISBN 0–85312–115–X (Ellis Horwood Ltd., Publishers, Library Edition)
ISBN 0–87055–356–9 (AVI Publishing Company)

Typeset in Press Roman by Ellis Horwood Ltd.
Printed in Great Britain by Fakenham Press Ltd.

Table of Contents

Foreword

From the Foreword to the first edition

Only a diet adequate in energy and nutrionally complete can form a secure basis for health necessary for the intellectual, cultural and ethical development in man.

This is recognised by present day food chemistry and technology, which must ensure that industrially produced foods are adequately supplied and maintained in quantity and quality. To do this, the processing techniques must be adapted to the chemical and physical properties and to the reactivity and the sensitivity of foods which themselves consist of complex mixtures of natural materials.

The historical development of food chemistry is closely bound up with analytical and organic chemistry and with physiology. There is constant fruitful interaction between these sciences. Food chemistry, with food technology and food microbiology (biotechnology) has now become a separate scientific discipline.

Modern food science and technology is largely applied biochemistry, and embraces kinetic and thermodynamic aspects. The processes of food science and technology, then, take their place in the general biochemical scheme. Food chemistry is an important part of the science of food and nutrition which, with physiology, examines the metabolism and health of man as it is influenced by the diet. Progress, especially in the field of natural materials, and therefore in the field of food chemistry and technology, is now only possible with close cooperation between the various other disciplines.

Thus, modern food chemistry has become a complex field interconnected with a whole series of scientific disciplines. This may confuse the beginning student, but may also offer a particular attraction to this branch of science. The fundamentals of food chemistry give an insight into the chemistry and physiological purpose of our food: furthermore they enable us to recognise the structure and reactivity, of the physical and chemical interactions of the building blocks and components of food in both its unprocessed and processed state.

This book aims to serve as a basic introduction for the student of food chemistry and food science and technology and to initiate him into the problems of this complex field. It is hoped it will lead him, with the help of telling examples, to independent critical scientific thought. I was therefore primarily concerned to make the student fully aware, over and above the factual material presented, of the intellectual connections and to indicate how research in this field is vitally important for the further development of food chemistry.

Because of the tremendous extension of the various scientific disciplines, it is not possible, in the compass of this work, to treat all fields and individual facts equally. On the basis of my long experience of teaching food chemists and technologists at the University of Karlsruhe, some fields are treated more extensively to give the student examples of general principles of chemistry and technology of foods. However, I have had to use my judgement about the extent to which the various branches of science must be considered in these 'fundamentals' and to what extent new developments can be brought in. I hope that this book will succeed in giving the food chemist and food scientist and technologist at the beginning of his studies, a basic understanding for the necessary cooperation between the chemist, the process technologist, the scientist, the microbiologist and the nutritionist.

The IUPAC rules have been followed for the names of compounds but the old names, for instance of enzymes, are frequently also given in the text.

Karlsruhe, Autumn 1968.

From the Foreword of the second edition.

A literature list is now included at the end of this book to stimulate the student to increase further his basic knowledge. It should also help to give colleagues indications of new literature of the subject. The list consists mainly of a choice of reviews indicating new research directions as well as monographs.

The sections on amino acids and proteins (biological value, new sources of protein), sections in the chapters on carbohydrates, enzymes, preservation of food, have been rewritten and made more complete.

Karlsruhe

W. Heimann.

In translating and editing the first part of the comprehensive book by Professor Dr. W. Heimann, we have tried to adapt the book to English conditions. As far as possible food tables and recommended daily allowances of nutrients are based on the most recent publications. The literature list has been completely revised and represents in our view the review articles and monographs which should be readily available to an English reader.

We should like to express our thanks to our colleagues and especially Dr. R. Scott for his invaluable help and encouragement in an editorial capacity.

London 1979,

C.M.

I.D.M.

Introduction

According to where and when it took place, the industrial revolution led to vast changes in man's diet. The various human diets which developed empirically over the centuries, were only subjected to systematic scientific consideration and investigation from the middle of the nineteenth century*.

Food chemistry is therefore a relatively new branch of science. In spite of the comprehensive and basic knowledge gathered over the last decades in modern scientific discovery, this complex field is still in the process of development. There are always new problems to be tackled, both in research and in practical application.

Food chemistry has been helped by the development of biochemistry. Biochemical problems are to a large extent concerned with the activities of enzymes which influence the type and extent of desirable or undesirable processes in the production, processing and storage of animal and vegetable foods. From this point of view, food chemistry is an important part of 'applied biochemistry'.

Modern food chemistry and its aims are based on the physiological view that man's nutrition is not static, and cannot be regarded just as a matter of analytical composition and of the energy value of food. Rather, this view has now been superceded by a dynamic one, which sees the diet as part of the whole metabolic process in the healthy and sick man.

Food chemistry can be regarded as being closely connected to nutritional physiology – an important field of modern medicine.

Food science and technology has become increasingly important as world population increased by leaps and bounds, because more and more it has to supply processed and industrially produced foods of a high quality. In this task the food technologists face a great variety of problems, in view of the chemical heterogeneity of most foods, which are complex mixtures of naturally

*Among the early successful pioneers of food chemistry and nutrition the following may be mentioned: J. v. LIEBIG, M. v. PETTENKOFER, C. v. VOIT, M. RUBNER.

occurring materials. The variety of processing possible and the ways of carrying them out, must be suited to often sensitive foods.

Based on this concept of food science, which includes biochemical, nutritional and technological aspects, this book covers the basic principles of nutritional science and the necessary background for understanding them.

Part 1

Nutrition

CHAPTER 1

The Function of Food

1. THE FUNCTION OF FOOD

Life on earth is made possible by the sun. Its radiation allows plants to grow. A colossal surge of energy – the process of photosynthesis – creates the conditions necessary of the synthesis of carbohydrates (sugar and starch). The plant contains the materials and the energy needed for chemosynthesis of all other organic necessities from carbohydrates. The synthesis of the seed oils and plant protein, are examples of this (the latter using the nitrogen compounds from the soil). Because man and animals have to use as food the materials manufactured by the plant as nutrients for its own metabolic processes of growth and maintenance, the whole biological process depends on photosynthesis and carbon dioxide assimilation. Photosynthesis provides the only possibility so far available, for converting the sun's radiation into chemical energy and reconversion of CO_2, (a useless product of decomposition) into an energy-rich state. In this way the photochemical reaction of the assimilation of carbon dioxide in the atmosphere (via the chlorophyll of the green plant), becomes the most important process which yields material and energy on earth.

It is necessary to digest food to cover the organism's needs for growth and maintenance.

Potentially energy-rich organic nutrients, provided in food, are usually broken down chemically, into less energy-rich products of metabolism (CO_2, H_2O, urea etc.). The energy which is set free by this reaction is used by the body for its life functions: as heat energy to maintain the body temperature, as mechanical energy for the work of organs and muscles, as chemical energy for the synthesis of its own protein, fats, glycogen, polynucleotides, energy-rich phosphate compounds, for the development of electrical (static and redox) potentials, for osmotic work, and so on.

However, obtaining energy is only one of the main functions of food, the second aims to provide and ensure the supply of basic material (of the most varied chemical and physiological nature) for the growth and maintenance of the body substance, for instance, soft tissue, bones, protoplasmic substances, formation of hormones, enzymes and certain vitamins. For this reason, the necessity of a nutritionally complete diet, both from an energy and a material point of view, must be emphasised. A diet is only of complete nutritional value when

1. it is calorifically (or joule) adequate and therefore covers the organism's energy requirements, and when
2. all necessary nutrients are present in sufficient quantity, in suitable proportions to each other and in usable form.

Quantitative and qualitative criteria must therefore be taken into account. Only food of full nutritional value provides an optimal diet from a physiological point of view.

Part 2

The Chemistry of Food

CHAPTER 2

Constituents of Food

2. CONSTITUTENTS OF FOODS

Our bodies are made and maintained from the food we eat together with oxygen we breathe in from the air.

Commonly we eat such foods as milk, fish, egg, bread, fruit and vegetable produce. Although we have a large variety of foods available, they contain only a limited number of constituents which occur repeatedly and are described as nutrients; proteins, fats, carbohydrates, vitamins, minerals and water. These substances allow our organism to grow, to maintain itself and to function normally.

Many foods contain, as well as the essential nutrients, other substances which are of little importance for human nutrition or may even have harmful effects. Examples of these are the toxic proteins present in eggwhite, fish and legumes; the natural plant substances, such as antitoxins which affect the thyroid, oxalic acid and gossypol.

Only the nutrients supply our nutritional needs and not the whole of the foodstuffs themselves. The latter are only important because they are carriers of the nutrients and therefore our food can be made up from a great variety of foodstuffs and yet be of full nutritional value.

Over the centuries people have obtained their food requirements from very different foods. Country, climate and culture are all determining influences. In spite of their different diets, these peoples remain healthy.

Eskimos, for instance, eat an almost entirely animal diet in which protein and fat are predominant. The main food of the Chinese, rice, is vegetable and is rich in carbohydrate; milk and milk products are almost unknown to them. The inhabitants of Central Europe and North America prefer a varied diet of animal and vegetable foods. These few examples show that it is possible to supply nutritional needs of man from very different foodstuffs, and that it is not a matter of the foodstuffs themselves but rather the supply of nutrients which is important.

These so-called **foods** which contain none or very few nutrients must be distinguished from the main group of nutritious foods. For centuries man has used them at all stages of civilisation to produce pleasant or stimulating effects on the stomach and intestines and also on the brain or the heart, via the nerves from organs of taste and smell. In the case of stimulants in such foods as tea, coffee, tobacco, alcoholic drinks, meat extract, spices etc., not only are specific ingredients involved such as caffeine in coffee but also a large group of aromatic substances. Their effect is mainly to stimulate the physiological functions of the body but they are not in any real sense, foods.

Foodstuffs containing cellulose and hemicellulose, such as fruit and vegetables, have a mechanical stimulating effect on peristalsis of the gut.

All constituents of food which our organism needs for energy and body building are considered to be nutrients. They can be listed as follows.

proteins	minerals and trace elements
fats	vitamins
carbohydrates	water

RUBNER, in his law of **isodynamics** (1883) states that the three main nutrients necessary to cover the energy needs of the body – proteins, fats and carbohydrates – can replace each other calorically, but by widely varying amounts, because they are subject to the same final oxidative decomposition (metabolic cycles, citric acid and respiratory cycles). The specific material importance of the individual groups of nutrients need not be considered because the question of energy comes first. In practice however the groups of nutrients listed above as suppliers of **energy** and **nutrients** possess, individually, essential and important functions, which are indispensable to the organism for its ordered metabolic processes. They are all of vital importance for nutrition. Although the carbohydrates and fats (as opposed to the proteins) carry the main burden of supplying energy (calorie suppliers), they are also used in the structural framework of the body to form structural elements in the cell. In the case of proteins, the main function is tissue formation. The organism does not normally make much use of them for energy. However it is possible for the body to use the protein from food (or from the body itself) as a source of energy in times of hunger when the body is compelled to do so because of inadequate supply of fat and carbohydrate from food or body sources. All those components of foods which (by RUBNER's definition) cannot be replaced by other nutrient's are called **essential** (= necessary for life). They are not synthesised by the body and must therefore be supplied as 'exogenous factors' in the food to build up the body and regulate matabolism.

Essential **fatty acids**, essential **amino acids**, **vitamins**, **minerals** and **trace elements** are all essential food constituents without which life is unable to continue.

They can be divided into three groups, according to their function in the organism;

1. **Suppliers of energy**
2. **Materials for growth and maintenance**
3. **Regulatory materials**

The suppliers of energy provide the energy necessary to maintain the chemical processes in the organism, to produce body heat. Their energy value is calculated in calories or joules because they can be considered as physiological 'fuels'. The three main foods, carbohydrates, fats and proteins in RUBNER's restricted definition belong to this group because they can be suppliers of energy.

However proteins and fats, in addition to simply supplying energy, also have important functions as building blocks of living tissue. As constituents of the various organs die or perish during the metabolic process, materials are constantly needed for the building up of new body substance and for the renewal of body building blocks. A nutrient such as protein therefore has to fulfil a number of functions which go far beyond simply supplying energy. Thus the **building blocks for our body are taken from the building blocks in the nutrients contained in our food.**

Vitamins are a conspicuous group of exogenous nutritional factors. Some can be converted by the organism from so-called provitamins into the active form (see p. 202). In the smallest concentration, vitamins act as powerful catalysts in specific essential metabolic reactions in the organism and are also important factors in intermediate metabolism as regulatory substances. When they are absent, an otherwise completely adequate supply of all other nutrients will not be able to keep the organism healthy.

Minerals function partly as building materials and partly as regulatory substances. They do not function as a substrate for energy production in the body. Their importance lies in their specific chemical function. The same is true for all the other essential factors in nutrition (essential amino acids and essential fatty acids, see pp. 44 and 81).

It is clear that a dynamic view of food components is necessary, since the whole process of metabolism combines within itself an exactly balanced inter-relationship between the individual nutrients. Recent physiological information about active agents such as catalysts in living processes (biocatalysts, regulators of metabolism), allows an extended concept of nutrition.

Our food must also contain other substances in addition to the nutrients described above. These are not necessary for life, but rather act as a beneficial aid to digestion. They are mainly bulky fibrous materials and aromatic substances.

Among the bulking materials can be included a series of undigestible components of food, such as cellulose, hemi-cellulose and lignin. Although they are not digested, they are important in normal digestion, because they promote peristalsis, absorption of nutrients and elimination of undigested material. Foods such as wholemeal bread, fruit and vegetables are relatively rich in bulking materials.

Aromatic substances (see odour and taste) are important agents influencing the secretion of the digestive juices and stimulating the appetite. Aromatic materials are present in many natural foods and can be added to food by the use of seasonings, spices and herbs. Combinations of these materials also arise during the preparation of food, in roasting, frying and baking.

In summary: the diet must have a sufficiently high energy value in order to be adequate for nutrition. In addition there must also be an adequate content of all 'essential' nutritive factors, as well as tissue building materials.

CHAPTER 3

Protein Materials

2.1 PROTEIN MATERIALS

Among nutrients which the organism needs, three groups stand out: proteins, fats and carbohydrates. The protein group is the most complicated because of chemical and physiological function in the cell. Along with lipids, proteins participate in membrane formation and tissue building. Apart from the structural framework of cells, proteins are also components of enzymes and hormones and therefore take part in most functions.

Proteins derive their name from the Greek protos = the first, which indicates that they are the material basis of life. Even if this original view is only true to a limited extent, proteins are among the most important constituents of the body for growth and maintenance. Today particular attention must be paid to protecting and preserving them during the processing of food to retain their full nutritive value (see p. 66).

Elementary analysis of protein shows that the main components are carbon, hydrogen, oxygen, nitrogen, sulphur, sometimes phosphorus and in some cases also iron, manganese, copper, iodine, zinc and so on. The most characteristic element of protein is the nitrogen, both in quantity and as a structural component. The proteins are termed the nitrogenous material – as opposed to N-free carbohydrates and fats. The characteristic smell of burnt hair, horn or wool, which is given off when protein is burnt in contrast to burning carbohydrates and fats – derives from the nitrogen content of protein.

The nitrogen content in proteins is fairly constant varying from 15-18%, and on average is about 16%. The simplest methods for the determination of protein in our foods are based on this fact.

Animal foods contain considerable amounts of protein, but many vegetable foods also have useful amounts: meat 15-25%, egg white 12%, egg yolk 16%, bread 6-10%, flour 10-15%, milk 3-4%, vegetables and potatoes 1-4%.

Until the turn of the century little was known about the detailed structure of proteins. The complicated structure of macromolecules which make up proteins was known, but details of their composition and detailed structure only became clear from the classical work of Emil FISCHER and his co-workers and through the researches of many other scientists such as PAULING and PERUTZ. Building on this work, further questions about chemical composition, physical and biological behaviour have been more fully studied in recent decades, using more modern and exact methods.*

Despite the results of research it is still difficult to define all details of the structure of this group of natural substances and their reactive capabilities.

Proteins, like fats and polysaccharides, have a compound structure. If a protein is boiled for some time (about 20 hours) with concentrated hydrochloric acid (20%) or sulphuric acid (35%), total hydrolysis occurs and the individual amino acids, which are the building blocks of the protein, are set free. Proteins are also broken up into mixtures of amino acids by alkalies. Enzymatic action (pepsin, trypsin) breaks down the proteins by cleavage at certain points to give peptides (i.e. groups of amino acids). Amino acids also sometimes appear during autolysis of tissue, or by break down of protein through microbial action as in the decomposition of milk casein (cheese ripening) by microorganisms. So far over thirty different amino acids have been found in protein hydrolysates and their structure ascertained. A number of amino acids are already produced by technical synthesis.

Research on proteins has shown that, apart from amino acids, other chemical groups, which can be split off unchanged, may occur within their structure. When proteins contain, as well as amino acids, the so-called prosthetic groups, they are called compound proteins, as opposed to the simple proteins which contain only amino acids.

3.1 AMINO ACIDS

Amino acids are compounds having both an amino group and a carboxyl group in the molecule and therefore they have both basic and acid characteristics.

The simplest general scheme for formulating these compounds is:—

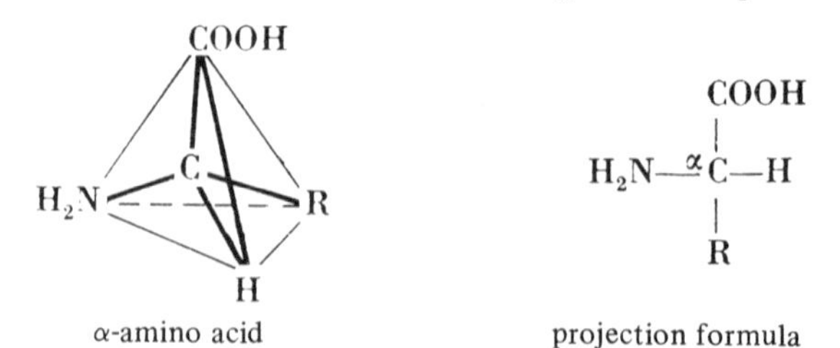

α-amino acid projection formula

*The use of physical and physico-chemical methods of investigation: X-ray analysis, sedimentation, ultraviolet and infrared spectroscopy, optical methods such as polarimetry, refractometry, colorimetry, nephelometry, photometry, as well as dialysis chromatography, electrophoresis, combined chemical-enzymatic and microbiological methods, isotope dilution method etc.

In the simplest case R=H(glycine). In all other cases R is an aliphatic or heterocyclic residue which can also contain other functional groups (see the formulae in Table 1).

All amino acids in proteins are α-amino acids, that is, the basic NH_2 group is attached to the –C atom directly linked to the acid carboxyl group. In the case of proline and hydroxyproline the α-amino group takes part in the formation of the ring (see p. 40). Amino acids are therefore amphoteric (ampholytes) and can react either as acids (proton donators) or as bases (proton acceptors) independently of the pH. They form two kinds of salts:

$$\left[R{-}\underset{\displaystyle NH_2}{\underset{|}{CH}}{-}C\begin{matrix}\nearrow O\\ \searrow O^{\ominus}\end{matrix}\right]^{-} Na^{+} \qquad \left[R{-}\underset{\displaystyle NH_3^{\oplus}}{\underset{|}{CH}}{-}C\begin{matrix}\nearrow O\\ \searrow OH\end{matrix}\right]^{+} Cl^{-}$$

sodium salt — hydrochloride

In solution three forms are possible, varying with the pH: cations, zwitterions and anions.

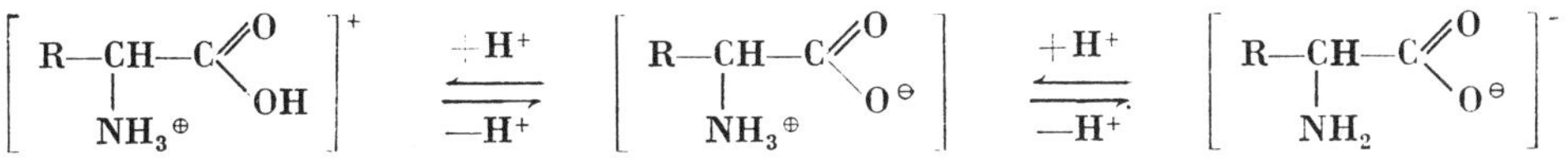

$$\left[R{-}\underset{\displaystyle NH_3^{\oplus}}{\underset{|}{CH}}{-}C\begin{matrix}\nearrow O\\ \searrow OH\end{matrix}\right]^{+} \underset{-H^+}{\overset{+H^+}{\rightleftharpoons}} \left[R{-}\underset{\displaystyle NH_3^{\oplus}}{\underset{|}{CH}}{-}C\begin{matrix}\nearrow O\\ \searrow O^{\ominus}\end{matrix}\right] \underset{-H^+}{\overset{+H^+}{\rightleftharpoons}} \left[R{-}\underset{\displaystyle NH_2}{\underset{|}{CH}}{-}C\begin{matrix}\nearrow O\\ \searrow O^{\ominus}\end{matrix}\right]^{-}$$

cation — zwitterion — anion

In the zwitterion form the acid nature of the amino acid depends on the NH_3+ group and the basic character on the COO^- group.

The zwitterion is an 'internal salt', i.e. anion and cation are in the same molecule. This stable form of an amino acid, which carries a positive as well as a negative charge, does not behave as an ion, but is electrically neutral.

An equalisation of the charge cannot take place within the molecule, because both charges are firmly fixed in the molecule. The true state of an α-amino acid, in aqueous solution as well as in crystal form, is more accurately represented by the 'zwitterion' form, the 'neutral' form does not exist there.

The salt-like character disappears with esterification of the carboxyl group and also when an acetyl group replaces an amino hydrogen atom, because in this reaction the acid character of the carboxyl group or the basic character of the amino group are neutralised.

As a result of their amphoteric nature the amino acids migrate with the electric field as cations to the cathode in acid medium, and as anions to the anode in alkaline medium. In both cases there is a sharply defined pH value where practically the whole amino acid has gone over to the zwitterion form which is strongly attracted to both poles in the electrical field. Here no amino acid is transported by the current. This pH value, specific for each amino acid, is called the isoelectric point.

The dissociation constant of the carboxyl group is called K_1, that of the amino group K_2. When a number of carboxyl or amino groups are present in the molecule, several K_1 or K_2 values will be obtained.

If $[^+AS]$ equals the concentration of the amino acid cation, $[^+AS^-]$ that of the zwitterion and $[AS^-]$ that of the amino ion, then the law of Mass Action is as follows

$$(1) \qquad \frac{[^+AS^-]\,[H^+]}{[^+AS]} = K_1 \qquad \frac{[AS^-]\,[H^+]}{[^+AS^-]} = K_2$$

In the same way as for pH values, negative logarithms to the base 10, are usually given for the dissociation constants K_1 and K_2. According to the definition the isoelectric point is defined as $^+AS = AS^-$. By transforming the equations (1) one obtains

$$(2) \qquad \frac{[^+AS^-]\,[H^+]}{K_1} = \frac{[^+AS^-]\,K_2}{[H^+]}$$

$$(3) \qquad [H^+]^2 = K_1 \cdot K_2; \quad [H^+] = \sqrt{K_1 \cdot K_2}$$

$$\log\,[H^+] = {}^1\!/_2 \log K_1 + {}^1\!/_2 \log K_2$$

$$(4) \qquad IP = {}^1\!/_2\, pK_1 + {}^1\!/_2\, pK_2$$

$$(5) \qquad IP = \frac{pK_1 + pK_2}{2}$$

Amino acids in aqueous solution react as ampholytes in either an acid or an alkaline manner according to the size of their pH value. Monoaminocarboxylic acids are strong acids (e.g. glutamic acid). At the other end of the scale diamino-monocarboxylic acids (e.g. lysine) react in a distinctly alkaline manner.

As the macromolecular proteins are ampholytes, because of the presence of acid and basic groups in the molecule, they also (e.g. casein, gelatine, albumin, globulin) have a characteristic 'isoelectric point'. The same is true for enzymes (see p. 250), which represent a group of proteins particularly important in scientific research and practical food technology. The buffer action is due to the ampholytic nature of the amino acids and proteins respectively. Buffer action means the interception of added hydroxonium or hydroxide ions so that the pH of a solution is not changed towards either the acid or alkaline side by this addition.

Due to their salt-like character the amino acids are stable, crystalline, non-distillable substances, which are mostly readily soluble in water. In alcohol they are difficult to dissolve, in other organic solvents e.g. ether, they are insoluble.

The C atom, to which the NH_2 group is attached in the amino acids, is always asymmetrical (except in the simplest amino acid, glycine). Therefore all the natural amino acids except glycine are optically active.

The rotation of the natural amino acids is sometimes to the right (+) sometimes to the left (–). But account must be taken of the difference between the actual rotation of the plane of polarisation of light and the spatial (steric) structure (configuration). The natural amino acids have the same structural configuration as dextrorotatory sarcolactic acid [(+) lactic acid]. The (+) lactic acid for its part can be traced back configuratively to L-glycero-aldehyde as general source material. The naturally occurring amino acids are the L-forms. The D-configuration only occurs to a limited extent (e.g. in certain bacteria: D-glutamic acid in Bacillus subtilis, a bacterium which is partly responsible for putrefaction of meat, the 'swells' in tinned goods and red rot in eggs among other things). Therefore the correct description of naturally occurring dextrorotatory alanine must be: L-(+) alanine, i.e. it derives configuratively from the L series (L-glycero-aldehyde), but is dextrorotatory (+).

$$
\begin{array}{ccc}
\begin{array}{c} COOH \\ | \\ HO—C—H \\ | \\ CH_3 \end{array} &
\begin{array}{c} COOH \\ | \\ H_2N—C—H \\ | \\ CH_3 \end{array} &
\begin{array}{c} CHO \\ | \\ HO—C—H \\ | \\ CH_2OH \end{array}
\end{array}
$$

L-lactic acid — L-alanine — L-glycero-aldehyde

When the molecule contains several asymmetric carbon atoms, the spatial arrangement of the configuration in the amino acids generally follows the position of the α-amino group. In this it deviates from the convention which applies to the carbohydrates, where arrangement follows the configuration of the C-atom with the highest number. Thus I is called D-threonine: according to the carbohydrate designation it should be regarded as an L-derivative.

$$
\begin{array}{cccc}
\begin{array}{c} COOH \\ | \\ H—C—NH_2 \\ | \\ HO—C—H \\ | \\ CH_3 \end{array} &
\begin{array}{c} CHO \\ | \\ H—C—OH \\ | \\ HO—C—H \\ | \\ CH_2OH \end{array} &
\begin{array}{c} COOH \\ | \\ H—C—NH_2 \\ | \\ H—C—OH \\ | \\ CH_3 \end{array} &
\begin{array}{c} CHO \\ | \\ H—C—OH \\ | \\ H—C—OH \\ | \\ CH_2OH \end{array}
\end{array}
$$

D-threonine I — L-threose — D-allothreonine II — D-erythrose

The diastereomers, which are given the prefix 'allo', belong to the II group according to the amino acid classification and are called allothreonine.

Reactions of amino acids

A series of characteristic derivatives can be formed from the two functional groups in the amino acids.

Reactions of the carboxyl group.

The carboxyl groups can form esters as well as salts. Esterification follows the introduction of HCl into a solution of the amino acid in absolute alcohol, when the hydrochloride of the ester will be obtained. The free esters are strongly basic substances. Earlier the separation of amino acids took place (E. FISCHER) by fractional distillation of the esters. Reduction of the esters with lithium aluminium hydride $LiAlH_4$ produces amino alcohols. Other derivatives of the carboxyl group, such as acid chlorides, anhydrides and azides, are important in peptide synthesis.

Decarboxylation can occur by thermal, oxidative or enzymatic means. Enzymatic decarboxylation of certain acids gives the biogenic amines, such as putrescine, cadaverine, histamine and serotonin. Oxidative decarboylation with ninhydrin (see p. 69). Reaction to determine the presence of . . .) is used for quantitative determination by collecting the CO_2 which is formed. In the same way the blue-violet colour formed from ninhydrin and ammonia, resulting from the STRECKER decomposition, can be used for estimation of amino acids. The formation of this blue-violet dye occurs with all amino acids with a free $-NH_2$ group (with the exception of proline and hydroxyproline) through the oxidative decomposition of amino acids in aqueous solution with ninhydrin. This ninhydrin reaction is used as a test for the presence of amino acids in natural hydrolysates and also for the detection of small amounts of certain amino acids after their separation e.g. by paper chromatography.

Quantitative determination of amino acids in absolute ethanolic (rather than aqueous) solution by WILLSTATTER's method depends on the repression of the zwitterion form in favour of the neutral form, so that the amino acid is now like any other acid, and the proton of the carboxyl group can be titrated with an alkali (phenolphthalein as indicator):

$$^{\ominus}OOC—CHR—NH_3^{\oplus} \rightleftarrows H_2N—CHR—COO^{\ominus} + H^{\oplus}$$

SORENSEN's well known **formol titration**, which is important in food analysis, depends on the fact that the basicity of the amino group is virtually eliminated because of the influence of formaldehyde. The freed proton H^+ can now be titrated with alkali:

$$^{\ominus}OOC—\underset{\displaystyle R}{\underset{|}{C}H}—NH_3^{\oplus} + 2HCHO \rightarrow {}^{\ominus}OOC—\underset{\displaystyle R}{\underset{|}{C}H}—N\begin{matrix}\diagup CH_2OH \\ \diagdown CH_2OH\end{matrix} + H^{\oplus}$$

Reactions of the amino groups

The reaction of amino acids with nitrous acid leads via the unstable diazonium salts to hydroxy acids:

$$\text{R—CH(NH}_2\text{)—COOH} \xrightarrow{HNO_2} [\text{R—CH(N}_2^{\oplus}\text{)—COOH}] \xrightarrow{HOH} \text{R—CH(OH)—COOH} + N_2 + H^{\oplus}$$

This reaction is used for the quantitative determination of amino acids (volumetric or monometric determination of N_2 by VAN SLYKE's method).

N-alkylisation can be achieved by reaction with alkylhalogenides or diazoalkanes (e.g. diazomethane). Depending on the method and conditions of the reaction, mono- di- and trialkyl derivatives are obtained. N-tri-methylglycine, and betaine occur in sugar beet (0.3%).

The N-dinitrophenyl derivatives (DNP-amino acids) are the most important of the N-aryl compounds. 1-fluoro-2,4-dinitrobenzene reacts at room temperature under weakly acid conditions:

$$\text{R—CH(NH}_2\text{)—COOH} + \text{F—C}_6\text{H}_3(\text{NO}_2)_2 \longrightarrow \text{R—CH(HN—C}_6\text{H}_3(\text{NO}_2)_2\text{)—COOH}$$

The yellow DNP derivatives crystallise well and are stable to hydrolysis. They are used to determine end groups of peptides and proteins as well as for the quantitative determination and chromatography of amino acids (see p. 46). N-acyl compounds of amino acids are used in peptide synthesis. The acyl residue blocks the amino group. The carbobenzoxy compounds (M. BERGMANN) are often used, because after the synthesis the carbobenzoxy residue can be split off by catalytic hydrogenation (H_2/Pt) with the formation of CO_2 and toluene.

$$\text{R—CH(NH}_2\text{)—COOH} + \text{C}_6\text{H}_5\text{—CH}_2\text{—O—C(=O)—Cl} \longrightarrow \text{R—CH(HN—C(=O)—O—CH}_2\text{—C}_6\text{H}_5\text{)—COOH}$$

N-trifluoracetyl compounds of amino acids, which can be split with alkali under mild conditions, are used to separate the esters (F. WEYGAND) by gas chromatography. For enzymatic transamination see pp. 46 and 47.

Classification of amino acids

The 20 or so amino acids which occur in the hydrolysates of proteins can be divided into two main groups as follows:

(a) amino acids where the residue R contains an aliphatic chain,
(b) amino acids where R is an aromatic or heterocyclic residue.

Both main groups can be further subdivided:

1. amino acids where the residue R is a pure hydrocarbon.
2. amino acids where R contains polar groups such as –SH or –OH.
3. amino acids which have in their residue R, a second carboxyl group (monoamino dicarboxylic acids, acid amino acids).
4. amino acids which contain in their residue R, a second amino group (diaminocarboxylic acids, basic amino acids).

All these R groups in the amino acid molecule, with their sometimes acid, sometimes basic, sometimes neutral grouping, (compare serine, valine, leucine) determine the chemical, physical and biological character of the individual amino acid (and also contribute to the protein molecule) in which they are incorporated; compare primary, secondary and tertiary structure of the proteins p. 49, denaturation p. 56, immunological (serological) behaviour p. 70.

The subtle nature of the amino acid molecule and of the proteins built up from them have far-reaching consequences for the food chemist and food technologist analytically, biochemically and technologically.

The most important amino acids

Table 1 gives a summary of this classification of the most important amino acids and the formulae for their structure and physical details.

Acyclic amino acids

Monoaminomonocarboxylic acids. Glycine, the simplest amino acid, is optically inactive because of the lack of an asymmetric C atom. It is frequently present in structural proteins (collagen), but is absent in milk proteins and in most albumins. The human organism can synthesise glycine, so it is not essential for nutrition. Glycine and serine can be transformed enzymatically into one another (see p. 119).

Alanine occurs in all protein cleavage products and is optically active, as are all the natural amino acids apart from glycine: L-(+)-alanine.

β-alanine (β-aminopropionic acid) is an isomer of α-alanine and occurs in the bound form in the vitamin pantothenic acid (see p. 219). β-alanine is especially important physiologically and so far is the only β-amino acid found in natural substances.

Valine is found in lupin seeds and cereal products. During the alcoholic fermentation of yeast, potatoes and cereals, valine arises as a cleavage product and in turn is fermented to isobutyl alcohol. The presence of isobutyl alcohol in the form of fusel oil is well known (in the distillation of spirits from potatoes and cereals).

Leucine, isoleucine and **norleucine**. Of these three isomers, leucine is the most abundant in nature and is found in all proteins. It is only slightly soluble in water, in protein hydrolysates and in the gut contents leucine separates out in the form of small fat globules. Free leucine is also produced in the ripening of cheese by the activity of bacteria.

In alcoholic fermentation the three leucines are transformed to the corresponding amyl alcohols (fusel oils). Isoleucine produces the optically active amyl alcohol by fermentation.

Other amino acids present in the fermentation mash, such as tyrosine and tryptophan, yield corresponding alcohols. The formation of these alcohols by living yeasts can be explained as follows: the amino acids serve as a source of nitrogen for the yeast in building up other protein building blocks, whereas the alcohols formed (fusel oils) are not used further by the organism.

The conversion of amino acids to so-called fusel oils is now understood and occurs mainly by an enzymatic transamination (transaminase, pyridoxal-phosphate as co-factor) to the corresponding α-keto acids. These are decarboxylated by specific decarboxylases and the aldehydes formed are enzymatically reduced to the alcohols by an alcohol-oxido-reductase (NADH as co-factor).

Monoaminodicarboxylic acids. The most important representatives of this group are **asparaginic** acid and its homologue **glutamic acid**.

Both acids are present with only one amino group (see p. 34). The two semi-amides of the two amino acids are called asparagine NH_2–CO–CH–CH(NH_2)–COOH and glutamine NH_2–CO–CH_2–CH_2–CH(NH_2)–COOH. Asparagine is found mainly in plants as a nitrogen containing reserve material; glutamine in plants and animal organisms. It is thought that nitrogen derived from the decomposition of protein is stored in the form of asparagine and glutamic acid for future use in the plant [metabolite (glucogenic) reserve]. Asparagine is present in large amounts especially in the germ, presumably because part of the protein is bound by asparagine during the process of germination. In animals glutamic acid is of the greatest physiological importance as the product of intermediate nitrogen exchange with the carbohydrate metabolism (transamination: keto acids $\rightleftharpoons$ amino acids).

A large percentage of the protein in cereals consists of glutamic acid; wheat gluten (gluten protein) contains about 40% glutamic acid. Recently glutamic acid in the form of its monoglutamate salt, which enhances the existing aroma of various foods, has become important in the preparation of foods; it is now manufactured on a commercial scale and wheat gluten, maize gluten, sugar beet syrup and soya are good sources from which it may be manufactured.

Diaminomonocarboxylic acids. In these amino acids, two amino groups are present with only one carboxyl group and they have a basic reaction (see p. 34). This is why the basic amino acids with 6 C atoms (lysine, arginine and histidine) are called 'hexobases' (KOSSEL).

Ornithine, **arginine** and **lysine** are universal building blocks of proteins. Ornithine and urea are formed in the organism by hydrolytic cleavage of the basic guanidine residue in arginine.

Table 1 Composition of important amino acids (according to classification p. 33)

amino acid	internationally accepted abbreviation†	formula	isoelectric point	at t (°C)	$[\alpha]_D$	at t (°C)	solvent
(a₁) glycocoll, glycine (amino-acetic acid)	Gly	$CH_2(NH_2)-COOH$	6.064	25	–	–	–
L-alanine (L-α-amino-propionic acid)	Ala	$CH_3-CH(NH_2)-COOH$	6.107	25	+ 2.7	22	H_2O
L-valine (L-α-amino-*iso*valeric acid)	Val	$CH_3-CH(CH_3)-CH(NH_2)-COOH$					
L-leucine (L-α-amino-*iso*butylacetic acid)	Leu	$CH_3\ CH(CH_3)-CH_2-CH(NH_2)-COOH$	6.036	25	− 10.57	25	H_2O
L-isoleucine (L-α-amino-β-methylethylpropionic acid)	Ile	$CH_3-CH_2-CH(CH_3)-CH(NH_2)-COOH$	6.038	25	+ 11.20	20	H_2O

(a_2) L-serine (L-α-amino-β-hydroxypropionic acid)	Ser	NH_2 | CH_2–CH–COOH | OH	5.68	25	− 8.8	16	H_2O
L-threonine (L-α-amino-β-hydroxybutric acid)	Thr	NH_2 | CH_3–CH–CH–COOH | OH			+ 28.4	26	H_2O
L-cysteine (L-α-amino-β-mercaptopropionic acid)	CyS	NH_2 | CH_2–CH–COOH | SH	4.44	30	− 212.9	25	1nHCl
L-cystine (bis[β-amino-β-carboxyethyl] disulphide]	CyS-SCy	NH_2 | CH_2–CH–COOH | S | S NH_2 | | CH_2–CH–COOH	6.064	25	− 214.4	24	1.02n HCl

†Three letters of the name are used, following BRAND and EDSALL

Table 1 Composition of important amino acids (according to classification p. 33)

amino acid	internationally accepted abbreviation†	formula	isoelectric point	at t (°C)	[α]D	at t (°C)	solvent
L-methionine (α-amino-γ-methylmercaptobutyric acid)	Met	$CH_2{-}CH_2{-}CH(NH_2){-}COOH$, with CH_2 bonded to S–CH_3	5.74	25	− 8.11	25	H_2O
(a₃) L-aspartic acid (L-amino-succinic acid)	Asp	$HOOC{-}CH_2{-}CH(NH_2){-}COOH$	2.98	25	+ 4.7	18	H_2O
L-asparagine (L-amino-succinic acid monamide)	Asn	$H_2N{-}C({=}O){-}CH_2{-}CH(NH_2){-}COOH$			− 5.3	20	H_2O
L-glutamic acid (L-α-amino-glutamic acid)	Glu	$HOOC{-}CH_2{-}CH_2{-}CH(NH_2){-}COOH$	3.08	25	+ 12.0	20	H_2O
Hhydroxyglutamic acid (L-α-amino-β-hydroxyglutaric acid)	Hyglu	$HOOC{-}CH_2{-}CH(OH){-}CH(NH_2){-}COOH$			+ 3.7–4		H_2O

L-glutamine (L-α-amino-glutaric acid monoamide)	Gln	$H_2N-C(=O)-CH_2-CH_2-CH(NH_2)-COOH$			+ 7.3	23	H_2O
(a₄) L-lvsine (L-α, ε-diamino-caproic acid)	Lys	$CH_2(NH_2)-CH_2-CH_2-CH_2-CH(NH_2)-COOH$	9.47	25	+ 14.6	20	H_2O
L-arginine (L-α-amino-δ- guanidyl-valeric acid)	Arg	$CH_2(NH-C(=NH)-NH_2)-CH_2-CH_2-CH(NH_2)-COOH$	10.76	25	+ 12.5	20	H_2O
L-ornithine (L-α, δ-diamino-valeric acid)	Orn	$H_2N-CH_2-CH_2-CH_2-CH(NH_2)-COOH$		25	+ 11.5	25	H_2O
(b) L-phenylalanine (L-α-amino-β-phenylpropionic acid)	Phe	$C_6H_5-CH_2-CH(NH_2)-COOH$	5.9		− 35	20	H_2O

† Three letters of the name are used, following BRAND and EDSALL

Table 1 Composition of important amino acids (according to classification p. 33)

amino acid	internationally accepted abbreviation†	formula	isoelectric point	at t (°C)	[α]D	at t (°C)	solvent
L-proline (L-pyrrolidine-α-carboxylic acid)	Pro	CH_2–CH_2 / CH_2 CH / N(H) COOH	6.03	25	− 84.9	20	H_2O
L-tryptophan (L-α-amino-β-indolylpropionic acid)	Trp	(indolyl, NH)–CH_2–CH(NH_2)–COOH	5.88	25	− 35	20	H_2O
L-tyrosine (L-α-amino-[4-hydroxyphenyl]-propionic acid)	Tyr	HO–(phenylene)–CH_2–CH(NH_2)–COOH	5.63	25	− 8.5	20	HCl 21%ig

L-hydroxyproline (L-4-hydroxy-pyrrolidine-2-carboxylic acid)	Hy Pro	HO–CH—CH$_2$ / CH$_2$ CH / N(H) COOH (ring)	5.8		−75.2	22	H_2O
L-histidine (L-α-amino-β-imidazolylpropionic acid)	His	CH═C–CH$_2$–CH(NH$_2$)–COOH / N NH / CH (ring)	7.64	25	−38.5	25	H_2O

†Three letters of the name are used, following BRAND and EDSALL

A large percentage of arginine is present in the protamines (see p. 58). The free amino groups of proteins arise partly from lysine which is why these amino acids are more involved than any others in the bonding of acids by proteins (acid buffering of protein e.g. for maintaining the pH value of blood).

When foods (meat, game, poultry) are improperly stored diamines, which belong to the group of biogenic amines, can be formed from these amino acids by microbial decarboxylation (putrefactive bacteria). Ornithine yields **putrescine** $H_2N-(CH_2)_4-NH_2$; **cadaverine** $H_2N-(CH_2)_5-NH_2$, is formed from lysine. These amines are also formed in the ripening of certain cheeses during an extensive decomposition of the protein.

Lysine is an indispensable 'essential' amino acid for man (see p. 44). Plant proteins, especially cereal proteins which are particularly important for human nutrition, contain little lysine. Therefore one tries to make up for this deficit with complementary foods (meat, mixed diet) from other nutritive groups (e.g. diet) (see p. 19).

The free NH_2 group of lysine can easily be bound by other materials; for instance in foods containing protein with carbohydrates, carbon-nitrogen compounds (like SCHIFF's bases). Sometimes high molecular browning products containing humic acids (melanoidins) may be formed* as the result of poly-condensations between carboxyl and amino groups, which are no longer, or only to a limited extent accessible to the proteolytic enzymes of digestion. A reduction in the biological value of the protein or food takes place.

Such a loss of lysine in flour proteins may occur in baking (biscuit baking), under some conditions it may also occur during lengthy storage of dried milk (loss of casein nitrogen, reduction of solubility), in dried egg and flours. Lysine and S containing amino acids may also be lost through heat treatment (auto-clave) of legumes (e.g. peas, lentils, beans, soya). When meat is pickled lysine is destroyed by the added nitrite. Unsuitable technological methods of food processing must be avoided (e.g. thermal treatment at too high a temperature or for too long), in order to prevent a serious reduction in the effective nutritive value due to loss of lysine and other amino acids in food (see p. 287).

Hydroxy amino acids. Serine and threonine are the important amino acids in the alicyclic series.

Serine occurs in many proteins, including those in silk, but it is the compounds of serine with phosphoric acid which are more important in the chemistry of food than serine itself.

In phosphoproteins, such as **casein** in milk and vitellin in egg, serine is bound to the phosphoric acid as the prosthetic group. In certain phosphatides also e.g. in wheat and rye germ, serine is bound to H_3PO_4 (see p. 120).

Threonine belongs to the essential amino acids (see p. 44), it cannot be synthesised by the human body. Whey protein, meat and brewer's yeast are

*So-called MAILLARD reaction, see p. 139.

particularly rich in threonine. The typical 'bouillon smell' of gravy mixes and soups (protein hydrolysates) comes from α-ketobutyric acid, which is formed from threonine.

Hydroxyglutamic acid has been found in milk casein.

Amino acids containing sulphur. **Cystine**, and its dimer **cystine** which readily arises from it by oxidation, have a particularly important function in the cell processes in the oxydation and reduction (=redox) system. The tripeptide, glutathione, has a similar function as a hydrogen carrier. It is made up from cysteine, glutamic acid and glycine and is synthesised in the human body. Cystine is found in quantity in hair, wool, nails, skin and in milk.

Methionine, as essential amino acid (see p. 45) is by far the most important sulphur containing amino acid. Its particular importance lies in the fact that in metabolism it acts as a methyl donor. The transference of methyl (transmethylation) occurs after activation with ATP (p. 240), which with methionine forms the so-called 'Active methyl' or S-adenosylmethionine, which functions as co-substrate for many methyl transferases (p. 239).

Choline (p. 119) is formed from cholamine via methionine. It is necessary for the phosphatides (p. 118), which are present and essential in all organs and transport fat around the body among other physiologically important functions. Lack of methionine and with it disturbance of choline and phosphatide synthesis, interferes with the removal of the fatty acids synthesised in the liver, which are normally given up to the blood as phosphatides. Methionine therefore belongs to the so-called 'lipotropic' materials, which counteract the laying down of fat (fatty degeneration) of the liver.

Methionine on its own and within the natural protein structure is sensitive to heat (heat decomposition of legumes to inactivate the trypsin inhibitor) and is also damaged when heated in the presence of carbohydrates e.g. in the baking process, (cystine, however, is not). This sensitivity is explained by the loss of methionine on heating and during the inappropriate storage of foodstuffs (dried egg, milk, dried milk, flour, see p. 291).

Lanthionine:

$$\begin{array}{lcl} CH_2 & \text{---S---} & CH_2 \\ | & & | \\ CH\text{---}NH_2 & & CH\text{---}NH_2 \\ | & & | \\ COOH & & COOH \end{array}$$

is a building block from the polypeptide subtilin, which occurs in Bacillus subtilis and has antibiotic properties.

Cyclic amino acids

The various amino acids in this group have iso-cyclic (benzene) or heterocyclic rings (pyrrolidine, imidazole and indole rings).

Aromatic series. **Phenylalanine** and **tyrosine**: Phenylalanine is an important protein cleavage product, which, like tyrosine, is found in almost all proteins.

Tyrosine (like leucine and cystine) can be dissolved only with great difficulty in cold water and therefore sometimes crystallises out of protein in the form of bundles of hydrolysates in long needles and is also found in very ripe cheese. Tyrosine is present in food and is the mother substance of the biologically important hormones (thyroxine, adrenaline).

Heterocyclic series. **Histidine** is an α-aminopropionic acid, which carries an imidazol ring in the β-position. Because the imidazol ring is strongly basic, histidene belongs to the hexobases with lysine and arginine (KOSSEL).

Tryptophan, important to life and therefore an essential nutritional amino acid, is destroyed in the acid hydrolysis of protein and can only be obtained from protein through fermentation or with alkali. Skatole, indole, and sometimes indoxyl are formed during bacterial putrefaction of protein. Tryptophan is necessary for the synthesis of haemoglobin; if it is lacking, anaemia results. (Tryptophan can be considered as provitamin PP, p. 214).

Proline and **hydroxyproline** (the imino acids) also belong to the heterocyclic amino acids. They contain a pyrrolidine ring, so that the amino group takes part in the formation of the ring, see table 1, p. 33.

Proline and hydroxyproline are present in most proteins (prolamine p. 59). It is possible for them to be formed in the body from glutamic acid, as has been shown by experiments with labelled substances (use of isotopes). On the other hand the transformation of proline to glutamic acid also occurs in the kidneys.

Essential amino acids

In its daily food, the organism is provided with a supply of protein containing varying amounts of amino acids. Many of them can be built up by the organism itself for its own nutritional needs. On the other hand, those others which the organism cannot synthesise, must be provided for the body in its food regularly and in sufficient quantities. These are grouped together under the heading 'essential (indispensable) amino acids', because their action cannot be replaced by other nutrients in food. Our organism needs a daily supply of a certain minimum amount of each essential amino acid, so that the nutritional **biological value** of a protein is limited by those amino acids present in a concentration below normal i.e. the **limiting essential amino acids**. In protein metabolism, the law of the limiting amino acid is of fundamental importance, and affects nutrition and nutrients in particular.

When there is a shortage or lack of even one of the essential amino acids in the nutritional protein, the organism reacts eventually with symptoms of deficiency, disturbances or growth, skin changes, and general disturbances of the protein and general metabolism. Biosynthesis from protein in general is reduced and certain fermentation systems of the body cells can no longer act because vital protein building blocks are missing.

These facts make it clear that, as has already been mentioned, the old system employed by chemists and physiologists of taking only calories into consideration, must be replaced by a more comprehensive biological approach, in which the supply of all parts of the diet should be optimal from a physiological point of view (see p. 19).

From a physiological point of view, the most important amino acids can therefore be divided into two groups, the essential and the non-essential amino acids:

Essential amino acids		Non-Essential amino acids	
valine	methionine	glycine	asparagine
leucine	threonine	alanine	glutamic acid
isoleucine	phenylalanine	norleucine	hydroxyglutamic acid
lysine	tryptophan	serine	tyrosine
arginine	histidine*	cystine	proline
			hydroxyproline

When the protein composition of a food is known, it is possible to draw certain conclusions about its biological value as a measure of the extent to which the organism can make use of the nutritional proteins (see p. 275). For example varieties of rice are being bred which are rich in lysine. Important research is also taking place in enriching cereal proteins with the limiting amino acids which can be made by purely chemical or enzymatic means.

When cereals are made into flour, there is no technological problem in enriching the flour during the milling process. On the other hand the enrichment of whole grains e.g. rice grains, is difficult, particularly because of the solubility of amino acids during the washing process. It has therefore been suggested that simulated grains should be used, that is, the amino acids should be added to the rice as a water soluble casing. In fact amino acid supplements are already added to various mixtures of proteins which have been developed in some countries which have a shortage of protein. In Central America, the lysine enriched Incaparina (INCAP) products have proved acceptable (see also p. 54), and are used in baked goods, sauces, soups or drinks. Fortifex, a Brazilian preparation, is fortified with methionine.

When enriching with amino acids, care must be taken to avoid creating an amino acid imbalance, as this can lead to considerable reduction in protein utilisation. The body can only use amino acids satisfactorily in favourable ratios (see also p. 273). Plant proteins which are of low biological value because of limiting amino acids can be made biologically complete by combining them

*The adult can do without arginine and histidine, whereas the young organism needs them for growth.

with other proteins (see p. 45). In countries with a predominantly vegetable diet, a combination of cereals and legumes can be used which is physiologically satisfactory, e.g. maize and soya, and complement each others limiting amino acids. The Incaparina products mentioned above are based on this idea.

Separation and determination of amino acids

Physical and chemical methods

A series of methods, which are important in the techniques involved in analysis of foods, have been developed for qualitative and quantitative determination of the amino acids which form part of a protein.

Hydrolysis of a protein may be by acid, alkali or enzymatic means.

In **acid hydrolysis** 6N HCl is usually used. The protein is boiled in a 10-100 fold excess of acid under reflux, or heated in an ampoule. The time taken for hydrolysis varies from 10 to 100 hours according to the protein. Tryptophan is destroyed completely under acid hydrolysis while serine, theonine and cysteine are partly destroyed. The peptide bonds of valine, leucine and isoleucine are hydrolysed with difficulty because of the hydrophobic nature of the amino acids.

The humin-like deposits which frequently occur on acid hydrolysis, are formed by the reaction of the decomposition products of tryptophan with aldehydes (carbohydrates, decomposition products of serine, cystine etc.) leading to the formation of melanin.

Alkaline hydrolysis (2-4 N NaOH) is most frequently used for the determination of tryptophan. Alkaline hydrolysis usually proceeds faster than acid hydrolysis. Arginine, cystine and cysteine are destroyed by it and most amino acids are racemised and deaminated in part.

The total enzymatic hydrolysis of a protein is a very time consuming process. Enzymes are therefore mainly used for the production of partial hydrolysates. Here enzymes have the great advantage over H^+ and OH^- ions, as by the use of suitable enzymes it can be decided which peptide bonds are to be broken.

Chromatography is generally used for the separation of the amino acids in the hydrolysate. Ion exchange chromatography is very important. Artificial resins, which are insoluble high polymers containing acid (–COOH; $-SO_3H$) or basic groups (primary, secondary, tertiary and quaternary ammonium salts) act as ion exchangers in water and many organic solvents.

The most commonly used acid exchangers for the separation of amino acids contain sulphonic acid groups (Amberlite IR 120, KPS200, Dowex 50). Here an exchange of H+ and amino acid cation occurs, so that the sulphonic acid salts of the amino acids are formed on the column. The amino acids are then eluted in succession from the column with buffer solutions of increasing pH and increasing ionic strength. The eluted material is collected fraction by fraction in aliquots. In the individual fractions, the amino acids can be determined by a quantitative ninhydrin reaction or by other methods. Automatic amino-acid analysers work on this principle (MOORE and STEIN) and the intensity of the ninhydrin colour reaction of the eluate fraction, is recorded directly by a recorder pen. In this way curves can be obtained, which make it possible to read off the qualitative and quantitative composition of the hydrolysate with great precision (Ninhydrin reaction, see p. 69).

Paper and thin layer chromatography of amino acids is simpler (but less exact). A two dimensional method is usually adopted. The fully developed chromatogram is sprayed with ninhydrin or other colouring reagent. The colour spots which develop are analysed, after elution or directly, by photometric methods. The chromatography of the dinitrophenyl derivatives of amino acids is frequently used (see p. 71). After two dimensional chromatography, the clearly visible yellow spots can be eluted and the absorption at 360 nm and 390 nm measured.

Microbiological determination of amino acids

Many microorganisms need amino acids for growth or for metabolism, but are not able to synthesise them. The amino acids must therefore be supplied to the microorganism with the nutrient medium. If the nutrient medium contains all the necessary materials with the exception of the amino acid to be determined, the metabolism and growth of the microbes will cease. If the missing amino acid is now added, relationship can be found where the renewal of growth or the production of a metabolic product is proportional to the amount of the added amino acid. A surplus of this amino acid will lead to no further increase in the metabolic processes once the maximum has been reached.

In this way it is possible to detect the presence of amino acids and to determine them quantitatively. The measurement can be based on the density of the microbial suspension or may be by the determination of a metabolic product. These methods have been considerably extended by the newly discovered possibility of altering the metabolic type of the microorganism by chemical or physical means. It was found that Neurospora, a common bread mould, in particular, can be altered by X-rays and by some chemical materials, so that it lacks certain enzymes and its intermediate metabolism is disturbed. Then these materials, which are normally involved in the intermediate metabolism, become substances which have to be supplied from outside. These artificially created mutants of microorganisms make it possible to determine a great variety of materials quantitatively. Microbiological methods have the great advantage that they are easy to carry out, especially in the case of a series of experiments.

3.2 PROTEINS

Constitution of Proteins

E. FISCHER demonstrated the linkage of amino acids to form peptides and proteins, by means of acid amide bonds. These bonds are also called peptide bonds.

The carboxyl group of an amino acid reacts with the α-amino group of a second amino acid, with the loss of water:

$$H_2N\text{—}\overset{\overset{\displaystyle R_1}{|}}{\underset{\underset{\displaystyle H}{|}}{C}}\text{—}COOH + H_2N\text{—}\overset{\overset{\displaystyle R_2}{|}}{\underset{\underset{\displaystyle H}{|}}{C}}\text{—}COOH \xrightarrow[-H_2O]{} H_2N\text{—}\overset{\overset{\displaystyle R_1}{|}}{\underset{\underset{\displaystyle H}{|}}{C}}\text{—}\overset{\overset{\displaystyle O}{\|}}{C}\text{—}\underset{\underset{\displaystyle H}{|}}{N}\text{—}\overset{\overset{\displaystyle R_2}{|}}{\underset{\underset{\displaystyle H}{|}}{C}}\text{—}COOH$$

Peptides are called di-, tri-, or tetrapeptides according to whether 2, 3, 4 or more amino acids are bonded together. The nomenclature indicates that the amino acid linked to the carboxyl group of the next amino acid is given the ending 'yl'. e.g.

$$CH_3\text{—}\underset{\underset{\displaystyle CH_3}{|}}{CH}\text{—}CH_2\text{—}\overset{\overset{\displaystyle NH_2}{|}}{\underset{\underset{\displaystyle H}{|}}{C}}\text{—}\overset{\overset{\displaystyle O}{\|}}{C}\text{—}\underset{\underset{\displaystyle H}{|}}{N}\text{—}\overset{\overset{\displaystyle CH_3}{|}}{\underset{\underset{\displaystyle H}{|}}{C}}\text{—}COOH$$

leucyl-alanine

$$CH_3\text{—}\overset{\overset{\displaystyle NH_2}{|}}{\underset{\underset{\displaystyle H}{|}}{C}}\text{—}\overset{\overset{\displaystyle O}{\|}}{C}\text{—}\underset{\underset{\displaystyle H}{|}}{N}\text{—}\overset{\overset{\displaystyle CH_2\text{—}CH(CH_3)_2}{|}}{\underset{\underset{\displaystyle H}{|}}{C}}\text{—}COOH$$

alanyl-leucine

If up to 10 amino acids are linked or bonded together, they are called oligopeptides, when they are more than 10, polypeptides, and with more than 100 amino acid residues, macropeptides and proteins.

Usually only the α-amino acid group $H_2N\text{–}CH\text{–}COOH$ takes part in the formation of peptide chains. It forms the backbone of a peptide or protein. The R residues function as short side chains. These side chains are very important for the chemical, physical and biological behaviour of a protein, as shown by the following formula:

$$\cdots NH\text{—}\underset{\underset{\displaystyle CH_2OH}{|}}{CH}\text{—}CO\text{—}NH\text{—}\overset{\overset{\displaystyle CH_2\text{—}CH(CH_3)_2}{|}}{CH}\text{—}CO\text{—}NH\text{—}\underset{\underset{\displaystyle CH_2\text{—}C_6H_4\text{—}OH}{|}}{CH}\text{—}CO\text{—}NH\text{—}\overset{\overset{\displaystyle CH_2\text{—}SH}{|}}{CH}\text{—}CO\text{—}NH\text{—}\underset{\underset{\displaystyle CH_2\text{—}CH_2\text{—}COOH}{|}}{CH}\text{—}CO\text{—}NH\text{—}\overset{\overset{\displaystyle (CH_2)_4\text{—}NH_2}{|}}{CH}\text{—}CO \cdots$$

Section of a peptide chain.

Macromolecular proteins are built up from hundreds of amino acids, of which each may be included in the molecule only once or many times. When the number of individual amino acid building blocks, from which the peptide or protein is built up, is known, we still know nothing about its structure. For that, it is necessary to determine firstly the order (sequence) in which the individual building blocks are linked together.

Primary structure: Various methods have been developed in the last two decades to determine the sequence (also called primary structure). The analysis of the sequence of insulin by SANGER was a milestone. The primary structure of ribonuclease is given in Fig.1.

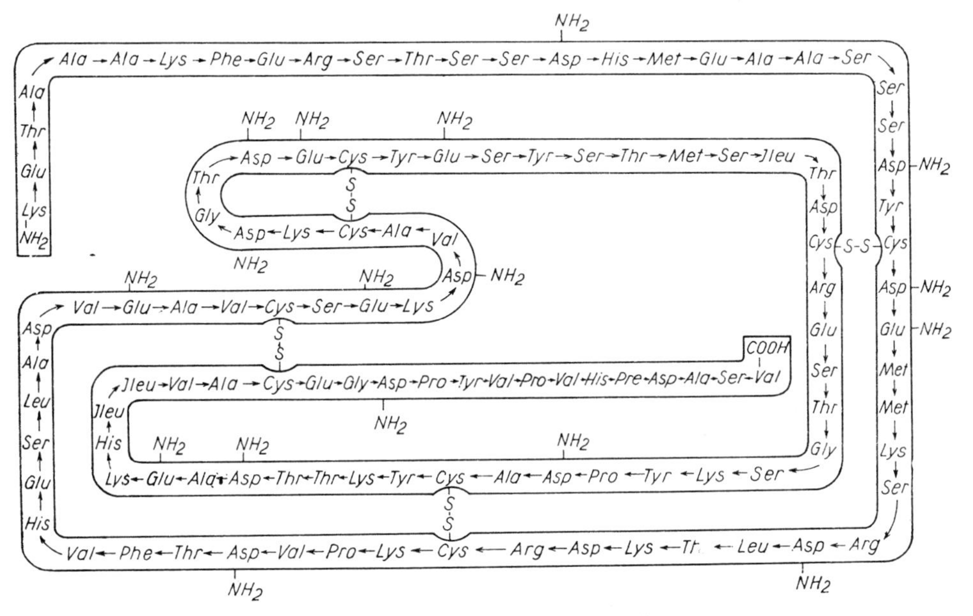

Fig. 1 Primary structure of ribonuclease

The abbreviations of BRAND & EDSALL are used here for the amino acids. The chain is drawn in such a way that the amino acid with the free amino group is shown at the left hand end, whereas the free carboxyl group is at the right hand end. In the case of longer chains the direction in which it should be read can be shown with arrows.

Secondary structure: If the primary structure of a protein is known, then the question of the spatial arrangement of the molecule arises. It is easy to see that there are in principle many billions of possibilities for spatial arrangement in the case of a fibre molecule which consists of 100 or more amino acid residues. However, the fibrous structure of certain proteins (scleroproteins) and the tendency to crystallise of many globular proteins, permits the conclusion that in nature only a few of these possibilities actually occur.

For the analysis of the secondary and tertiary structure, physical methods, especially X-ray structure analysis are used. Here the simplest materials to study are the scleroproteins, where the fibrous structure can be seen macroscopically. Groupings which repeat regularly within a chain at regular intervals, the so-called identity periods, can be calculated on the basis of X-ray interference patterns. An identity period of 7.23 Å (= $7.23.10^{-7}$mm) can be expected for the simplest example of a stretched chain, see Fig. 2.

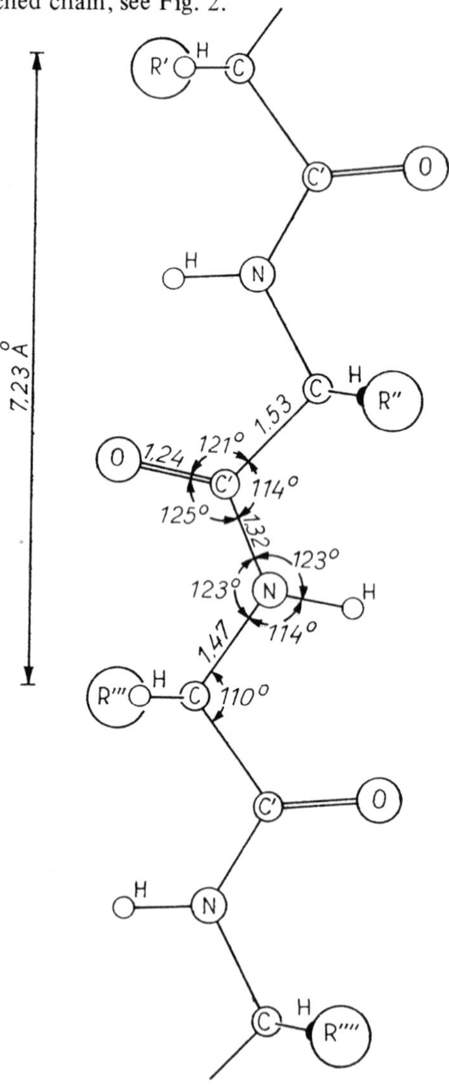

Fig. 2 Measurement of a stretched polypetide chain.
After PAULING AND COREY.

Scleroproteins which occur in nature can be divided into 3 groups according to their measured identity periods:

1. silk fibroin-β-keratin group 6.5 – 7.0 Å
2. α-keratin-myosin-fibrogen group 5.1 – 5.4 Å
3. collagen group 2.8 – 2.9 Å

It can be seen from these values that a stretched chain is not possible. In the formation of the secondary structure only those forms are likely which are stabilised by a maximal number of hydrogen bonds. In addition the atoms of the peptide bond must lie in the same plane, so that tauomerism as illustrated is possible:

$$\mathrm{C{-}C(=\overline{\underline{O}}){-}N(H){-}C} \longleftrightarrow \mathrm{C{-}C(-|\overline{\underline{O}}|^{\ominus}){=}N^{\oplus}(H){-}C}$$

The C=O and HN–C groupings must be around 2.8Å so that the formation of a hydrogen bond is possible. Various models comply with these conditions e.g. in the 'folded sheet structure', where two peptide chains lie next to each other in the same direction (parallel) or opposite directions (antiparallel) in such a way that in each case the C=O group of one chain lies opposite an NH group of the other. A band is then obtained in which the atoms of the peptide bonds which lie opposite one another are in the same plane, with kinks at the α-C-atoms. See Fig. 3.

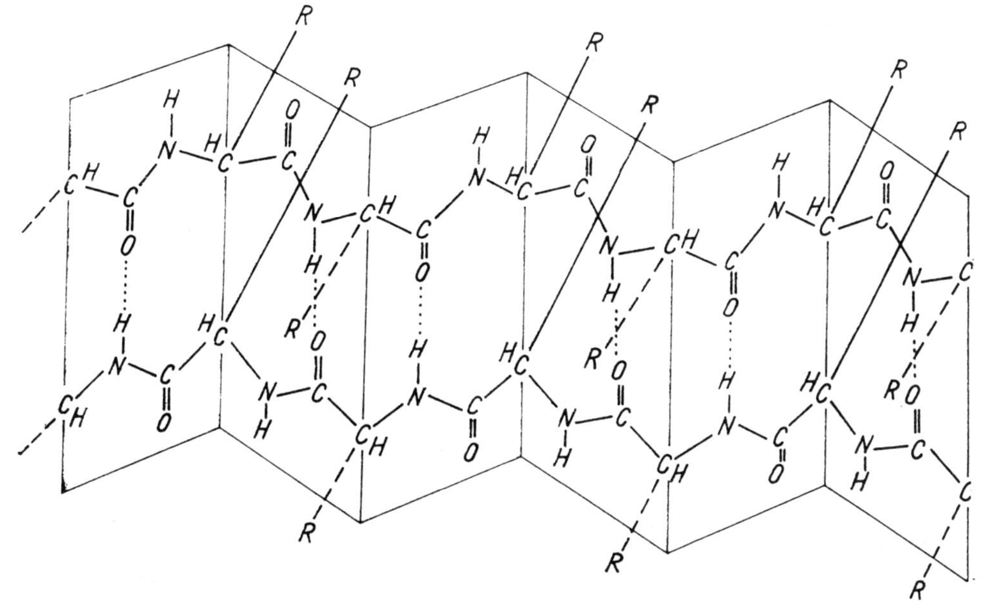

Fig. 3 Concertina structure of a fibrillar protein (silk fibrin). After PAULING AND COREY.

The folding of the band brings about a shortening of the identity period of 5-10% compared with the stretched form. This kind of secondary structure was probably involved in the silk fibroin-β-keratin group.

The α-helix (α-screw) structure calculated by PAULING is much more favourable than the concertina structure. In it the peptide chain, in the form of a spiral, is wound up to that the C=O– and NH groups are opposite each other at suitable intervals. The hydrogen bonds can now develop inside their own chain and stabilise the structure. See Fig. 4. Each revolution involves 3.7 amino acids; the atoms of the peptide bonds lie in planes, which are transposed by 80° each time. The identity period measures 5.44 Å. The α-helix can be constructed as a left or right winding screw. The side chains of the amino acids stretch outwards from the screw. The helix formation is thermodynamically favourable because of the stabilising H-bonds. Both forms can change into each other.

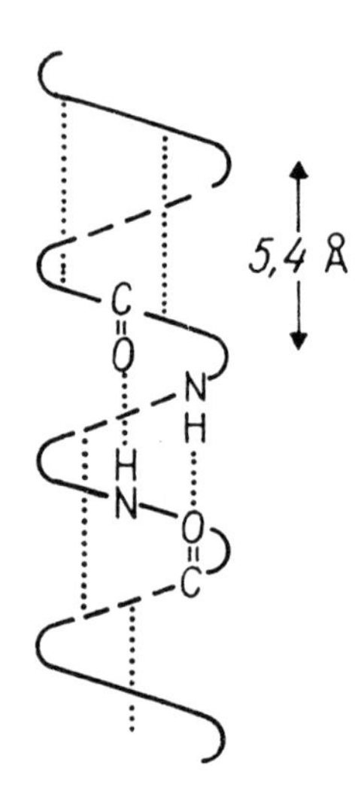

Fig. 4 α-helix. Screwlike conformation of the peptide chain.
(. . . = hydrogen bonds)

Tertiary structure: The tertiary structure of a protein reproduces in space the distribution of the ordered system described in the secondary structure. So far it is only known for a few proteins because the evaluation of the X-ray interference spectra is extraordinarily difficult. The arrangement in space of a peptide chain for myoglobin and haemoglobin is well known. The secondary structure consists of α-helix sections which envelop the prosthetic group in an irregular double loop. In myoglobin the protein section consists of a single peptide chain, in haemoglobin there are four peptide chains of which each pair is identical. So far as is known at present, it is probable that the tertiary structure of a protein is already determined by the primary structure, that is by the sequence of the amino acids in the chain. The proline residues usually have the effect of causing a kink in the α-helix. Cysteine residues which are linked together to form cystine by S–S bonds, determine a particular spatial arrangement of the α-helix sections between the proline residues, and so on. Of interest here are experiments with ribonuclease, where it has proved possible to split the four S–S bonds and then to recombine them again in the correct way (among many other possibilities). Such occurrences can also play a part in the spontaneous regeneration of certain enzymes after previous inactivation. See p. 245 (Kinetics of enzymes – influence of temperature). Secondary and tertiary structure are now included together under the heading of chain formation.

Quaternary structure: a concept introduced by BERNAL describes the association of certain tertiary structures into stable aggregates without these being linked covalently. For instance, haemoglobin consists of four peptide chains. Certain enzymes, e.g. aldolase, a trimeric complex with molecular weight 150,000, can dissociate in strongly acid conditions into three monomers of 50,000 molecular weight. The monomer has a different tertiary structure in the association from that in the monomeric condition. Both glutamic dehydrogenase and lactic dehydrogenase have several stages in their structural arrangement. In glutamic dehydrogenase the complex separates into four enzymatically equally active monomers when it is diluted or when $NADPH_2$ is added as co-substrate. In lactic dehydrogenase, the sub-units have different molecular weights and they differ in their enzymatic activity. This can be linked with isoenzymes.

Protein sources and biological value

Proteins are available to us in many different forms – in meat, egg, milk, legumes and cereals. From the nutritional aspect it is important to know the biological value of the food proteins.

The concept of 'biological value' was developed to describe the extent to which a protein provides nitrogen for the body. In an ideal case, when all the nitrogen available from the protein is retained in the body, the biological value would be 100%, (expressed by the number 100). The biological value, expressed as a number, indicates how many parts of body protein can be replaced by 100 parts of the protein fed in the test. The figure will therefore be lower, the lower the value of the food protein. According to THOMAS, whole egg, milk and meat are almost 100, whole potatoes 79, peas 56, wheatflour 40 and maize 30.

When determining the value of food protein the so-called protein efficiency ratio (PER) (growth value) is today in common international use. PER is expressed by the following ratio:

$$\frac{\text{bodyweight increase in grams}}{\text{protein consumed in grams}}$$

The most valuable proteins make up an average of 21% in the flesh of warm blooded animals and average 12–14% in fishes. Important sources of animal protein are eggs (11–16%), milk (3–4%) from which various protein-rich products such as curd cheese, yoghurt, dried skim milk (35%) and hard cheese (32–37%) are made.

Among vegetable foods, the protein content e.g. legume (27–41%), cereals (7–14%), is of quantitative importance particularly when cereals are the staple food. However, vegetable proteins are not nutritionally perfect due to specific limiting amino acids. See p. 53.

A low value vegetable protein can be transformed into a nutritionally complete protein not only by adding the missing limiting amino acids but also by the addition of other proteins, particularly those of animal origin or from yeast. According to LANG the following combinations have a good reciprocal value: legumes + milk or meat or eggs: potatoes + milk: cereals + meat or milk or eggs: cereals + soya or yeast.

For optimal nutrition, a mixed animal and vegetable protein diet, such as is normally available to us, is desirable. In most of the developing countries in Asia, Africa and Latin America, where animal foods are still to a large extent in short supply, the aim should be a useful combination of those cereals and legumes which are available.

Protein mixtures are of considerable help today in countries with a protein shortage. They are usually made from local cereals and legumes with the addition of minerals, vitamins, amino acids and sometimes also dried milk. Examples are the Incaparina products of Columbia made from maize, de-fatted soya flour, cottonseed flour, vitamin A lysine, $CaCO_3$, with a protein content of 27.5%.

Protein concentrates from plants which can be used to fortify such products have become of great importance recently. The main source of these are de-fatted oil seeds particularly soya beans, peanuts and cotton seed. Concentrates with up to 70% protein can be made from de-fatted soya flour. The extraction of protein concentrates from leaves e.g. rape, clover, wheat, maize, bananas, sugar cane, grass seems promising but better economic technical processes for extraction need to be developed.

Potential protein sources for the future

1. **Food yeasts**, for instance dried *Torula utilis* is obtained by normal protein synthesis by growing varieties of *Torula* on carbohydrates (molasses, sulphur waste lye, wood, reed or straw hydrolysates), N-containing inorganic nutrient salts, phosphates and potassium salts. See also pp. 195, 196. A very small part of the yearly growth of timber would be sufficient to ensure the increase in protein required by the population of the year 2000 (6–8 Billion) using the transformation of wood carbohydrates into yeast.
2. **Microorganisms.** (*Torula and Candida* yeasts), can be grown on the gas oil fractions of mineral oil with inorganic salts added as above. In addition, the huge

natural gas reserves (containing 90% methane), can be used as a source of carbon for protein synthesis with certain microorganisms (bacteria, yeasts) and so could provide an important potential basis for supplying world protein needs.

3. **Algae** (*Chlorella*).

Physico-chemical properties of proteins

Like other macromolecular combinations formed by condensation, proteins differ from each other in many of their physico-chemical properties, including their building blocks, the amino acids. The **molecular weight** of proteins can be determined by various methods including osmotic pressure, diffusion, chemical analysis, sedimentation and viscosity. Cryoscopy and ebullioscopy are inadequate because of the very small measurable effects and also because of denaturation. See p. 56.

Chemical analysis gives only the minimum molecular weight (the number of amino acid residues or prosthetic groups must always be an even number), the actual molecular weight can be a multiple of the minimum molecular weight. Nowadays the determination of molecular weight is frequently carried out by measuring the rate of sedimentation in the ultracentrifuge (SVEDBERG).

The following table summarises the molecular weights of various proteins:

lactalbumin (bovine)	17,400	egg albumin	44,000
myoglobin	16,000	serum albumin (bovine)	68,900
ribonuclease	12,700	haemoglobin (human)	64,000
insulin	6,000	globulin (human)	156,000
β-lactoglobin (bovine)	35,400	catalase	250,000
pepsin	35,000	urease	480,000

It is not always easy to determine whether the weights obtained are the actual molecular weights. It is well known that proteins very easily form associates, Insulin, readily associates in the presence of metal ions in solution and 'forms' molecular weights of 12,000, 16,000 or 48,000. Sedimentation in the centrifuge is often taken as a criterion of purity for proteins, but the existence of a single symmetrical 'peak' in the sedimentation diagram is a necessary but by no means adequate criterion for purity.

The **tendency** of many proteins to **crystallise** is extraordinarily high. Proteins with globular or elliptical tertiary structure in particular crystallise readily because of their spatial arrangement. Some haemoglobins even crystallise inside the red blood corpuscles, if they are cooled down to 0°C.

The **solubility** of proteins depends on the number of polar (hydrophilic) and non-polar (hydrophobic) groups. Only polar solvents such as H_2O, HCOOH, dimethyl formamide, formamide and glycerine can be used. Small concentrations of salt often increase the solubility (salting-out effect). The strong hydration of inorganic ions which occurs in water solutions forces the protein out of the solution. This is frequently used in fractional precipitation of protein mixtures, e.g. with ammonium sulphate used in the purification of enzymes, (p. 238).

The solubility of proteins, related to pH depends strongly on their **amphoteric character**. According to its pH value, a protein can exist as a polyvalent cation, anion or zwitterion. One distinguishes between the total charge of a protein (= the sum of the positive and negative charges) and the excess charge, which can be positive or negative according to the pH. Like their building blocks, the amino acids, proteins have an isoelectric point. Here the total charge reaches a maximum (maximum formation of zwitterions); the surplus charge is zero, i.e. the total number of positive and negative charges in a protein molecule is exactly equal. Viscosity and solubility are at a minimum at the isoelectric point. The tendency to crystallise and to precipitate out are therefore at their highest here. At the isoelectric point the electrophoretic mobility is zero.

The **isoionic point** is different from the isoelectric point. The former is defined as the pH value at which the number of protons which are taken up by the basic groups is equal to those given by the acid groups. The isoelectric and isoionic point may differ from each other by a whole pH unit according to the salt content of the solution.

Optical Rotation. In aqueous solution, proteins rotate polarised light to the left (by $-30°$ to $-60°$ according to the protein concerned). The magnitude of the rotation is apparently dependent on the secondary structure of the protein. Statistically, convoluted protein chains show a higher rotation ($-80°$ to $-120°$) than a pure α-helix. The high rotation of collagen (from calves' hide: $\alpha D = -415°$) should be noted.

The measurement of **light absorption** of proteins shows strong bands in the UV area at 180-230nm for the CO–NH bonds and at 250-300nm (maximum at 280nm) for the aromatic rings of phenylalanine, tyrosine and tryptophan. UV spectroscopy permits quantitative determination of tyrosine and tryptophan in proteins. Infra-red spectroscopy of peptides and highly organised proteins, gives indications of their configuration.

Denaturation

Certain changes in proteins, which are collectively included under the heading of denaturation, are of importance for the nutritional and technological aspects of food. Denaturation involves structural alteration of proteins leading, among other things, to the loss of certain biological properties (enzymatic or serological activity, hormonal action and such like). Denaturation of proteins means destruction of their secondary and tertiary structure, where in general co-valent bonds are not broken. The highly organised structure of native protein is broken up in favour of a purely fortuitous and unorganised one. The unfolded peptide chains then develop hydrogen bonds and internal salt formation in a haphazard way. This is clearly seen in an X-ray diagram which resembles that of β-keratin (laminar structure). S–S bonds can be broken in denaturation, as shown by the increase in SH groups. The associated high positive change in entropy (entropy is the measure of the likelihood of a state) is characteristic of the process of denaturation. At the same time the physical properties of the protein also change (rotation, viscosity, solubility, diffusion constants etc.). Denatured proteins can no longer be crystallised. Chemically an 'unmasking' of certain, not previously identifiable –SH, –S–S imidazolyl groups etc. occurs. Denatured proteins are more rapidly digested enzymatically.

Denaturation can be achieved by various methods: by heating in the vicinity of the isoelectric point (coagulation), by strong acids and alkalies, concentrated urea or guanidine solutions, organic solvents, aromatic acids (sulphosalicyclic acid), detergents (dodecylsulphate among others), ionizing radiation (α–, β– and γ-rays), other methods include the use of precipitants, such as phosphotungstic acid, and large surface areas (surface denaturation in foams) and others. Denaturation is usually irreversible, but can under certain conditions be made reversible. (See also p. 248 and under Regeneration of enzymes p. 260).

Classification of proteins

It is still necessary to classify the numerous and varied proteins empirically, since chemical criteria are inadequate. The classification is based mainly on external criteria such as origin, solubility, precipitation conditions, coagulation properties and other physical properties. They are divided into two major groups.

1. Simple proteins.
2. Compound proteins.

Simple proteins yield only amino acids on hydrolytic decomposition. Compound proteins on the other hand, yield other substances among the products of breakdown, such as carbohydrates, nucleic acids, colouring agents, certain metals and vitamins which were bonded to the protein. These are called prosthetic groups.

The subdivision of compound proteins can be made according to the characteristics of the phosthetic groups. In the differentiation of simple proteins, chemical, physical and physico-chemical differences are decisive, including composition based on salting-out, biological characteristics, source etc. This leads to the following division of proteins.

Simple proteins:

albumins
globulins
glutelins
histones
protamines
prolamines (gliadines)
scleroproteins

Compound proteins:

chromoproteins (metalloproteins)
lipoproteins
mucoproteins (glycoproteins)
nucleoproteins
phosphoproteins

Simple proteins

Globulins. Globulins are the commonest simple proteins. They act as important reserve proteins in plants, for example in the luguminosae (legumin in peas, lentils and beans, phaseolin in white beans) and in grain and potatoes. They are also widely distributed in the animal kingdom, in blood, milk, meat and eggs (ovoglobulin).

One of the special characteristic of globulins is their faintly acid character*, which is due to the increased content of amino-dicarboxylic acids. Another is their insolubility in pure water and ready solubility in dilute neutral salt solutions. From such salt solutions, some globulins are precipitated by the addition of water (dialysis against pure water) or by acidification (some even by the introduction of CO_2). Globulins are completely precipitated out by half saturation with $(NH_4)_2SO_4$. These properties distinguish them sharply from the albumins, so that it is easy to separate these two groups, which commonly occur together†.

Albumins. Albumins form the second main group of proteins and nearly always occur with globumins. Characteristically albumins differ from globulins in their solubility in distilled water as well as in dilute acids, and in their ready and more complete precipitation at full saturation by ammonium sulphate. Albumins contain no glycine but are rich in sulphur. They are found in organ fluids, in tissue, in the blood serum (serum-albumin) in milk, (lactalbumin), and in egg white (ovalbumin). In plants, albumins are mainly present in the seeds. Some plant albumins are very poisonous, for instance ricin in castor beans (inactivated by heating) crotin, phasin and abrin.

Protamines. Protamines occur in the spermatozoa of fishes (bound to the nucleic acid by a salt-type bonding: clupein is the protamines of the herring sperm, salmin that of the salmon sperm. They are characterized by a high hexone base (including ornithine, arginine and histidine, see p. 38) among which arginine predominates. They are therefore of a basic character. Protamines contain no sulphur (hence no cystine or methionine) and also lack the aromatic amino acids. Protamines are not split by pepsin, but by the protein-splitting enzymes of the intestinal tract.

Histones. The histones, like the protamines show a basic reaction because of their hexone base content. They contain no tryptophan and little cystine and methionine. They occur in the cell nucleus of nearly all body cells, probably in the form of nucleoproteins. They combine with other proteins and acid glycolipids to form compounds which are difficult to dissolve. They are split by trypsin and pepsin.

†The so-called euglobulins are, according to the latest research, lipoglobulins, that is lipoproteins = symplexes. See symplexes p. 65.

*Isolectric point between pH5 and 6.

Heat does not coagulate them. Most research has been done on globin from haemoglobin and histone from the thymus, lymphocytes and liver.

Gliadins. The gliadins (prolamines) occur only in the endosperm of grain in seeds: gliadin in wheat and rye, hordein in barley and zein in maize. They are insoluble in pure alcohol and water, but can be extracted from the flour with 50–90% alcohol. The name prolamine only indicates the high content of proline and the evolution of ammonia on acid hydrolysis. Another characteristic of prolamines is their high glutamic acid content.* Because of the lack of lysine and the small arginine and histidine content, gliadins have a low biological value (see, however, the section of glutelin) Coeliac disease (intolerance to gliadin) can lead to chronic disturbance of growth and a gliadin-free diet has a very favourable effect.

Glutelins. The glutelins occur with the gliadins in grain seeds, e.g. **glutenin** in wheat. The combination of glutelins and gliadins represents the most valuable protein, called **gluten**. Glutelins are estracted from cereals by means of dilute acids and bases and can be salted out from alkaline extracts with small amounts of ammonium sulphate. In contrast to the gliadins, glutelins contain the amino acids lysine and tryptophan and therefore complement the gliadin, so that the gluten of the cereals as a whole has full biological value. Gluten composition plays an important part in the baking process.

Skeletal proteins. The main representatives of the skeletal proteins (scleroproteins) are **collagen** and **keratin**. The first is found in the connective tissue, the sinews, ligaments of the face, other ligaments, bones and cartilages. When slightly acidified and boiled with water, it yields water-soluble hydrolysis products known as **glutin, glue** and **gelatine**.*

Gelatine of varying composition according to the place of origin, contains much glycine, proline, hydroxyproline and arginine, but no tryptophan and no (or only traces of) tyrosine. It does not have full biological value and it is therefore important, from the nutritional point of view, to determine the gelatine content of many foodstuffs (determination of gelatine nitrogen in meat and meat products). Elastin, a kind of collagen, the basic matrix substance of elastic fibres, does not yield any gelatine on boiling. The most obvious characteristic of the keratins, frequently called horny substances, is their high sulphur content due to the high cystine content (up to 17%). This group of proteins is to be found mainly in hair, nails, horns, wool and feathers. These skeletal proteins perform mainly mechanical functions. Their resistance to chemical reagents means that their solubility is poor in water, acids and alkalis, even dilute acid

*See p. 35, preparation of glutamic acid from wheat gluten etc.

*Their N-content (18%) is higher than that of other proteins. Therefore, to determine the gelatine content, the nitrogen found analytically is multiplied only by 5.55 instead of the usual 6.25.

hydrolysis under heat occurs only very slowly and no gelatine is produced. Hydrolytic splitting yields considerable cystine, tyrosine and glutamic acid. Keratins are not attacked by protein splitting enzymes.

Compound proteins

These proteins – unlike the simple proteins – yield other substances in addition to amino acids when they are broken down. Many of these compounds are linked to the protein by salt-like bonding because of their acid nature (carboxylic groups or other acid groups). These groups are non-protein and are generally called prosthetic groups. Compound proteins occur more widely in nature than simple proteins, and it seems likely that, with further research in this difficult area, many new ones will be discovered. In very many proteins (casein, collagen, serum protein) small amounts of other accompanying substances are regular constituents, particularly carbohydrates, and we do not yet know for sure whether they are true **prothetic** groups, or parts of **symplexes** (see p. 65).

Phosphoproteins. In this group of proteins, ortho-phosphoric acid (H_3PO_4) functions as the prosthetic group and is usually bound as an ester to the alcoholic hydroxyl groups serine and threonine (see p. 31), forming phosphoric acid and threonine phosphoric acid as a mono- or di-ester:

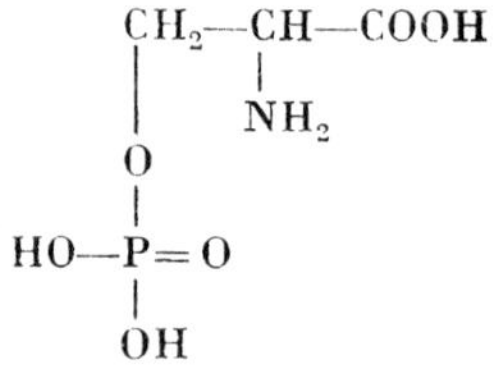

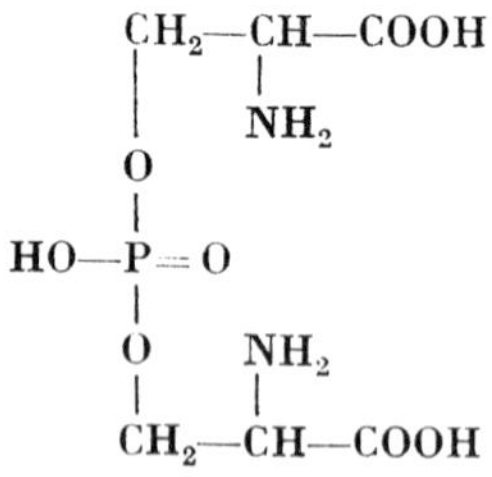

```
      CH2—CH—COOH              CH2—CH—COOH
      |    |                   |    |
      |    NH2                 |    NH2
      O                        O
      |                        |
  HO—P=O                   HO—P=O
      |                        |
      OH                       O    NH2
                               |    |
                               CH2—CH—COOH
```

This group of proteins (H_3PO_4 content 0.7%) behaves in an acid manner and is soluble in dilute alkalis and ammonia, forming salts and on acidification is precipitated out again. Phosphoproteins are almost insoluble in water. Phosphatases and other splitting enzymes (esterases, see p. 259) do not, as far as is known split off phosphoric acid.

The most important representative of this class of phosphoproteins is **casein**, the principal protein in milk. It is present in milk as a colloidal dissolved opalescent or milky calcium salt (calcium caseinate). Solutions do not coagulate on heating under neutral conditions; the other milk proteins (albumin and globulin) when heated are also protected from coagulation by calcium caseinate.

In **acid coagulation** of milk, the ubiquitous lactic acid bacteria produce lactic acid from lactose which extracts the calcium from the calcium caseinate. As soon as the process has reached a certain stage, the free casein precipitates out. In rennet coagulation the dissolved calcium caseinate is transformed into paracasein an insoluble calcium salt of casein*.

The **ovovitellin** of egg yolk is another important phosphoprotein. With lecithin, it is the main phosphorus containing reserve material of egg yolk. It occurs together with lecithin to which it is firmly attached and from which it can only be separated by denaturation. Possibly ovovitellin and lecithin are chemically loosely bound together, although strongly connected symplexes in the form of lecithin albumin (see p. 260) are also present.

Nucleoproteins. Nucleoproteins are one of the main constituents of the cell nucleus. We find them in cell plasma, in the organs and in the body fluids (gall, milk, kidney, thymus gland). They are also widely distributed in vegetable matter, especially in yeast cells. For prosthetic groups they have nucleic acids, high molecular compounds of strong acid character due to the phosphoric acid residues. The bond between the nucleic acid and the protein is in many cases salt-like. (See Symplexes p. 65).

Nucleic acids and their building blocks. Nucleic acids are polynucleotides of high molecular weight, made up from low molecular mononucleotides or 'simple nucleic acids'. These mononucleotides have three components arranged in the following manner:

base – pentose – phosphoric acid

According to the type of pentose, two groups of nucleic acids are recognised: ribonucleic acids (RNS) with ribose (p. 147) as the sugar, and desoxyribonucleic acids, with desoxy-ribose (p. 152) as the carbohydrate.

Pyrimidine or purine derivatives act as base*.

The following heterocyclic ring formulae are the basis for these compounds.

```
       1  6                     1  6
       N==C--H                  N==C--H
       |  |                     |  |
  H--C2  5C--H             H--C2  5C--N7
       |  |                     || ||    \\ 8
       N--C--H                  N--C--N   C--H
       3  4                     3  4  9 \
                                         H

     pyrimidine                   purine
```

The following purine bases have been found in nucleic acids:

cytosine: 2-hydroxy-6-amino-pyrimidine
uracil: 2,6-dihydroxy-pyrimidine
xanthine: 2,6-dihydroxypurine
adenine: 6-aminopurine
guanine: 2-amino-6-hydroxypurine
hypoxanthine: 6-hydroxypurine

*Often, in the literature, casein (= calcium caseinate) is called caseinogen, and paracasein on the other hand is called casein.

*The stimulants present in our coffee, tea and cocoa: caffeine, theophylline, theobromine, are purine derivatives.

The purine bases, hypoxanthine and xanthine, derived from nucleic acids are present in meat juices and meat extracts (see p. 27).

Inosinic acid, hypoxanthineribose-5-phosphoric acid, is an example of a mononucleotide and is an integral constituent of meat extract†.

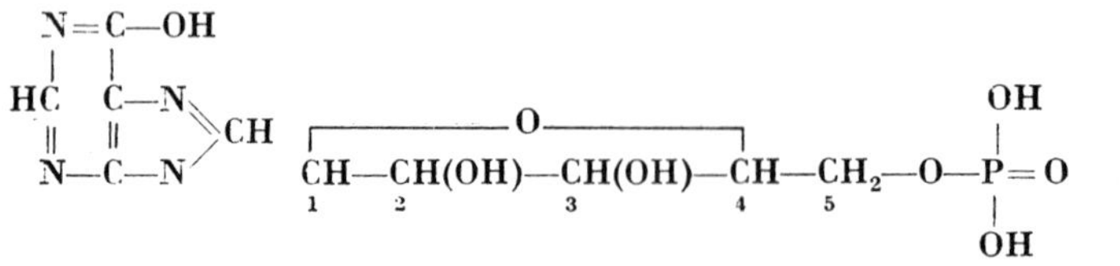

```
 N=C—OH
 |  |
HC  C—N                                                    OH
 ||  ||   \CH   ┌───────────────O──────────────┐           |
 N—C—N/        CH—CH(OH)—CH(OH)—CH—CH2—O—P=O          inosinic acid
               1     2        3      4    5        |
                                                   OH

 hypoxanthine            D-ribose                phosphoric acid
```

Adenylic acid, which is built up from adenine ribose and H_3PO_4 and its pyrophosphoric acid compound, **adenosinetriphosphoric** acid (ATP), are physiologically important mononucleotides. ATP plays an essential role as co-substrate in the transfer of energy in the cell, giving up one or two molecules of phosphoric acid readily, so that a multiplicity of phosphorylations may occur, or **energy-rich** phosphate compounds may be formed (see p. 43). Similar processes also take place in alcoholic fermentation. Not only phosphate residues, but other groups (e.g. methyl groups) can be transferred with the help of energy-rich ATP; compare the formation of choline from cholamine via D-adenosylmethionine ('active methyl'). See pp. 43, 119, 239.

Mononucleotides can be broken down into their three building blocks by chemical and enzymatic hydrolysis. One component alone, for instance phosphoric acid, may be split off, so that a compound of pentose and base, the so-called nucleosides remain:

```
              mononucleotide
 ┌──────────────────┴───────────────────┐
 base – pentose – phosphoric acid
 └──────┬───────┘
     nucleoside
```

The end product of the pyrimidine nucleotide is uric acid, a 2, 6, 8-trihydroxypurine, which is normally precipitated out from urea*.

The molecular size of the polynucleotide molecule cannot be given with any degree of certainty, because isolating it can easily lead to decomposition. However molecular weights of several millions seem likely to be the rule (calves' thymus desoxyribonucleic acid DNA 8.10^6). In the Coliphage T_2 (virus) 45.10^6 was found. Linkage between the individual mononucleotides is by the formation

†Inosinic acid is formed through oxidative deamination of muscle adenylic acid.

*The formation of uric acid from the pyrimidine bases present in the nucleotides occurs through progressive deamination and oxidation in a series of steps.

of esters. Here the phosphoric acid on the 3 and 5 OH-groups of the pentose form the central point of linkage. This can be observed in **yeast nucleic** acid, a tetranucleotide.

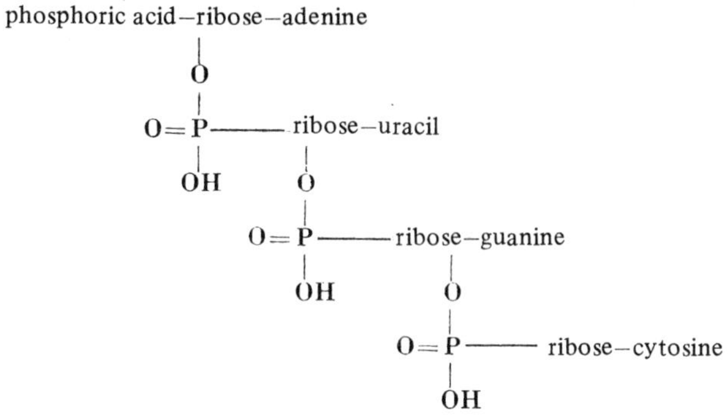

```
phosphoric acid–ribose–adenine
                 |
                 O
                 |
             O=P———ribose–uracil
                 |          |
                OH          O
                            |
                        O=P———ribose–guanine
                            |          |
                           OH          O
                                       |
                                   O=P——— ribose–cytosine
                                       |
                                      OH
```

The physiological importance of nucleic acids (polynucleotides) in the wider sense, should be emphasised, because all the changes, which can be recognised morphologically in the cell nucleus of animals and plants (division of the nucleus, behaviour of chromosomes etc.) take place essentially in the nucleoproteins. Desoxyribonucleic acids and ribonucleic acids play a decisive role as carriers of genetic information and in the synthesis of species specific proteins and enzymes.

The prosthetic (active) groups of certain enzymes (see p. 238) are also built up in a similar way to nucleotides. The active groups of the 'yellow respiratory ferment', lactoflavin phosphoric acid, a nucleotide whose synthesis has been carried out using an iso-alloxazine derivative as base, a carbohydrate (ribose) and phosphoric acid.*

The **codehydrases** (see p. 238) are important dinucleotides. Codehydrase I (co-enzyme I, diphosphopyridine nucleotide DPN) is identical with the cozymase of alcoholic fermentation. The exact chemical description of codehydrase I is di-phospho-pyridine-adenine-dinucleotide (see p. 239).

Glycoproteins. The glycoproteins (mucoids, mucoproteins) are widely distributed in nature. The carbohydrate components are co-valently linked to the protein and can make up 4–70% of the whole molecule according to the source of the glycoprotein. Glycoproteins are found in skin, in cartilage, in bones, in connective tissue, in eggs, blood, urine, saliva, gastric juice, cerebrospinal fluid, eyes and synovial fluid and in pathological cysts. The prosthetic group contains principally N-acetyl-hexosamines (glucosamine, galactosamine) as well as other sugars, hexuronic acids and sialic acid (see p. 57).

* Lactoflavin phosphoric acid is the phosphoric acid ester of Vitamin B_2 (see p. 212).

As the name mucoid (mucos = slime) indicates, many of the glycoproteins are mucous substances, the mucous membranes of the mouth, nose, throat, stomach etc., which have a protective action against chemical, mechanical and infectious attack. So far little is known about the structure of glycoproteins. The submaxillary mucus of the sheep has, as prosthetic group, a disaccharide from N-acetyl galactosamine and sialic acid, linked via the reducing group of galactosamine with the second carboxyl group of a glutamic acid residue.

Chromoproteins. Chromoproteins are compounds of proteins with low molecular weight prosthetic groups similar to pigments, which nearly always but not necessarily contain a metal (Fe, Mg).

The pigment component may be a porphyrin compound* (haem, chlorophyll), but carotenoids, which are fat-soluble pigments (see p. 122) can also function as prosthetic groups. The protein part is a simple globulin.

The most important and widely distributed chromoproteins containing porphyrin pigments are: haemoglobin, (respiratory pigment), myoglobin, (muscle pigment) (see p. 57), a few oxidation enzymes with a haem-like active group such as Warburg's respiratory ferment (cytochromoxidase), cytochromes, catalase and peroxidase.†

The chromoprotein chloroplastin, which is responsible for assimilation in plants, belongs to this group and contains chlorophyll as a prosthetic group. Chlorophyll, in contrast to porphyrin pigments of haemoglobin and haem in enzymes which contain iron in a complex bonding, has magnesium as its central atom.

In the shell of crustaceans, e.g. in lobster and crab shells, there is a brown-black chromoprotein whose prosthetic group astaxanthine is a red polyene pigment (carotenoid). The red colour of **astaxanthine** only appears when the lobster is cooked when the protein components of the chromoproteins are denatured.

Symplexes

The proteins of the protoplasm of animal and plant cells, as well as the biological fluids (e.g. blood) are frequently combined with other cell building blocks in the form of complex entities to form symplexes or cenapses. Symplexes between protein and lipids, so-called **lipoproteins** have been isolated, but there are also compounds between proteins and carbohydrates **(glycoproteins)** which are as widespread as compound lipids and carbohydrates **(liposaccharides)**.

The existence of such protein types is shown by the well known fact that, with lipophilic solvents (ether, light petroleum) there is incomplete extraction of the fat from animal and vegetable material (for example, from serum lipids, fresh

*Porphyrin pigments have porphyrin as the basic framework.
†For more about these oxidising enzymes see p. 252.

milk, milk products, bakery goods, yeast, chocolate, liquid egg, starch). Other examples are the partial extraction of sucrose in the determination of fat in chocolate, and the presence of carbohydrate-containing phosphatides in fat extracts. These few examples show that chemical and physical interactions occur between the individual components leading to the formation of complexes (symplexes) such as lipoproteins, liposaccharides and glycoproteins. The reciprocal interaction between active compounds (e.g. vitamins) and proteins with the formation of certain enzymes should also be noted, as for example the carboxylase in alcoholic fermentation, which has vitamin B_1-pyrophosphate, as its co-enzyme (see p. 239). In addition there are indications that the formation of liposaccharides is aided by the presence of proteins, so that in this case, all three main groups of nutrients take part in symplex formation.

Relatively little is known about the nature of the forces which link the proteins with the lipids, the carbohydrates or the active compounds. Covalent bonds, i.e. homopolar valency bonds, are not common. Generally the bond is similar to that in salt formation, i.e. of an electrovalent nature (ion formation; Coulomb forces); however molecular forces (intermolecular hydrogen bonds, adsorption forces, di-polar forces, van der Waals forces) act as binding forces between the individual basic components of the symplexes and so bring about the most varied degrees of bonding strength in the arrangement of the various parts of the symplex. As an example we can take the bonding of carbohydrates to the polar guanidino groups of arginine and to the hydroxyl groups of tyrosine as well as the bonding of fats, fatty acids, phosphatides, sterols, waxes, carotenoids and other lipochromes to the aliphatic side chains of leucine in the formation of a lipoprotein.

Compounds such as alcohol, sodium chloride, urea, invert soaps, dimethyl formaldehyde, acids and caustic alkalis, can break the bonds between the symplex partners, so that in many cases the components of the symplex can be determined. However, the protein components of the protein symplexes are often irretrievably denatured.

It is understandable that the symplex, because it is a system consisting of various components, exhibits clear differences in its chemical, physical and biological behaviour compared with its individual components. For instance a water soluble lipid (fat, fatty acid, carotenoid, sterol etc.) becomes water soluble when the corresponding lipoprotein is formed and therefore can be transported through the cell wall of the organism. The larger part of the serum proteins are present in the form of lipoprotein molecules.

The nature, peculiarities and behaviour of the symplexes, as biocolloidal systems containing a number of different materials, have to be considered in the technology and procedures for isolating processing and preserving foodstuffs. These include extraction, concentration, condensing, drying, heating, pasteurising, sterilising, fermenting, chemical preservation, irradiation in the analysis of foodstuffs. This also applies, for instance, in the determination of fats in protein and sugar containing foods, and in the breakdown or separation of the lipoprotein and liposaccharide symplexes by means of acids, caustics and invert sugar. Compare also the production of bakery goods, egg powder, milk powder, chocolate, alimentary pastes, fruit juices etc.

Behaviour of proteins during food processing

As, in general, the native protein from eggs, meat and cereals (with the exception of milk), is not eaten raw, man has for a long time tried to use artificial means to make the most of and increase the acceptability and enjoyment of his food. This has been done by mechanical reduction (grinding of seeds into flour), by heating, (boiling, baking, frying, roasting, browning) and by chemical methods (salting, acidifying, pickling, smoking etc.) and sometimes also by using the enzymatic action of microbes (sour milk, yogurt, cheese). One of the aims of the preparation of food at home is to make the protein more readily accessible to the digestive enzymes by loosening the protein structure, that is by converting it into a more easily digestible form. Meat protein is in any case only pleasant and easily digestible when it has been subjected to a ripening process by hanging during which some (enzymatic) decomposition of the natural meat protein takes place.

For example an **increase in the nutritive value** of soya protein, also occurs when the toxic protein in the raw soya beans, the so-called trypsin inhibitor, is destroyed (p. 263). Then the digestive enzyme trypsin can digest the protein without any interference.

Changes which increases the nutritive value must be set against those processes, not fully understood, where the protein is damaged and its biological value reduced. Although cooking in the home as, far as is known, produces practically no reduction in value, changes which lower the nutritive value have been observed during many technological processes in production, processing, storage and preservation of proteins.

Recent research has shown that damage can result from long storage, excessive heating, from the effect of chemical agents as in the bleaching of flour and from enzymatic action. For example, in soya protein on storage, the sulphur containing amino acids (methionine, cystine) are altered and the arginine, histidine and lysine reduced. Milk and egg proteins are especially sensitive; frequently a considerable decrease in the essential amino acid content is found when dried milk and dried egg are stored for a long time. When foodstuffs which contain both protein and carbohydrates are heated, amino acids of the protein component (especially lysine, but also arginine, tryptophan, histidine, methionine, cystine, asparagine and glutamic acid) can be bonded to the carbohydrate or changed into humin-like materials in the so-called Maillard reaction (see p. 139). Then they will no longer be fully digested because of reduced absorption from the gut. Sterilisation can also cause a loss of amino acids in meat. Chemicals may affect proteins (e.g. the damage to gluten when flour is bleached with nitrogen trichloride). In pickling, the added nitrite salts affect mainly the amino acid lysine. Current knowledge can increase the number of examples. In any case it is clear that these processes affect the nutritive value of protein and result in reduction of its biological value and its nutritional effects.

There are obvious problems here for food chemistry and technology from the nutritional point of view. It is necessary to protect and conserve the protein present in the various foodstuffs during technological processing: for instance by appropriate handling of foodstuffs during sterilisation, evaporation, drying and storage, avoiding heating for too long or at too high a temperature in sterilisation and in drying, by keeping to fixed rates of drying. Sometimes the exclusion of air, as in vacuum drying and storage of foodstuffs rich in protein (flour, dried milk, dried egg) is desirable, as well as control of the moisture in the storage area and the relative humidity during packaging.

Reactions for the presence and determination of proteins

We must distinguish in principle between those reactions of which involve proteins as macromolecules, and those in which particular amino acids react in their capacity as protein building blocks in the sidechains (R) of the protein molecule.

Flocculation processes (coagulation) and many precipitation reactions belong to the first group. As amino and carboxyl groups are always present in the end groups of the peptide chains, and frequently also in side chains, each protein gives characteristic reactions for these groups: salt formation with heavy metals, binding of acids and alkalis without change in the pH (buffer action). The so-called Biuret reaction (p. 68) is specific for the peptide bond. Serological tests also involve the whole protein complex.

A whole series of reactions are not specific to protein itself, but rather to individual amino acids in the whole bundle of protein macromolecules. These are the mainly well known colour reactions (see p. 69). In tests for protein groups characteristic of amino acids, these groups only react when they are not structurally fixed inside the peptide chain and chain bundles.

Proteins can only give reactions for those amino acids present. For instance gelatine (gelatinous substance) does not give the tryptophan reaction because tryptophan is lacking, as shown by ADAMKIEWICZ (p. 69). The xanthoprotein reaction characteristic of tryptophan and tyrosine (p. 69) is very weak in gelatine, showing a lack of tyrosine as well as the absence of tryptophan.

Precipitation reactions

Many reagents precipitate protein from water solutions. They are usually divided into two groups, according to whether or not the precipitation effect is accompanied by an irreversible denaturation.

Precipitation reactions accompanied by denaturation. The following precipitating reagents give precipitates from dissolved protein which cannot be redissolved i.e.

are irreversible. Their use allows the partial or total separation of proteins from solutions containing protein, even in the presence of non-protein substances.

Acid precipitation: Mineral acids, such a hydrochloric acid, sulphuric acid, nitric acid (Heller's Ring test) as well as acetic acid (coagulation test), precipitate proteins from water solutions; when concentrated acids are used, there is complete denaturation. The precipitation is often accelerated by heating. An excess of acid may cause the resulting precipitant to remain in solution with the formation of acid albuminates which precipitate on neutralisation. On the other hand, when alkalis are used, soluble alkali albuminates are formed when protein solutions are treated with alkalis. This is accompanied, as with acid albuminates, by the formation of salts, which precipitate out when the neutral point is reached.

More sensitive protein precipitation can be achieved with sulphosalicylic acid, metaphosphoric acid (orthophosphoric acid does not cause precipitation) trichloracetic acid, phosphotungstic acid, phosphomolybdic acid, phenol picric acid, etc.

Heavy metal salts precipitation: The following salts are commonly used, mercury ($HgCl_2$), lead (especially lead acetate), copper, silver etc. However, some of these protein precipitates easily dissolve in an excess of the precipitant. TANRET's reagent consists of an acetic acid solution of potassium mercuric iodide and precipitates all proteins as well as alkaloids. But these latter are again dissolved by heating (peptones also to some extent) whereas the protein precipitate remains insoluble.

Precipitation reactions not involving denaturation. Protein is reversibly precipitated from water solutions by alcohol and acetone or by neutral salts such as ammonium sulphate, sodium sulphate and mixtures of alkaline phosphates. Salt precipitation is strongly dependent on the pH and the salt concentration. Usually the protein is precipitated near the pH of the isoelectric point (see p. 29); at this hydrogen ion concentration solubility in water and salt solutions is at its minimum. By varying the salt concentration, fractional precipitation of certain proteins can be carried out, e.g. albumins and globulins. The use of $(NH_4)SO_4$ is preferred in enzyme chemistry for the precipitations, separation and purification of enzymes.

Colour reactions

Among the colour reactions of proteins, only the Biuret reaction is specific to protein; it is given by the integral structural elements of the peptide bonds (–CO–NH–) present in all proteins.

The other colour reactions refer to specific amino acids present in the protein molecule.

Biuret reaction. Proteins, from polypeptides down to tetrapeptides (and sometimes even tripeptides) produce a characteristic blue-violet colour in a strongly alkaline solution with dilute copper sulphate solution (0.1–1.0%). This reaction is specific to peptide bonds. Dipeptides and free amino acids, with the exception of histidine, give a negative reaction. The reaction depends on the formation of protein-alkali-copper-complexes and occurs in all compounds which have two or more peptide bonds in suitable opposing position (e.g. as also in oxalamide, malonamide). The reaction is suitable for the determination of the degree of hydrolysis of proteins (e.g. in the production of seasonings).

ABDERHALDEN's ninhydrin reaction. Proteins, polypeptides (e.g. peptones, albuminoses) and also amino acids, give a blue colour (or sometimes a red-purple colour) when boiled with ninhydrin triketohydrindine hydrate. This reaction must be carried out in a neutral solution. It is unsuccessful with the amino acids, proline and hydroxyproline. It is however not very specific because it is also produced by NH_4 salts and amines.

Xanthoprotein reaction. Concentrated nitric acid produces a yellow colour with protein which turns orange with the addition of ammonia or alkali. The yellow colour is due to the formation of coloured nitro-derivatives of the aromatic amino acids phenylalanine, tyrosine and tryptophan which are present in nearly all proteins.

MILLON's* reaction. Gives (white) precipitate in a protein solution, which becomes red to dark red on heating. The colour reaction is caused by the oxidation of tyrosine: hence gelatine, collagen and certain keratins do not give this reaction as tyrosine is absent.

PAULY's diazo-reaction. A sodium hydroxide solution of the protein is mixed with a freshly prepared solution of diazotised sulphanilic acid. A red colour results from coupling with the histidine and tryosine present.

ADAMKIEWICZ's tryptophan reaction. Solutions of protein containing tryptophan mixed with glacial acetic acid and with an underlayer of concentrated sulphuric acid, give a beautiful intense violet colour. The reaction is negative for collagen, gelatine, elastin and silk fibroin because tryptophan is absent.

Lead sulphide reaction. Most proteins in strongly alkaline solution yield a black to brown colour when they are boiled in the presence of lead sulphide. The colour derives from the splitting off of H_2S (PbS formation) from the sulphur amino acid cystine.

Nitroprusside reaction. The nitroprusside reaction (blue-red colouring of cysteine with sodium nitroprusside in weak alkaline solution) depends on the presence of

*MILLON's reagent: Acidified nitrite containing a solution of mercuric(II) nitrate in water.

amino acids containing sulphur, and on the SH grouping of cysteine. The reaction is positive for cystine also if KCN or sulphite are added, because the S–S bond is broken with the formation of cysteine.

Less frequently used are the SAKAGUCHI reaction for arginine, EHRLICH's reagent (p-dimethylaminobenzaldehyde-HCl for tryptophan), KNOOP's reaction for histidine, LANG's reaction (for hydroxyproline), FOLIN's reaction (for amino acids).

Serological distinction of proteins

Serological testing, in the so-called 'precipitin reaction' is also of importance to the food chemist.

When proteins (e.g. remains of a drop of blood, milk or vegetable protein) are introduced into the bloodstream 'parenterally' (subcutaneously, intravenously or intraperitoneally), by-passing the stomach and intestinal tract, these proteins cause the body to defend itself by forming 'anti-bodies', because proteins are species specific and foreign to the body or blood. The foreign protein is somehow bonded and made harmless.

A serum – the so-called antiserum – which contains the antibodies (precipitins) in more than adequate amounts from the blood of an animal treated with a particular protein (animal or vegetable) is used. These precipitins form a deposit (precipitate) *in vitro* with the type of protein used for immunisation. A rabbit treated (immunised) with human blood produces a serum which only yields a deposit (precipitate) in the test tube with human blood: a rabbit treated with ox blood produces a serum which only gives a deposit with ox blood serum and a rabbit treated with vegetable protein, a serum which only gives a deposit with the appropriate vegetable protein.

The reaction is generally specific, so that only closely related animals, such as horse, donkey and mule, or sheep and goat react in the same way.

The precipitin reaction offers a biological method of differentiating between animal and vegetable proteins, which is not possible by purely chemical methods. Precipitating sera are therefore indispensable for forensic chemistry and for the testing of foodstuffs, e.g. for distinguishing between types of blood and meats (also for raw, dried and pickled meat; but not for cooked or heated meat because of denaturation), for testing honey, alimentary pastes, various kinds of milk. They are used also for testing margarine for dried egg yolk, for the detection of foreign nuts in marzipan etc.

Quantitative determination of proteins

The simplest quantitative method measures protein as 'nitrogenous material' or raw protein because of its almost constant nitrogen content. In KJELDAHL's method the protein containing foodstuff is decomposed by means of concentrated sulphuric acid together with a catalyst. In this oxido-reduction process, nitrogen is reduced to NH_3 and then determined by titration. The approximate protein content (expressed as raw protein) is calculated by multiplying by 6.25 the N value determined from the ammonia content.*

The factor 6.25 is commonly used in analytical practice for the calculation of raw protein, although this is somewhat inexact.* Small variations in nitrogen content of proteins are not taken into account, nor is the fact that non-protein nitrogen compounds may be present. These may include free amino acids, amides (e.g. asparagine, glutamine), betaine, biogenic amines, purine bases and alkaloids.

This method therefore determines total nitrogenous matter. Various precipitation reactions are used to obtain the actual protein content, the so-called pure protein. (Common methods: copper hydroxide method, uranium acetate method, alcohol method). For quantitative determination the Biuret reaction, the ninhydrin reaction, the UV-absorption and FOLIN's reaction can be used.

If the digestability of the protein has to be considered, which may be of importance for the nutritional characteristics of the protein-containing foodstuff, the digestible protein can be determined *in vitro* with the help of the enzyme pepsin, obtainable from calves stomachs.

To obtain a general biew of the composition of the proteins, a determination of the amino acids should be carried out. By hydrolysing the protein with hydrochloric or sulphuric acid or by enzymatic hydrolysis, the amino acids can be obtained. They can then be identified and determined by a variety of methods. A quantitative measurement of the free amino acids is also necessary for the control of technical processes, for instance the determination of the degree of hydrolysis in the preparation of seasonings from protein and other protein containing materials.

Colour reactions frequently permit a quantitative measurement of the various amino acids by colorimetric methods (e.g. MILLON's reaction, nitroprusside reaction etc.). However, sometimes only the total amount of free amino acids needs to be determined. Analytical methods which can be used here are based on the ability to react of the amino and carboxyl groups in the amino acids (c.f. SOERENSEN's formol titration; WILLSTATTER's titration; VAN SLYKE's nitrite procedure; see the chapter on amino acids). Modern methods for the determination of protein and amino acids can only be mentioned here: chromatography (chromatographic adsorption analysis, paper chromatography, partition chromatography, ion exchange chromatography, electrophoresis, isotope dilution method and microbiological methods.

*From the formula protein:nitrogen = 100:16, protein is calculated as $N \times \frac{100}{16} = N \times 6.25$.

†A more accurate figure for calculating the casein content of milk is 6.37, for gelatine 5.55 and for egg yolk protein 6.67.

CHAPTER 4

Fats and Associated Substances (Lipids)

Our nutritional fats are entirely of animal and vegetable origin. They are formed during the lifecycle of all plants and animals. Fats are nitrogen-free, organic compounds present in very variable amounts. They have many different physiological functions. Seen from the physical and mechanical respect – as heat insulators and organ protectors (c.f. kidney fat), they are a group of high caloric nutrients with a combustion value of 9.3 Kcal per g., the most important **source of energy** for the body and an **energy store**.

In addition to their function as energy reserve, fats also act biologically as building blocks in the cell, for growth and maintenance of the body structure. In this way they fulfil the most important vital functions in the dynamic physiological and chemical processes occurring in the cell. They are built into the cell structure combining with specific proteins as lipoprotein membranes. Together with their associated materials, they also make available to the body the fat-soluble vitamins A, D and E, (see Vitamins).

Included among the lipids are biogenetically closely related materials of the animal and vegetable kingdom. Like fats they are insoluble in water but soluble in one another as well as in such organic solvents as benzene, light petroleum, carbon tetrachloride, chloroform, and carbon disulphide, but are insoluble in water (although short chain acids are soluble in water).

The chemical composition of the individual lipids is not uniform. Sometimes their make up is similar to that of fats (triglycerides), in other cases it is quite different. Because of their similar solubility, lipids nearly always occur together with fats (glycerides) and are extracted along with them during solvent extractions. They are also found in the so-called 'crude fats'.

Fats are frequently included under the heading 'lipids', which should really be 'total lipids' since both fats and associated materials are included. In this wide classification, the following are among the most important components of lipid mixtures of our natural fats (solids) and oils (liquids):

triglycerides	esters of glycerol with fatty acids (usually accompanied by 0.1–0.4% of mono- and di-glycerides)

free fatty acids	
phosphoglycerides	glycerophosphates (e.g. lecithin and cephalin)
saccharoglycerides	phosphoinositides, phospherine: galactoglycerides
sphingosides	sphingomyelin), cerebrosides, gangliosides etc.
higher alcohols and their esters and ethers*	Long chain alcohols (e.g. cetylalcohol) and their esters (waxes). Glyceryl ether (e.g. glyceryl butyl ether)
sterols*	cholesterol and phytosterol
hydrocarbons*	squalene, pristane, gadusene
lipochromes*	carotene, lycopene and other carotenoids; chlorophyll
lipovitamins*	vitamin A, vitamin D_2 and D_3, vitamin E (tocopherols)
antioxidants*	tocopherols, sesamol, sesamolin and gossypol
flavour and aroma materials†	e.g. unsaturated hydrocarbons, lactones, unsaturated aldehydes, methylketones

*These accompanying substances are all found in the unsaponifiable matter
†They are also found principally in the unsaponifiable matter

Water, trace metals and fat oxidation products are present – mainly in crude fats – but also in processed fats.

The greater part of fats and oils consists of triglycerides, but they are found in conjunction with a large amount of associated material, according to their type and source. Phosphatides are the largest fraction, followed by sterols, tocopherols, lipochromes, hydrocarbons and the aroma and flavour materials. The mixtures of lipids which occur in nature fulfil a biological role. The 'fat associated materials' have an indispensable role in the total metabolism of the human body. Although compared with the quantity of the fat itself, they are often present only in small amounts, they are very reactive in biochemical metabolism. If the fat is to be a biologically essential food and not merely a carrier of calories, then these materials should be preserved during the chemical and technological processing of edible fats. Raw materials containing the fats, the intermediate products and the finished fat products should be protected as far as possible. Particular care should be taken to retain them, especially during extraction processes and storage of fats and fatty foods. (Compare also virgin oils p. 93).

Fats and oils are also important as the basis for a variety of industries which manufacture oils, varnishes, paints, lacquers, linoleum, soaps, emulsions, textile improvers, pharmaceutical products, surface active materials (e.g. emulsifiers, washing powders and wetting agents).

4.1 CLASSIFICATION OF FATS

Fats can be classified in various ways, by source, physical properties, chemical composition, use or physiological function.

The most common classification by source divides natural fats into two major groups:

1. **Vegetable fats**, characterised by the presence of phytosterols. Fruit and seed fats can be distinguished from each other.
2. **Animal fats**, characterised by their zoosterol content, especially cholesterol. The animal fats can be divided into land and marine types; the fats of land animals further divided into milk and depot fats; those of marine animals into the fats from mammals (whales, seals) and those from fish.

The difference between vegetable and animal fats, which depends on the sterols present is particularly important in fat analysis and in research on these two types of fat (see p. 97). Cholesterol has been found in the occasional higher plant and in red algae.

Fats obtained from microorganisms e.g. yeasts, mycelium, fungi and algae, form a third group (see p. 98). Their inclusion in human food has so far been of no practical importance because of their high unsaponifiable content and the presence of branched chain fatty acids.

The classification of fats according to their physical properties – state of aggregation and consistency – divides them into: hard fats (e.g. tallow, bone grease), semi-solid fats (e.g. lard, palm oil) and liquid fats (oils).

This classification is unsatisfactory because the liquid or solid state is only a question of the ambient temperature. Coconut fat and palm oil, for instance, are liquid in tropical temperatures, but solid in temperate ones. It would be better to discontinue this distinction between fats and oils, and only speak of fats. The crude classification by consistency does express differences in composition, because the liquid or semi-solid character of a fat reflects a relatively large content of either unsaturated or short chain saturated fatty acids.

Liquid fats (oils) can be further subdivided industrially according to their 'drying ability', into non-drying, semi-drying and drying oils.

The chemical classification of fats and their accompanying materials is based on their 'lipid' structure. In English and American literature they are classified as follows:

1. Simple lipids
   (a) Neutral fats
   (b) Waxes (true waxes, cholesterol esters, vitamin A esters, vitamin D esters

2. Compound lipids
   (a) Phospholipids or phosphatides (glycerophosphatides, sphingomyelins, phosphatidic acids
   (b) Cerebrosides (saccharolipids)
   (c) Sulpholipids

3. Derived lipids
   (a) Fatty acids
   (b) Alcohols (straight chain alcohols, sterols, vitamin A)
   (c) Hydrocarbons (aliphatic hydrocarbons, carotenoids, squalenes)
   (d) Vitamin D, E and K.

In the German language literature, the classification is as follows:

1. Simple lipids: glycerides, waxes
2. Phosphorus and nitrogen containing lipids
3. Combined lipids
4. Associated substances.

4.2 BUILDING BLOCKS OF FATS (NATURAL FATTY ACIDS)

Chemically fats are esters of the trivalent alcohol glycerol with fatty acids (see p. 86). Because the properties and the behaviour of fats are mainly dependent on the structure and configuration of the fatty acids, these are considered according to their occurrence and importance:

Saturated fatty acids

The saturated fatty acids included here have an unbranched C chain. Branched chain fatty acids are only rarely found in vegetable and animal fats used as food e.g. in mutton and butter fat in small amounts (1%), see p. 83. They have been found in ruminant body fat and are derived from the metabolism of the rumen microorganisms.

Palmitic acid (C_{16}) and stearic acid (C_{18}) are preponderant among the saturated fatty acids, but all the other saturated even numbered fatty acids from C_4 to C_{20} are found in natural glycerides. Certain milk fats are notable as there is no gap in the series of even numbered saturated fatty acids between C_4 and C_{20}.*

Saturated fatty acids with less than ten carbon atoms are liquid at room temperature, the longer chain ones are solid. The shorter the fatty acid chain (with the exception of acetic acid) the lower the melting point.

The solubility of saturated fatty acids increases with increase in the hydrophobic character of the solvent and with temperature. In addition solubility of hydrophobic solvents increases with the increase in the length of the fatty acid chain, whereas it decreases in hydrophilic solvents. The solubility of the even numbered fatty acids is less than that of the next highest odd numbered ones.

At room temperature in vitro, saturated fatty acids are somewhat inactive and are resistant to oxidation, although they are broken down and oxidised with little difficulty during metabolism. At higher temperatures the stability to oxidation decreases.

Details are given here of the most important fatty acids in edible fats*:

*The composition of the milk fat of ruminants with its unbroken series of fatty acids from C_4 to C_{20} gives a useful indication of the synthesis of milk fat from low molecular source material (C_2 fragments compare 'activated acetic acid').

*The figures denote % of the individual fatty acids to total fatty acids (100%)

Table 2 Saturated fatty acids

trivial name	formula	molecular weight	melting point °C	acid value	source
acetic acid	$C_2H_4O_2$	60.03	18	934	in seed fats of Rosaceae
butyric acid	$C_4H_8O_2$	88.10	−8	636.8	2.5–4.5% in butter from cows
caproic acid	$C_6H_{12}O_2$	116.16	−3.4	483.0	1-2% in butter from cows, traces in palm kernel oil
caprylic acid	$C_8H_{16}O_2$	144.21	16.3	389.0	1-2% in butter from cows 6-8% in cocoa butter
capric acid	$C_{10}H_{20}O_2$	172.26	31.3	325.7	in the milk fat of mammals and the seed fat of palms
lauric acid	$C_{12}H_{24}O_2$	200.31	43.5	280.1	in seed fats of Lauraceae, palms, and butter from cows
myristic acid	$C_{14}H_{28}O_2$	228.26	54.4	245.7	in many animal and vegetable fats, in milk fats, seed fats of palms
palmitic acid	$C_{16}H_{32}O_2$	256.42	62.9	218.8	in almost all natural fats
stearic acid	$C_{18}H_{36}O_2$	284.47	69.9	197.2	in body fat of terrestrial animals, in milk fats, in tropical seed fats
arachidic acid	$C_{20}H_{40}O_2$	312.52	75.4	179.5	traces widely distributed, in ground nut oil about 3%
behenic acid	$C_{22}H_{44}O_2$	340.57	79.6	164.7	in ground nut and rape seed oil, up to 2%
lignoceric acid	$C_{24}H_{48}O_2$	368.62	84.2	152.2	in ground nut and rape seed oil, under 3%
cerotic acid	$C_{26}H_{52}O_2$	396.68	87.7	141.4	traces in vegetable fats

n-Butyric acid. This acid occurs in small amounts (2-4%) in the milk fats of various animals. The unpleasant smell of rancid butter depends to a large extent on any free butyric acid present, derived from the hydrolysis of the glycerides. Even small amounts (traces) of the free acid render butter and foods containing butter unpalatable.

n-Caproic acid is present in milk fats in amounts up to 2%, in coconut fats (<1%) and in various palm kernel oils (<1.5%). Oxidative rancidity caused by microorganisms in butter fats gives it a scent due to ketone rancidity. Caproic acid is converted by microorganisms to n-propylketone which has an unpleasant smell. Caproic acid itself has a characteristic smell of the male goat.

n-Caprylic acid is found in milk fats in amounts (1-2%), in coconut oil (6-8%), in palm kernel oil (3-4%) and in some other oils.

n-Capric acid is mainly found associated with caprylic acid in milk fats, coconut oil (7-10%) and other palm kernel oils also contain this acid in the glycerides.

Lauric acid is one of the three most widely distributed saturated fatty acid to be found in nature. Practically all seed fats of the Lauraceae contain lauric acid (c.f. the name). A few of these contain up to 80% of total fatty acid as lauric acid. Palm kernel oil also contains large amounts of lauric acid (45-50%), partly in the form of trilaurin. In butter from cows, it is present up to 4–8% and is also found in the milk fat from other mammals.

Myristic acid is a component of almost all vegetable and animal fats, though it is usually only found in small amounts (1–5%). Milk fats contain 8–12%, palm kernel fats have up to 20% of total fatty acids in the form of myristic acid. This acid predominates in practically all the fats of the Myrtaceae (nutmeg plants) and amounts to 60–75% of the total fatty acids.

Palmitic acid is the characteristic saturated fatty acid of our natural fats. Although it is widely distributed it is not present in large amounts. In the usual culinary oils it is usually about 10% and in olive oil 15%. Palm oil 35–40%, Stilingia chinese tallow (mou tesu) 60–70%, shea butter (Bambok butter, Soudan, from *Buturospernum Parkii)* contains stearic acid 30–40%, oleic acid 49–50%, linoleic acid 4–5% and palmitic aicd 5–9%.

Stearic acid is the highest molecular weight saturated acid found in large amounts in natural fats and oils. Milk fats contain 5–15%, lard 10–12%, beef tallow 14–30%, cocoa butter and shea butter 30–35%, maize (2–4%), soya (2–6%), palm kernel (1–3%). Stearic acid is a characteristic component of commercial hydrogenated fats. Completely hydrogenated maize and soya oils may even contain up to 90% of stearic acid.

Arachic acid is widely distributed in nature, but in general is only found in small amounts. Ground nut oil (Arachis hypogaea) contains 5–7% C_{20} to C_{24} saturated fatty acids, of which about one third is arachic acid. It has also been found in the depot fat of some animals, in milk fats and in human milk fat.

Behenic acid is not found in commercial oils, except the Cruciferae seed oils. Among the Cruciferae oils, mustard oil contains 2–2.3%, radish seed oil 3.4% and rape seed oil 0.6–2.1% behenic acid. Behenic acid is also a component of wool fat. Surprisingly enough there are large amounts in hydrogenated fish and some vegetable oils, e.g. 7% in herring oil, 13% in cod and liver oil, 18% in sardine oil and 57% in rape seed oil.

Lignoceric acid is present in small amounts (under 3%) in ground nut and rape seed oil, whereas only traces of cerotic acid are found in vegetable fats.

Unsaturated fatty acids

Unsaturated fatty acids are characterised by having one to six reactive double bonds in the molecule. The most important ones found in fats are listed here.

palmitoleic acid	$C_6H_{13}CH{=}CH(CH_2)_7CO_2H$
oleic acid	$C_8H_{17}CH{=}CH(CH_2)_7CO_2H$
ricinoleic acid	$C_6H_{13}CH(OH)CH_2CH{=}CH(CH_2)_7CO_2H$
linoleic acid	$C_5H_{11}CH{=}CHCH_2CH{=}CH(CH_2)_7CO_2H$
linolenic acid	$C_2H_5CH{=}CHCH_2CH{=}CHCH_2CH{=}CH(CH_2)_7CO_2H$
arachidonic acid	$C_5H_{11}CH{=}CHCH_2CH{=}CHCH_2CH{=}CHCH_2CH{=}CH(CH_2)_3CO_2H$
erucic acid	$C_8H_{17}CH{=}CH(CH_2)_{11}CO_2H$

In natural fats these acids – with a few exceptions – are only found in the cis form. The all *cis* linoleic acid (*cis, cis*-Δ-9.10-12.13-linoleic acid) is particularly important because, as a so-called **essential** fatty acid it shows high biological activity.

In general the essential fatty acids can be distinguished from other unsaturated fatty acids by the number 2 to 4 and the position of their double bonds. The presence of double bonds in the Δ-6.7- and Δ-9.10 position, numbered from the methyl group at the end of the chain (w nomenclature) is the most important factor in determining their function as essential fatty acids.

Unsaturated fatty acids which occur in natural food oils and fats are generally found as isolenic acids with an interrupted methylene group between double bonds*. Traces may also be found in the conjugated acids, which can be formed in small amounts during the hydrogenation of fats (see p. 104). All unsaturated fatty acids show a strong tendency to oxidation, polymerisation and addition. They are therefore particularly prone to changes during technical processing and in the storage of foods. The term oxidative spoilage of fats is used when unsaturated fatty acids and their glycerides are attacked by the oxygen from air. (Compare p. 114–118).

Hydrogen and halogens can be added to unsaturated linkages (double bonds) to form saturated fatty acids (compare hardening of fats). This additional ability can be used analytically as the hydrogenation value, iodine number, thiocyanogen number or diene number. These numbers are characteristic both for unsaturated fatty acids and for the fats themselves.

Fatty acids with one double bond; monoenoic unsaturated acids

$$CH_3(CH_2)_nCH{=}CH{-}(CH_2)_nCOOH$$

The lower members (C_6–C_{14}) occur in very small mounts in the glycerides of various food fats; Δ-9.10-decenoic acid (caproleic acid) is only found in very small amounts in butter fat; Δ-9.10-tetradecenoic acid is found mainly in fish and marine mammal oils. The Δ-9.10-hexadecenoic acid (palmitoleic acid) is found in many fish oils and in numerous fats of warm-blooded animals (sardines 12%, whale fat blubber 14%, lard 3%, butter fat 4%).

The most widely distributed of these acids is oleic acid, which is liquid at room temperature (cis-Δ-9.10-octadecenoic acid) $CH_3{-}(CH_2)_7{-}CH{=}CH{-}(CH_2)_7{-}COOH$. It is the main component of vegetable oils and is found in all food fats.

*These acids have their double bonds in an isolated position: $CH{=}CH{-}CH_2{-}CH{=}CH{-}$; conjugated acids on the other hand have conjugated double bonds: $-CH{=}CH{-}CH{=}CH{-}$. Correspondong acetylenic acids are known. These include tariric acid (6-octadecynic acid, $C_{18}H_{32}O_2$) and isanic acid (with two acetylenic bonds) present in the isano and boleco oil, used for technical purposes. Analytically they can easily be distinguished by means of UV, IR or mass spectroscopy.

Table 3 Fatty acids with one double bond*

Trivial name	Systematic name	Molecular formula	Molecular weight	Acid value	Iodine number	Source
caproleic acid	Δ-9.10-decenoic acid	$C_{10}H_{18}O_2$	170.2	330	149	Milk fat
myristoleic acid	Δ-9.10-tetra decenoic acid	$C_{14}H_{26}O_2$	226.4	248	112	fish and sperm whale oils
palmitoleic acid	Δ-9.10-hexadecenoic acid	$C_{16}H_{30}O_2$	254.4	221	100	fish and whale oils, milk fat
petroselenic acid	Δ-6.7-octadecenoic acid	$C_{18}H_{34}O_2$	282.5	199	90	seed fats of Umbelliferae and Araliaciae
oleic acid	Δ-9.10-octadecenoic acid	$C_{18}H_{34}O_2$	282.5	199	90	most widely distributed fatty acid, principal fatty acid of many plant and animal fats
vaccenic acid	Δ-11.12-poctadecenoic acid	$C_{18}H_{34}O_2$	282.5	199	90	milk fat, animal body fat
	Δ-11.12-eicosenoic acid	$C_{20}H_{38}O_2$	310.5	181	82	Euphorbiacae
erucic acid	Δ-13.14-docosenoic acid	$C_{22}H_{42}O_2$	338.6	166	75	seed fats of Cruciferae (rape)
selcholeic acid	Δ-15.16-tetracosenoic acid	$C_{24}H_{46}O_3$	366.5	153	69	fish oils
ximenic acid	Δ-17.18-hexacosenoic acid	$C_{26}H_{50}O_2$	394.7	142	64	seed fats of varieties of Ximenia

*The symbol Δ denotes a double bond

The chemical properties of **oleic acid** correspond, as might be expected, to those of an unsaturated fatty acid. Reduction yields stearic acid, bromination gives threo-9.10-di-bromo stearic acid, whereas oxidation with diluted alkaline permanganate solution gives erythro-9.10-dihydroxy stearic acid.

Various naturally occurring isomers of oleic acid have been found. They can be classified according to the position of the double bond in the molecule: petroselinic acid (*cis*-Δ-6.7-octadecenoic acid) and vaccenic acid (*trans*-Δ-11.12-octadecenoic acid).

Vaccenic acid, surprisingly enough identified as a *trans* fatty acid, has been found in butter fat from 1-5%, in ovine fat and beef fat from 1-2%, and in lard around 0.2%. Its presence in the depot fat of animals is thought to be due to the feed, and in ruminants in particular to the activity of the bacteria in the rumen. During the hardening of fat by hydrogenation (see p. 104) oleic acid undergoes changes at the double bond the extent of which depends on the degree of hydrogenation. These can be seen in a spatial *cis-trans*-configuration change and also in a change in the position of the double bond. These stereo and positional isomers of octadecenoic (C_{18}) acids are grouped as '**isooleic acids**'. The development of the geometrical isomer form, that is the change of configuration to the *trans* form, is called elaidinisation. Elaidinic acid itself is formed from oleic acid. The *trans* configuration is more stable and therefore has a higher melting point (45°C). Elaidinisation of oleic acid itself or in the glyceride form occurs not only during hardening of fats with hydrogen, but also in the course of oxidation and in the process of thermal polymerisation.

Infrared spectroscopy has shown the presence of elaidinic acid and vaccenic acid in amounts from 5-10% in freshly rendered beef tallow, in oleostearin (pressed tallow) and in technical olein.

Erucic acid, *cis*-13.14-docosenoic acid is the main fatty acid in the seed oils from the Cruciferae (rape, mustard seed, water cress etc.). The *trans* form of erucic acid is brassidic acid. Rape oil contains 40-50% erucic acid, fatty oil from mustard seed about 40%, water cress 82%. The latter is therefore particularly suitable for obtaining pure erucic acid.

Nervonic acid, Δ-15.16-tetracosenoic acid ($C_{24}H_{46}O_2$) is found in brain cerebrosides.

Fatty acids with two or more double bonds; dienoic, trienoic and polyenoic acids
Denoic acids

$$C_nH_{2n-4}O_2 \text{ or } C_nH_{2n-3}COOH$$

The naturally occurring fatty acids possessing two double bonds almost always have a methylene group between the double bonds.

Linoleic acid, (Δ-9.10-12.13-octadecadienoic acid)

$$CH_3{-}(CH_2)_4{-}CH{=}CH{-}CH_2{-}CH{=}CH{-}(CH_2)_7COOH$$

is the important acid in this series. It is found in semi-drying and drying oils in considerable amounts and usually in much smaller amounts in semi solid and solid animal fats.

A structural conjugated isomer, Δ 9.10-11.12 of linoleic acid is found in grain germ oils in amounts up to 1-2%.

Linoleic acid (formerly called vitamin F) belongs, with linolenic acid and arachidonic acid, to the 'essential' fatty acids; they are indispensable and their absence leads to skin diseases in animal and in man. They are also involved as precursors for the prostaglandins.

Table 4 Unsaturated fatty acids with two or more double bonds.

Trivial name	Systematic name	Molecular formula	Molecular weight	Acid value	Iodine number	Source
linoleic acid	Δ-9.10–12.13-octadecadienoic acid	$C_{18}H_{32}O_2$	280.4	200	181	widespread in vegetable fats
linolenic acid	Δ-9.10-12.13-15.16-octadecatrienoic acid	$C_{18}H_{30}O_2$	278.4	202	274	linseed oil, soya oil etc.
elaeostearic acid	Δ-9.10-11.12-13.14-octadecatrienoic acid	$C_{18}H_{30}O_2$	278.4	202	274	Chinese heating oil
parinaric acid	Δ-9.10-11.12-13.14-15.16-octadecatetrenoic acid	$C_{18}H_{28}O_2$	276.4	203	367	seed fats of *Parinarium laurinum*
arachidonic acid	Δ-5.6-8.9-10.12-14.15-eicosatetrenoic acid	$C_{20}H_{32}O_2$	304.5	184	334	animal body fat, phosphatides
clupanodonic acid	Δ-4.5-8.9-12.13-15.16-19.20-docosapentenoic acid	$C_{22}H_{34}O_2$	330.5	170	384	marine animal oils
nisinic acid	Δ-4.5-8.9-12.13-15.16-18.19-21.22-tetracosahexenoic acid	$C_{24}H_{36}O_2$	256.5	157	427	Japanese sardine oil

Table 5 Average content of *cis, cis* linoleic acid (% of total fatty acids) in the most important food oils and fats

Name of fat	% linoleic acid	Name of fat	% linoleic acid
Safflower oil	70	Pork lard	1–10
Soya bean oil	60	Sheep tallow	5
Sunflower oil	50–70	Whale oil	0–6
Poppy seed oil	60	Herring oil	2-5
Cottonseed oil	45	Coconut oil	1-3
Sesame oil	50	Beef tallow	2
Maize oil	55	Butter	0.5-3
Ground nut oil	20	Palm kernel oil	0.5-2.5
Palm oil	5–10		
Olive oil	10		

Trienoic acids

$$C_nH_{2n-6}O_2 \text{ or } C_nH_{2n-5}COOH$$

Fatty acids with three double bonds are included in this series, and linolenic acid is the most important (Δ-9.10-12.13-15.16-actadecatrienoic acid),

$$CH_3–CH_2–CH=CH–CH_2–CH=CHCH_2–CH=CH–(CH_2)_7–COOH$$

Linolenic acid occurs in amounts in the glycerides of linseed oil (20–40%), but is also present in other oils (soya oil, Hemp oil, walnut oil).

Elaeostearic acid, a conjugated acid, is present in the fatty acids of wood oil or Tung oil at levels of 70–80%, but is not suitable for consumption because of unacceptable impurities, but is often used in the paint industry. It is described in the literature as d-elaeostearic acid, and the trans form as β-elaeostearic acid. Neither are used in food.

Polynoic acids. Arichidonic acid, an unsaturated C_{20} acid with four double bonds in the methylene interrupted arrangement, is found in fish oils as well as in the fat of terrestial animals and birds. An essential fatty acid, it is one of the mother substances for prostaglandin and is therefore of physiological importance.

Parinaric acid, has four conjugated double bonds. It is found (30–50%) in the seed oils of Balsaminaceae as an acetodiparinin.

In addition to **clupanodonic acid** ($C_{22}H_{34}O_2$), a methylene interrupted acid with five double bonds, other highly unsaturated acids have recently been found in whale and fish oils, e.g. nisinic acid ($C_{24}H_{36}O_2$) as in Japanese sardine oil.

Branched chain fatty acids

Although nearly all fatty acids found in nature are straight chain compounds, (see p. 76) some branched chain fatty acids do occur. They were first found in the cell wall fat of the tubercle bacillus and in the wool fat of sheep. They are also found in large amounts in the synthetic fatty acid mixtures from the Fischer-Tropsch-Gatsch oxidation of paraffins. The type and position of the branching determines the toxic action of the branched fatty acids. Because of this toxicity synthetic fats from technical grade fatty acids (paraffin oxidation) are unsuitable for use as food.

Isovaleric acid, which has long been known from dephinium oil and in valerian roots, is a branched fatty acid. Recently very small amounts of branched fatty acids have also been found in sheep tallow, in butter fat (milk fats) and in the fat from the uropygeal gland of the duck.

In some natural fatty acids ring branching has also been established, e.g. a cyclo-propane ring in the acid from *Lactobacillae,* a cyclo-propane ring in sterculic acid, a cyclo-pentene ring in chaulmoogric and hydnocarpic acid.

Sterculic acid, with a cyclo-propane ring, is found in small amounts in cotton seed oil. From its special chemical constitution, it has unusual properties, allowing it to polymerise rapidly at room temperatures and even to react by forming a gel at 250°C.

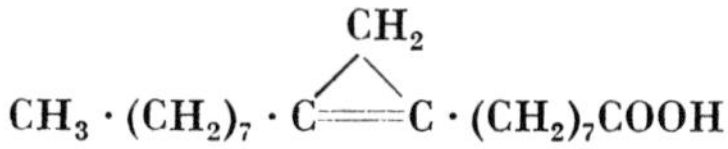

$$CH_3 \cdot (CH_2)_7 \cdot \overset{CH_2}{C{=}C} \cdot (CH_2)_7COOH$$

sterculic acid

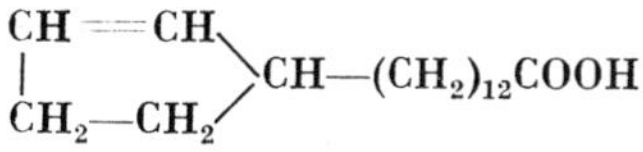

$$\begin{matrix} CH{=}CH \\ | \quad\quad \\ CH_2{-}CH_2 \end{matrix} \rangle CH{-}(CH_2)_{12}COOH$$

chaulmoogric acid

Chaulmoogric acid (13-cylcopent-2-enyl-tridecoic acid) belongs to the naturally occurring acids of the chaulmoogra group, which have 16–18 carbon atoms. Five of the carbon atoms are contained in the cyclo-pentene ring. The general formula shows that a cyclo-pentene ring can be attached to a saturated, or in the case of hydnocarpic acid to an unsaturated, side chain. Both these natural alicyclic fatty acids are optically active and have been used for hundreds of years in the treatment of leprosy and other skin diseases.

Hydroxy- and keto- fatty acids

Hydroxy- and keto- groups have been found as substituents in the fatty acids of natural fats, in addition to the methyl groups and carbon rings, which occur in branched fatty acids as substituents. Oiticica oil contains a keto- substituted fatty acid, licanic acid. The best known substituted fatty acid is ricinoleic acid with a hydroxyl group (=12–hydroxyoleic acid). Hydroxy- and keto- acids also occur in autoxidised fats (see p. 111).

Ricinoleic acid (12-hydroxy-cis-Δ-9.10-octadeneoic acid).

Up to 80–88% of the total fatty acids of castor oil consist of this acid. Castor oil is not used as food because of its pharmacological action, but is useful as a primary material for technical syntheses. Large amounts have been used pharmaceutically. Due to its ricinoleic acid content, castor oil, in contrast to the other oils, is soluble in alcohol, miscible with glacial acetic acid and is insoluble in mineral oils. Analytical proof of the presence of castor oil is easy via sebacic acid. The latter arises along with other products, from alkaline treatment of ricinoleic acid at 240°C. The chemical reaction is as follows:

$$CH_3(CH_2)_5CHOHCH_2CH{=}CH(CH_2)_7COOH \xrightarrow[\text{effect of alcohol}]{240\,^\circ C} CH_3(CH_2)_5CHOHCH_3 + COOH(CH_2)_8COOH \text{ sebacic acid}$$

Sebacic acid is also formed by dry distillation of castor oil.

Licanic acid (4-keto-Δ-9.10-11.12-13.14-octadecatrienoic acid).

Licanic acid is the only unsaturated keto- acid, which has been isolated from a natural fat. In addition to the keto- group, it has three conjugated double bonds. Oiticica oil, which can be obtained from *Licania rigida*, contains 70–78% of its fatty acids as licanic acid. The similarity of wood oil and oiticica oil is thoroughly exploited in their suitability for paints.

Polymerised fatty acids and epoxy fatty acids

When fats and oils are heated for some time at high temperatures, as can occur during roasting or frying, polymeric acids are formed in varying quantities. These arise from the polyunsaturated fatty acids by **oxidative** and **thermal poly-**

merisation depending on the amount of oxygen taken up from the air and the temperatures employed.

It is thought that **thermal polymerisation** takes place mainly between the polyunsaturated fatty acid radicals in the glycerides and occurs after previous partial conjugation by means of a DIELS–ALDER synthesis. However, the process of thermal polymerisation is by no means uniform; dimeric and polymeric products without a ring structure can also be formed by a radical mechanism.

In thermal oxidation, polymeric compounds are formed containing a variable amount of oxygen with hydroxyl groups, keto groups as ethereal oxygen.

The hydroperoxides and epoxides which are initially formed by thermal oxidation from the glycerides and fatty acids, are destroyed at temperatures above 180°C. Low molecular weight decomposition products can be present as well as dimeric compounds containing –C–C– bonds in addition to hydroxyl and carboxyl groups.

In fats and oils which have been subjected to oxidative thermal processes epoxy fatty acids are found among the large number of reaction products. They are formed by the addition of oxygen to a double bond with the formation of an oxirane ring:

$$2\text{—}\overset{\text{H}}{\overset{|}{\text{C}}}\text{=}\overset{\text{H}}{\overset{|}{\text{C}}}\text{—} + O_2 \longrightarrow 2\text{—}\overset{\text{H}}{\overset{|}{\text{C}}}\underset{\text{O}}{\diagdown\diagup}\overset{\text{H}}{\overset{|}{\text{C}}}\text{—}$$

In natural fats and oils epoxy fatty acids seldom occur, with the exception of vernolic acid, a 12.13-epoxy-Δ-9.10-octadecenoic acid from the plant *Vernonia anthelminthica.*

Epoxy acids are of particular technical importance in the manufacture of epoxy resins, which are produced by chemical means from the oxidation of unsaturated acids. These epoxy acids can be determined by their ability to react with a hydrogen halide (epoxy number).

Commercial production of fatty acids

Fatty acids, which play an important role in the soap and detergent industry, can be made either synthetically or from natural materials – i.e. fats.

The simplest method of obtaining fatty acids is by hydrolytic cleavage of the glycerides present in natural fats (see p. 90). Here the first product is a mixture of fatty acids of varying chain length and varying degrees of unsaturation.

A mixture of fatty acids is also obtained synthetically by **paraffin oxidation,** when the initial oxidation products (containing other oxidation products such as aldehydes, dicarboxylic acids and fatty acids) are reacted together.

With continuous methods of production fractional crystallisation, high vacuum distillation, chromatography, and the use of urea inclusion compounds, permits the separation of the complex mixture of fatty acids in a satisfactory technical manner. On the other hand the separation into saturated, simple unsaturated and polyunsaturated fractions by the use of polar and apolar solvents or their mixtures by low temperature crystallisation on an industrial scale is only being achieved in a limited number of factories.

4.3 CHEMICAL COMPOSITION OF FATS

Fats occurring in nature are esters mainly composed of the trivalent alcohol **glycerol** and three **fatty acid** molecules, and are therefore predominantly **triglycerides. Di-** and **monoglycerides**, in which only two or one OH group of the glycerol is esterified with fatty acids, are only found in small amounts (0.1-0.4%) in natural fats. However, mono- and diglycerides are produced industrially in large amounts and as their digestibility corresponds to that of the triglycerides, they are widely used in the food industry as emulsifiers.

Triglycerides

When, in a triglyceride, all three OH groups are esterified with the same fatty acid, they are called a single acid glyceride for instance tristearin (compare formula p. 88). In nature almost all glycerides contain a mixture of acids, esterified to a glycerol molecule. Glycerides with a mixture of fatty acids are bound to various carbon atoms of the glycerol, designated in order as 1, 2, 3 or α, β, α'. In this way, positional isomeric glycerides are formed:

(1) α	$CH_2O—CO \cdot R_1$	$CH_2O—CO \cdot R_2$	$CH_2O—CO \cdot R_1$
(2) β	$CHO—CO \cdot R_2$	$CHO—CO \cdot R_1$	$CHO—CO \cdot R_3$
(3) α'	$CH_2O—CO \cdot R_3$	$CH_2O—CO \cdot R_3$	$CH_2O—CO \cdot R_2$

The location and arrangement of the numerous, structurally different, natural fatty acids in the triglycerides determines the chemical diversity and the varying physical behaviour of natural fats.

The neutral fat glycerides, are synthesised in nature from *activated fatty acids* – the form of acyl-coA derivatives in biosynthesis from the esterification of α-glycerol phosphate. The first stage is the formation of a **diglycerol phosphate** and then a phosphatidic acid is formed enzymatically from two molecules of activated fatty acid and the diglycerol phosphate. Enzymatic dephosphorylation next yields a **diglyceride**, which is then transformed to a **neutral triglyceride** stored as depot or organ fat.

Although the distribution of fatty acids in the natural glycerides is not wholly understood, it is known that in most oils and fats it is not arbitrary and is probably determined by the cells own enzymatic system. It is clear that in animal fats the food eaten has an influence on the composition of the fatty acid mixture of the triglycerides and on the quality of the derived produce, for instance, in milk fat or lard. The influence of climate on the unsaturated fatty acid content can be observed in vegetable fats. Oils of the same kind of plant are correspondingly more saturated during colder growing conditions of the oil producing plant.

Oils and fats from different sources differ from each other considerably in the relative position of the saturated and unsaturated fatty acids in their triglycerides. In vegetable oils and fats the saturated fatty acid residues have a tendency to favour the 1-(α) position over the 2-(β) position; in animal fats the situation may be the opposite (lard). Tallow seems to take an intermediate position with both forms in a more or less arbitrary manner. This variability in the configuration of the glycerides has an influence on their digestibility and also in their melting behaviour during technical processing.

The composition of the triglycerides can be determined by splitting off the α, α'-positional (1, 3 positional) fatty acids with panarentic lipase and then saponification of the β-(2)-monoglyceride with determination of the fatty acids by gas chromatography.

Liquid triglycerides as well as mono and diglycerides have the property of crystallising in various crystalline or **polymorphic** forms. It is now thought that these forms exist for every glyceride. They are called α, β and β' (β prime) polymorphic forms. The β form is the most stable, the more unstable α and β' forms change into the stable β form with time. Rapid cooling of liquid glycerides results in the most unstable α form, whereas slow cooling gives the stable β form. These various polymorphic forms have varying melting points, the β form having the highest and the α form the lowest melting point.*

It is therefore necessary to ensure that the sample under test is in the most stable form. This can be achieved by storing the fat sample for the previous 24 hours or longer at low temperatures before the melting point is determined.

This polymorphic behaviour of the glycerides is of the greatest importance in industrial processes. It can lend flexibility to the fat product during processing and subsequent solidification, but on the other hand it can also lead to breakdown of the emulsion and crystallisation.

Nomenclature of the glycerides

The nomenclature of the glycerides is still not uniform, but the length of the fatty acid chain is used as the overriding principle in the description of glycerides, the lower fatty acids being placed before the higher acids, e.g. palmitodistearin (not distearopalmitin).

*The pure forms can be conveniently obtained by crystallising the glyceride from suitable solvents. The various polymorphic forms can be studied by X-ray crystallography. From the crystallographic and spectroscopic data, the differential packing of the molecules in the crystals can be derived.

When saturated and unsaturated fatty acids have the same chain length the saturated fatty acid comes first, e.g. palmitostearo-olein.

When unsaturated acids have the same chain length, but different degrees of unsaturation, the more saturated acid is placed first e.g. oleolinoleolinolenin.

The longest straight chain is taken as the determining one in branched chain fatty acids. In *cis-trans*-isomers the cis form is placed first; in optically active acids the L-form comes first. Substituted acids are placed last and the same applies to cyclic acids.

Mono- and diglycerides

The term mono- or diglyceride is used when only one or two of the OH groups of the glycerol are esterified with fatty acids. In natural fats they are normally only found in small amounts (0.1-0.4%).

When glycerol is esterified with fatty acids, not only are triglycerides formed but also mono- and diglycerides of varying chemical constitution. For example in the esterification of glycerol and stearic acid:

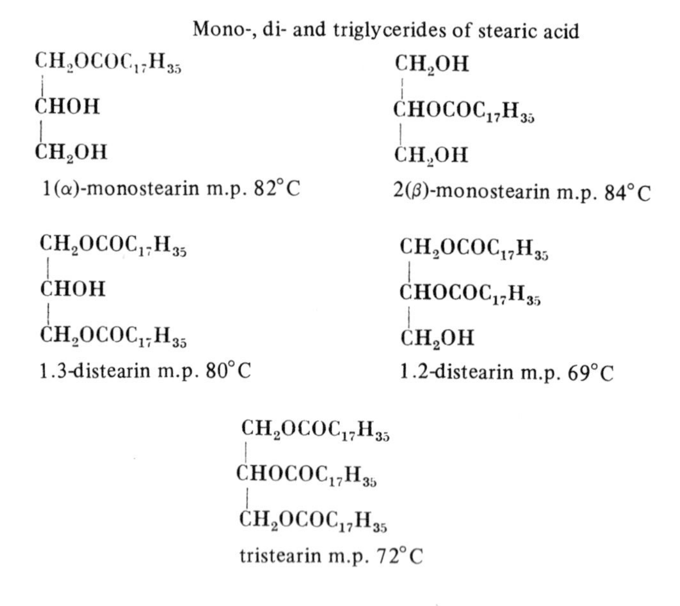

Two monoglycerides, two diglycerides and one triglyceride are formed. The four incompletely esterified compounds present in the mixture are called **partial glycerides**. When saturated and unsaturated fatty acids are used for the synthesis, mono-, di- and triglycerides are obtained which can be saturated and unsaturated.

Monoglycerides are surface-active (capillary-active) compounds as they contain both polar (water soluble) and apolar (fat soluble) groups. For this reason the monoglycerides of the higher fatty acids are of importance as emulsifiers in the food industry. They are particularly suitable in making water-in-oil emulsions. They also speed up crystallisation so that a fat which does not crystallise easily, quickly forms a micro crystalline structure when small amounts of monoglycerides are added. The monoglycerides are distinguished by this property and it is important in preventing oiling from materials containing fat, for example in the manufacture of margarine. The **diglycerides** are not so polar and are not that different from triglycerides in their general properties.

Nowadays mono- and diglycerides are obtained industrially by the controlled esterification of fatty acids or fat with glycerol or by the interesterification of fats with glycerol (glycerolysis) in the presence of a basic catalyst.

The reaction products from fatty acids or fat with an excess of glycerol consist of a mixture of mono-, di- and triglycerides together with any unchanged glycerol. The term 'monoglyceride' is used for the crude product. Commercial products described as 'monoglycerides' contain 50–60% monoglycerides, 30–45% diglycerides and a small amount of triglycerides. Molecular distillation is suitable for further purification of such commercial products; the mixture from the reaction above can be separated by this method and monoglyceride concentrations of 90–95% purity can be obtained*.

It is necessary to pay particular attention to these partial glycerides when analysing fat products manufactured from commercial monoglyceride preparations.

Compared with the triglycerides, the mono- and diglycerides are less soluble in lipophilic solvents, and the solubility of monoglycerides is less than that of the diglycerides. Glyceromonostearin is barely soluble in cold ether and light petroleum and almost insoluble in water. However, glyceromonostearin is soluble in other fatty solvents, although in lower concentration than the triglycerides. Partial glycerides, which contain unsaturated fatty acids, can be used, in many different ways and in many fields, as surface active compounds, for instance as emulsifiers because of their emulsifying effect. The small amounts of mono- and diglycerides present in natural fats can lead to the formation of unwanted emulsions during the refining of fats and so lead to losses of neutral oil.

From a nutritional point of view the digestibility of mono- and diglycerides is equal to that of triglycerdies.

4.4 PROPERTIES OF FATS

The ester linkages are reactive positions in the fat molecule. Hydrolytic processes play an important part in fat chemistry.

Hydrolytic cleavage – long called saponification in the soap industry – is

*Special laboratory methods are necessary to obtain pure mono- and diglycerides.

carried out commercially with the aid of water, dilute acids, caustic alkalies, fat splitting enzymes (lipases) and other additives (e.g. TWITCHELL reagent)*.

Under normal pressure or increased pressure (autoclave splitting). Recently 'simple water cleavage' in high pressure autoclaves (20-30 at 220-260°C) without catalysts has been used on a large scale. The fatty acids split off are used to produce soaps, tensides, disinfectants, and fatty acid derivatives. The glycerol obtained by distillation has a wide variety of industrial uses (food, medicine, cosmetics, textiles, explosives etc.).

Hydrolytic processes can also occur during the deterioration of fats, when enzymes in particular split the fats into their components (souring, rancidity, see p. 114).

Hydrolysis or saponification of fats is used analytically to characterise fats. The so-called saponification number indicates how much alkali (mg KOH) is necessary to saponify 1g fat completely.

The saponification number is higher, the higher the content of low molecular fatty acids in the fat; it is a measure of the average molecular weight of the fatty acids.

The double bond in the unsaturated fatty acids are the reactive positions. Addition and splitting reactions also occur at these positions. Some of these reactions may be controlled, for instance by hydrogenation, halogenation, preparation of stand oils (polymerisation) and blown oils (oxidation), while other reactions are undesirable, for example, autoxidation and deterioration of fats.

When fats are heated to about 300°C, pungent smelling fumes of acrolein (allylaldehyde, $CH_2{=}CH{-}CHO$), which sting the eyes are given off. They are formed by the splitting out of water from the glycerol molecule.

This reaction can be used to identify glycerol or fat; the fat is dried and heated with a dehydrating agent (e.g. $KHSO_4$).

The physical, chemical physiological and technological behaviour of a natural fat or oil is determined by the fatty acids present, except where it is influenced by naturally accompanying materials. The configuration of the glycerides is also important as well as the individual components, the saturated or unsaturated character and the molecular size. All the important questions in fat chemistry (processing, preservation, deterioration), in fat technology (hardening, polymerisation, preparation of stand oil, drying of oil, blowing of oil, artificial conjugation, preparation of lacquers and varnish) and in physiology and the physiological chemistry of fats are connected basically with the components of the fatty acids, with particular emphasis on the unsaturated fatty acids and their positions in the glyceride molecule.

*The advantage of TWITCHELL cleavage lies in its rapid saponification without excess pressure, so that the products of the reaction are not decomposed. The TWITCHELL reagent (a sulphonic acid) is formed by condensation of ricinoleic acid or oleic acid with naphthalene and sulphuric acid. This catalyst is soluble both in fat and water and can emulsify fats, so that the cleavage proceeds rapidly.

Purely external differences in fats (whether solid or liquid) are also determined by the nature of the fatty acids. In general the fats richest in saturated fatty acids are hard (tallow), and the longer the chain of the saturated acids (palmitic acid and stearic acid), the harder. In contrast the fats rich in unsaturated fatty acids (oleic acid, linoleic acid, linolenic acid) are softer and frequently liquid.

Table 6 Distribution of the most important fatty acids in edible fats

	Capronic C_6	Caprylic C_8	Caprinic C_{10}	Lauristic C_{12}	Linolenic C_{18}	Linoleic C_{18}	Oleic C_{18}	Stearic C_{18}	Palmitic C_{16}	Myristic C_{14}
coconut oil	0–2	6–10	5–11	45–51	16–20	4–8	1–5	2–10	1	
palm kernel fat (seed fat)	0–2	3–5	3–6	50–55	12–16	7–9	1–7	4–16	1	
babassu fat		6	3	46	20	7		18		
palm oil (from the flesh of the fruit)					1–3	33–45	3–7	40–50	8–11	
olive oil					1	7–10	2–4	64–86	4–12	
ground nut oil*						6–11	3–6	42–61	19–33	
rape oil† (rapeseed oil)						1–2	2	35	14	2–3
sesame oil						7–8	4–5	35–46	36–38	
soya oil					1	2–7	4–7	26–36	51–57	5–8
cottonseed oil					1	20–23	2	24–35	40–45	
butter‡	1–2	1–2	2–3	3–6	8–15	20–30	2–11	27–42	3–6	
beef tallow					2–6	25–30	14–28	38–50	2–5	
lard					1–2	25–32	8–15	50–60	0–10	
whale oil**					2–4	12	3	28–36		

*Ground nut oil contains 2–5% arachidic acid
†traditional rape oil contains 50–60% erucic acid
‡Butter contains 3–4% butyric acid
**Whale oil contains 10–20% unsaturated fatty acids of the C_{20}– and C_{22} series.
} as glycerides

Fats normally liquid at room temperature are called oils. Fish oils or liver oils are derived from marine animals, for example, whale oil, cod liver oil and so on. All liquid fats can be converted into a solid form by the process of fat hardening, that is by the addition of hydrogen to the double bonds of the unsaturated fatty acids (see p. 104).

As can be seen from Table 6, vegetable fats contain mainly saturated acids of the C_{16} and C_{18} series and unsaturated acids of the C_{18} series (oleic, linoleic acid). A few vegetable fats (coconut oil, palm kernel fat, babassu fat, muscat butter, laurel kernel oil etc., also contain fatty acids with shorter C chains (from C_6 to C_{14}). Cruciferae oils (e.g. rape oil) can be distinguished because of their content of the high molecular (C_{22}) mono unsaturated erucic acid.

The vegetable fats solid in our climate (coconut oil, palm kernel fat, palm oil, cocoa butter) are built up mainly of saturated fats, whereas vegetable oils contain predominantly unsaturated acids. Vegetable oils can be divided into the following groups corresponding with their drying capability depending on the content of polyunsaturated fatty acids.

Non-drying oils, in which oleic acid predominates: Iodine number* about 75-100: olive oil, groundnut oil.

Weak or semi drying oils in which oleic acid and linoleic acid predominate; iodine number about 100-150: sesame oil, sunflower oil, soya oil, rape oil, cottonseed oil.

Drying oils in which linoleic and linolenic acid predominate; iodine number 150-190: linseed oil, hemp oil.

The body fats of land animals contain principally palmitic acid and stearic acid, with oleic acid and sometimes linoleic acid (lard). Milk fats of ruminants also contain the whole series of even numbered, low molecular saturated fatty acids (from C_4 onwards).

The composition of the body fats of marine animals, blubber, fish oils and liver oils, must be treated separately. They are conspicuous by their content of more highly unsaturated fatty acids, specially in the range of C_{20}-, C_{22}- and C_{24} acids. The C_{18} acids (oleic acid, linoleic and linolenic acid) are unsaturated acids of the natural edible fat glycerides – excluding marine animal oils.

Modern refined physical methods* have recently made it possible to improve the analysis of fats, and hitherto totally unknown fatty acids have been identified.

In shark liver oils – which have been interesting for a long time because of their characteristic squalene and glycerol ether content – hitherto unknown fatty acids have been found, both saturated (C_{14} to C_{22}) and unsaturated (C_{14} to C_{26}) with 1-5 double bonds.

Recent research has also shown, suprisingly enough, that fatty acids with conjugated unsaturated (double bond) linkages are widely present in natural animal and vegetable glycerides, though usually only in very low concentrations†. Parinaric acid, known in plants, was recognised as a C_{18} acid with four conjugated double bonds. In the seed oils of European Balsaminacae it has been found linked with acetic acid, as in acetodiparinin, the first occurrence in a natural fat of the acetic acid residue.

*Iodine number (as measure of unsaturation = g iodine added per 100 g fat).

*Low temperature crystallisation, fractional distillation, ultra violet–infra red and raman spectroscopy, chromatography, radiography, polarography, partition chromatography, mass spectroscopy.

†For instance in butter fat 2.5% dienoic and 0.7% trienoic acids.

Conjugated acids are formed during the autoxidation of fats (conjugation of methylene interrupted acids).

The production of 'conjugated acids' on a technical scale from isolated unsaturated fatty acids or by the dehydration (splitting out of water) of the hydroxymonounsaturated ricinoleic acid from castor oil is of great importance in the varnish and paint industry. By this step naturally non- or slightly drying oils can be converted into drying oils.

The almost universal presence of conjugated unsaturated acids in those natural fats which have been investigated and also the presence of acetic acid as in acetodiparinin, makes it likely that fats are built up from acetic acid. Latest research in this field has proved that the synthesis of fatty acids in the organism is based on the two carbon system of acetic acid. In addition the biochemical breakdown of carbohydrates, proteins and fats proceeds via the 'activated acetic acid' stage. On the one hand, synthetic processes in the body proceed and on the other hand biological oxidation occurs to obtain energy giving as the end products CO_2 and H_2O via the so-called citric acid cycle (citrate cycle), see also p. 93.

The metabolic scheme (Fig. 5) showing that the breakdown of all three main groups of foods, fats, carbohydrates and proteins (after deamination takes place via 'activated acetic acid' (=acetyl co–enzyme A). This makes possible the synthesis of fatty acids for the body as well as other complicated compounds, for instance sterols in plant and animal organisms. See also p. 276.

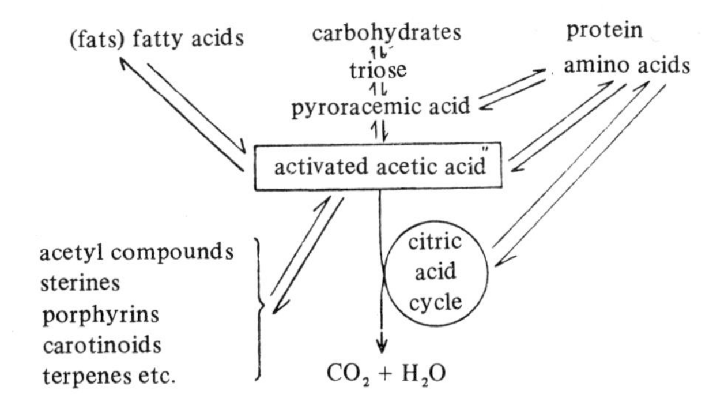

Fig. 5 Schematic diagram of the metabolic cycle.

4.5 PRODUCTION OF FATS

Technical processing of vegetable fats

Pressing and extraction are used for the production of vegetable fats on a commercial scale, taking into consideration the nature of the raw materials. Processing of oil seeds can be spread over the whole year, since the oil seeds can be stored for long periods under suitable conditions. Olives and the fruits of the oil palm are exceptions; the flesh of these fruits readily ferments.

Before edible oils are processed – by pressing or extraction – the seed must be cleaned using sorters, sieves, aspirators and electromagnets to remove dirt, dust stone and iron fragments, foreign seeds, glass fragments of textiles etc. When the seed is crushed in the rolling mils, the cells are broken open and the oil can escape.

Pressing procedure

Smaller amounts of oil are obtained by cold pressing e.g. of olives. Hot pressing produces a better yield, but gums, colour, odour and flavour materials as well as free fatty acids may pass into the oil. Oilseeds are frequently warmed to 70°C in warming pans or ovens before the actual pressing in screw extruders. The two processes – warming and pressing – are also combined when warmed screw extruders are used (on the principle of a mincer, see Fig. 6). Several pressings are carried out to obtain the highest possible yield of oil. Before each pressing the expeller cake remnants are further broken up by back breakers, fluted rollers or

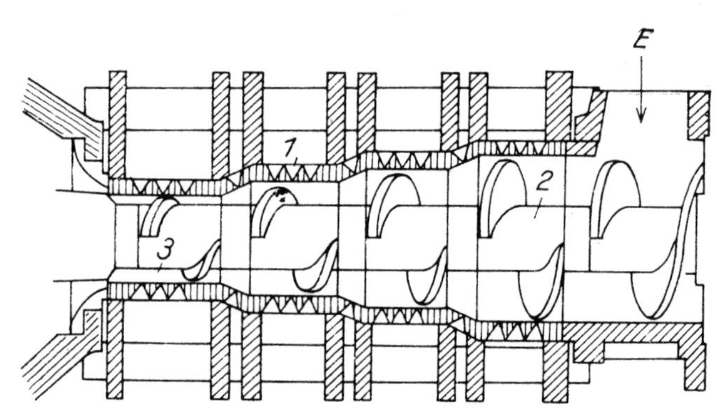

Fig. 6. Typical screw extruder
E seed entry, 1 step-tapered filter, 2 main body, 3 exit for pressed residue (from R. LUDE, Die Gewinnung von Fetten und fetten Olen; Steinkopff, Dresden und Leipzig 1948)

hammer mills. After the final pressing, the expeller cake, which is still relatively rich in oil (5-10% oil) is extracted with solvent in order to de-oil it further. The extracted cake contains only about 1% of oil at the end and is in demand as a valuable feed, rich in protein. Only the oil from the first pressing is usually suitable as edible oil, as the quality of oil deteriorates with the second and consequent pressings.

The seed used in pressing should have a desired moisture content (about 14-18% H_2O). Within these limits of humidity and at a suitable temperature, the oil can be readily forced out from the interior of the seed by pressure.

Standardisation of the oil bearing raw material before pressing under fixed mechanical and physico-chemical conditions (cleaning, crushing, grinding, control of water content, pre-warming) to obtain the highest possible yield of oil is called conditioning.

Modern methods, e.g. the SKIPIN process, force the oil present in the seed cells out directly with water.

There is a basic distinction, in the oil presses used today, between open and closed presses with the batch (interrupted) methods of production (e.g. packing presses, shelf presses, filter presses, ring presses) on the one hand and on the other continuous process screw extruders (e.g. ANDERSON expeller, KRUPP press, MIAG press.

Extraction process

Before extraction the oil seed must first be pre-dried to a moisture content of 10–12%, because a higher moisture content makes it more difficult for the lipophilic solvent to reach the cells (this is also a form of conditioning), with seeds rich in oil a preliminary pressing is frequently carried out to simplify the fine crushing and to save solvent (combined process).

Extraction of the crushed vegetable seeds is s diffusion (= enrichment) process. In batch extraction the solvent runs through a series of vertical extraction tanks (batteries) which have previously been filled with crushed oil containing material on the counter current principle, see Fig. 7. The individual extraction tanks can be switched in or out of the main circuit as required and can be refilled when exhausted, without interrupting the whole extraction process.

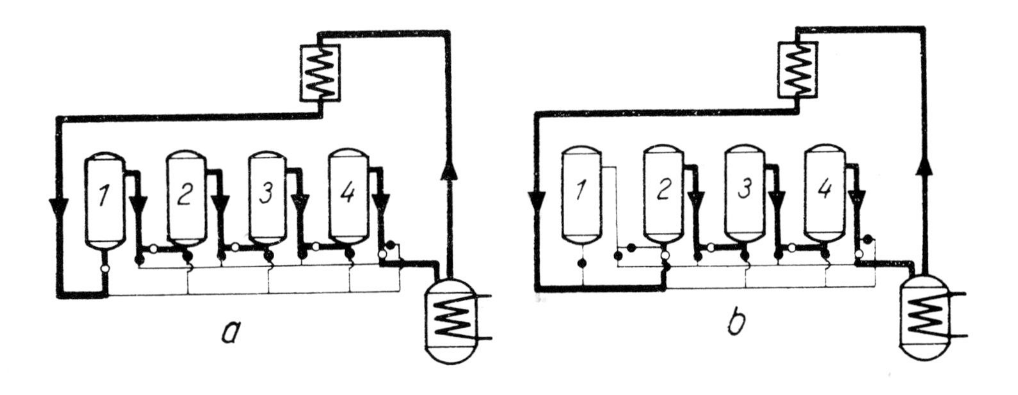

Fig. 7. Battery arrangement of a multiple extraction installation with four extractors (a) feeding in of the solvents to the first extractor, (b) feeding in of solvents to the second extractor after the first has been switched off. (According to VAULK–MULLER).

Nowadays extraction is usually carried out in a continuous process (again on the counter current principle). Two continuous process extractors are pictured in Figures 8 and 9.

The mixture of oil and solvents, called miscella, which flows from the extractors, is run through closed filter presses or centrifuges into distillation columns to remove the fine suspended matter and the solvent is carfully distilled off under vacuum. The last remnants of the solvent are removed by means of direct injection of slightly superheated steam.

The oil, now free of solvent, is suitable for refining (see p. 98).

The solvent, free from oil, is returned to the extraction circuit after condensing and cleaning.

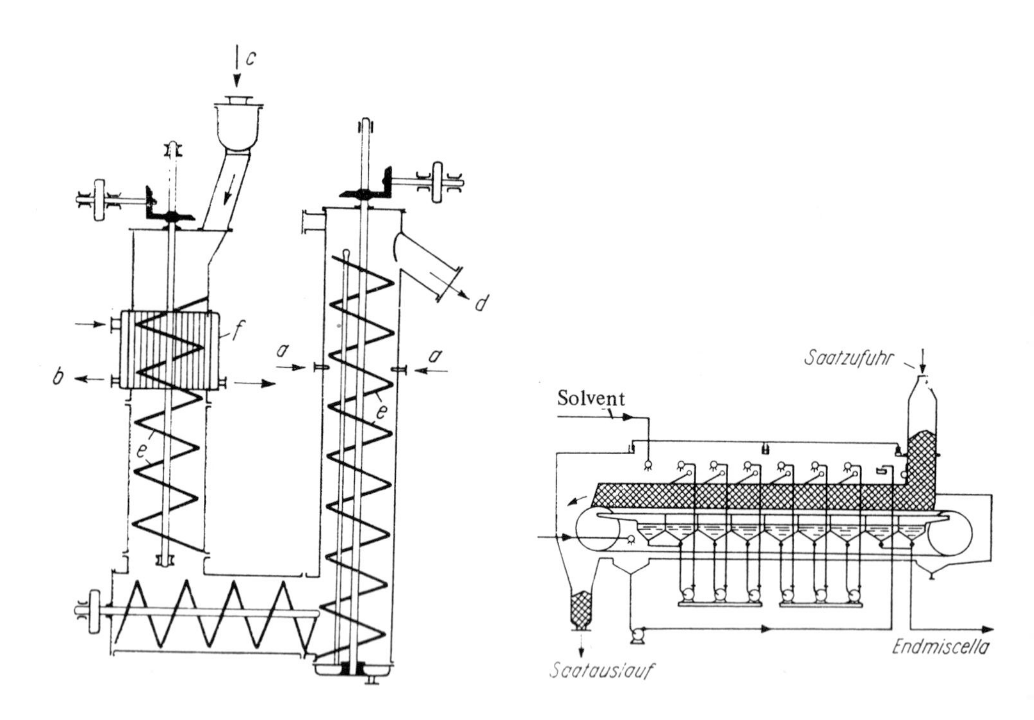

Fig. 8 Continuous extraction apparatus according to HILDEBRANDT, (a) solvent entry, (b) exit for miscella, (c) entry for oil-containing material, (d) exit for de-oiling materials, (e) continuous conveyor belt, (f) filter, strainer.
(From ULLMANN Encyclopaedia of Technical Chemistry, Vol. 7, p. 686, Urban and Schwarzenberg, Munich 1951).

Fig. 9 DE SMET extractor
(from ULLMANN Encyclopaedia of Technical Chemistry, Vol. 7, p. 491, Urban and Schwarzenberg, Munich 1956).

Nowadays light petroleum (cracked petrol, b.p. 70–100°C) is used instead of benzene 'tetra' (CCl_4), 'tri' (trichlorethylene), carbon disulphide and alcohol. It does not corrode the steel equipment and the non-fatty materials (starch, protein) have very low solubility in light petroleum. Suitable safeguards for its inflammability are essential. Soya beans are best extracted with alcohol.

Extracted oils are purer than pressed oils, because they are less contaminated with contaminated with protein, gums, etc., but as far as taste is concerned they are inferior to the pressed oils. They need refining before they can be used as table oil. Neutral flavoured oils are required today for the margarine industry (see also 'full oils' p. 98).

Processing of animal fats

Animal fats, especially from land animals, are usually obtained by the wet or dry rendering process. Impulse rendering by the CHAYEN process is a special type of cold rendering. The extraction process with solvent and the pressing procedure are only used for the production of bone fats and fish oils. Special centrifuge procedures are used to obtain the fat from milk.

When obtaining fats from land animals, it is necessary to process the fatty tissues of the slaughtered animals quickly after cleaning, washing and sorting, as they deteriorate rapidly.

Fats are obtained either by so-called dry rendering, in which the fat is melted out in boilers with direct heating, or more gently in boilers with a hot water or steam heater jacket. In wet rendering the fat is melted out by direct steam (fresh steam) or by hot water in open or in closed steam boilers (up to 6 atm. pressure).

Two groups of marine animals are important for fat production: fish and marine mammals (whales and seals).

In fish oil production (herring, sardine, menhaden), whole, or chopped fish are boiled to a thick mass and then pressed in a screw extruder. The fluid from the pressing is separated by centrifugal action into oil and water – previously this was done by settling – the oil is then clarified. The solid remains are made into fish meal. Fish oils are now more widely obtained by extraction. Continuous working wet rendering processes are used for whale – usually while still in factory ships at sea. The comminuted whale fat (blubber) is rendered under steam pressure in a continuously operating boiler. The separation of small pieces of tissue, fibre and glue liquor from the oil takes place in the fat separator. The oil is then freed from gums in special separators (e.g. Westphalia or DE LAVAL types).

Liver oils are obtained, after careful preparation of the raw materials, by rendering (50-90°C) or careful boiling and the oil and water phase separated in special centrifuges. The TITAN process is very gentle; here the oil is freed from the liver, after comminution in suitable mixing vessels, by direct steam at 5 atm and after removal of the tissue, the oil water mixture is separated continuously in gum separators.

The CHAYEN cold rendering process has been developed in England. Here, cold water is used instead of heat and solvents to remove the fat from the raw material (bones, oil seeds, fish liver, herrings etc.). The comminuted raw material is led continuously, with large amounts of water, into a kettle where mechanical vibrators send strong high frequency impulses through the water. The effect is similar to that of high pressure waves. The cell walls are ruptured and the contents are washed out immediately. The expelled fat collects on the surface of the water and is separated by centrifuge. This 'impulse rendering' is very quick and yields high quality fats. Bones (and also oil seeds) can be defatted up to 98.5%.

Fat production by microbiological synthesis

The principle of this method utilises the metabolism of the microorganism to change carbohydrates (sugar) or other C containing nutrients (organic acids, acid amides, hydrocarbons, CO_2) into fats. Inorganic nutrient salts (especially N and P compounds) need to be added. Attempts have been made since the first world war to use microorganisms for industrial fat production. Extensive research has shown that fats produced by various microorganisms may be of considerable use for nutritive purposes, since their composition is similar to that of fats from higher plant. Although the experiments carried out on a laboratory scale have shown good results, scaling up to commercial operation has problems which have yet to be resolved.

Recent work has shown that the moulds *Oidium lactis* and *Endomyces vernalis* (mould hyphae) are suitable for fat production as pentoses, hexoses and the disaccharide lactose can be utilised as carbohydrate sources. This raises the hope, that the lactose in whey and also wood sugar can be utilised. Both fungi can only grow on the surface and are therefore susceptible to foreign infection, which can affect the fat synthesis and make it uneconomic.

In general the object of all the experiments has been to culture microorganisms which will grow rapidly when submerged and give a good fat yield; from a commercial point of view, microbiological fat synthesis only looks promising with submerged cultures.

Microbiological fat synthesis can be developed on an industrial scale with Fusarium species (mycelium fungi) submerged cultures. In addition to the pure carbohydrates (sugars), commercial spent liquor (e.g. sulphite spent liquor) and synthetic compounds (organic acids and acid amides) can be used as nutrient substrates. The Waldhof process, in which both fat and protein are produced, has already been developed on a commercial scale and shows promise, this is a continuous aeration process in which species of *Torula,* such as *Torula utilis* or 'budding fungi' are used for biosynthesis. They could also be important for protein production in the future.

4.6 REFINING OF EDIBLE FATS

Fats from land animals to be used for human nutrition, such as lard, edible tallow, poultry fats, can be produced in an edible storable state without further refining by choosing suitable raw material (fatty tissue) and appropriate processing (mainly rendering processes. Butter from milk fat serves for human food without refining; its storage life can be increased by rendering the hot vacuum separation of the butter oil.

Those vegetable oils which are commonly used for human nutrition are so pure when they are produced shortly after harvesting, that they can be enjoyed by man immediately e.g. freshly pressed olive oil and seed oils.

In todays world market conditions longer transport times and storage are inevitable and the raw oils can be affected by air, heat, humidity and enzymes. This increases the content of undesirable matter and affects keepability, useability for immediate human consumption decreases considerably – as does also suitability for the production of margarine or mayonnaise. Crude vegetable oils *have* to be subjected to cleansing and refining before they can be used as food (table oils, fat products). In these stages, the insoluble (suspended) and the dissolved impurities, which are undesirable components of the oil, are removed.

The suspended materials are removed mechanically by settling (clarification) and filtration. Dissolved materials are removed by various chemical and physical methods.

The non-dissolved (suspended) materials, which can be removed by **clarification** and **filtration** are principally, dust, hairs from the press cloth, water, precipitated gums and parts of the seed casing which have been carried through. Lipases from the seed may cause the formation of free fatty acids which frequently have an unpleasant taste.

Soluble impurities are partly or wholly removed from the oils during the various stages of the refining process. Among these are the free fatty acids formed by hydrolysis, small amounts of mono- and diglycerides, natural colouring matter such as the carotenoids and chlorophyll and new pigments formed during storage. There are also unpleasant smelling materials of unknown composition, gums, (e.g. lipoprotein and liposaccharide complexes), traces of heavy metals (Cu, Fe), sterols, phosphoglycerides (e.g. lecithins, phosphoinositides), resins, offensive smelling amines, aldehydes, ketones, hydrocarbons. Finally further autoxidation, polymerisation and cleavage products may be present and natural antioxidants such as tocopherols, gossypol and sometimes mycotoxins produced by fungi.

Normally in crude vegetable oils we find a whole series of materials ~~which~~ affect or reduce keepability, and which may even be toxic (gossypol). These must be removed in the refining process. However, this also removes some nutritionally valuable lipids, such as carotenoids, and tocopherols. When commercial refining processes are carried out carefully edible vegetable oils can be produced with their content of natural nutritionally important and valuable accompanying materials (tocopherols, phosphatides) virtually intact. KAUFMANN calls these ‘complete’ oils.

Modern refining practice is adapted to the composition of the crude oils and their intended uses and takes care to preserve the oils.

The following are the stages:

(a) degumming
(b) neutralising
(c) bleaching
(d) deodourisation

Degumming is usually carried out with a dilute brine or phosphoric acid solution or acid alkaline phosphate, to remove the resin and gum particles, proteins and phosphatides from crude fats. The hydrated gum which is deposited often called 'crude lecithin' in the case of oils rich in phosphatides (soya, rape seed), is separated in centrifuges and can be purified with water (or with acetone). Degumming improves the keepability of the oils because the gum materials are a good nutrient base for microorganisms and increase the danger of microbial breakdown of the fat. The removal of the phosphatides and gummy materials is also important for the commercial processing of the fats, because during hydrogenation, gums and phsophates can poison the catalysts (nickel) or render them inactive. The extracted phosphatides are useful natural emulsifiers, e.g. in the manufacture of margarine and the making of pasta.

Neutralising is frequently combined with de-gumming and is usually carried out by spraying or stirring in dilute caustic solution, followed by washing (bubbling water through) the oil. This also, to a large extent, bleaches the crude oil and removes traces of heavy metals (Cu, Fe) which promote autoxidation decomposition of the fat. The soaps formed during neutralising, separate off as compact, dark coloured 'soapstock' which can be drawn off and used in the soap industry.

Neutralising can also be carried out, independent of the degree of acidity, by means of superheated steam (steam distillation) under vacuum, when the fatty acids are distilled off. After lengthy caustic treatment remnants of acid are removed.

Distillation cannot be used on crude cotton seed oil, because dark coloured oxidation products are formed from materials accompanying the oil (gossypol).

Neutralising may be carried out in batches, that is discontinuously, or as a continuous process. In the U.S.A., with their large production of soya and cotton seed oil, the soapstock is separated as a continuous process using high speed centrifuges. The oil is then passed through a continuous drier and transported to the bleaching process. As contact between the raw oil and caustic is relatively brief, the neutral oil is not greatly affected: emulsion formation is slight, loss through refining is relatively low and the quality of the oil remains standard and is easy to control. Continuous neutralising can be carried out without technical difficulty to yield an oil containing 0.1% or less of free fatty acids (calculated as oleic acid). (Continuous processes: DE LAVAL, SHARPLES, CLAYTON, short-mix process, Zenit process).

Bleaching is carried out with absorbants (bleaching earths, such as bentonite, Florisil or active carbon), which absorbs the soap and gum residues, traces of heavy metals and colouring agents both natural and those which have been formed during storage.

Bleaching also to a large extent removes oxidation products still remaining in the oil (hydroperoxides, peroxides) which might affect keepability of the oil and result in flavour changes due to decomposition and further oxidative reaction. The unsaturated, conjugated diene and triene fatty acids with a tendency to polymerisation and oxidation, which are sometimes formed in very small amounts during bleaching, are a natural result of reactions due to the primary formation of hydroperoxides in the fat. Oxidative changes during the bleaching of oil can be used to advantage commercially.

Bleaching can also be carried out continuously, (e.g. BAMAG-MAGUIN and GIRDLER processes). Chemical bleaching, which used to be carried out with peroxides, is now no longer permitted.

The final, but undoubtedly the most difficult stage of the whole process of refining is **deodorisation**. The aim is to remove those quality reducing flavour and odour materials which are naturally present in the fat or those which have been formed secondarily by oxidative or hydrolytic reactions (chemical, enzymatic or microbial).

In coconut oil and palm kernel oil – with their high content of fatty acids of medium chain length – considerable amounts of unpleasant smelling methyl ketones are formed. The oil also acquires a typical rancid taste from the oxidation of unsaturated fatty acids resulting from the decomposition of the hydrogen peroxides and from the unsaturated aldehydes, hydroxy-acids, keto-acids, low molecular fatty acids and alcohols which are formed.

During hydrolysis, free fatty acids are frequently formed; those of low and medium molecular weight have an especially unpleasant odour and taste. Amine type compounds, derived from the phosphatide residues in the oil, must also be removed because of their unpleasant smell. These materials, even in concentrations as low as 1-10 mg/kg spoil the smell and taste of an oil.

Deodorisation is usually carried out as a steam distillation, which removes the relatively volatile compounds which give an undesirable smell or taste to the oil.

As the vapour pressure of the odour and flavour materials to be removed is low, high temperatures would have to be used. This difficulty is overcome by using the highest possible vacuum (5-20 Torr). This protects the hot oil from oxidation by oxygen in the atmosphere and also prevents hydrolysis of triglycerides by steam.

In practice deodorisation is carried out by means of a series of steam injectors which produce a vacuum of 5-20 Torr and heat the oil to 170-230°C. This partial pressure permits the volatile materials to distil off. Steam which is injected directly into the oil must be deaerated (oxygen free) and dry. To prevent traces of metals having a catalytic effect (and the fact that oxygen from the air cannot be completely excluded in commercial installations). a small amount of dilute solution of citric acid in water (about 5 mg/100 g calculated on the oil) is added just before the beginning and end of the steam injection. This complexes traces of metal which could act as pro-oxidants.

When deodorisation is complete, the oil is cooled to about 135°C, after which it is cooled to below 60°C *under vacuum* to prevent oxidation by oxygen in the atmosphere. In Europe deodorisation is still mainly a batch process (Fig. 10), although continuous deodorisation equipment in the USA has reached a

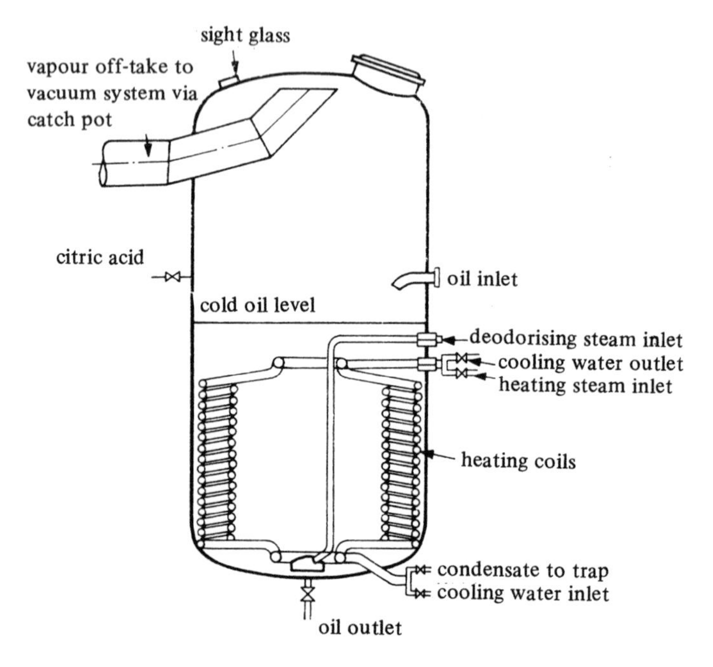

Fig. 10. Deodorisation apparatus
(from DEVINE and WILLIAMS, The Chemistry and Technology of Edible Oils and Fats, p. 41; Pergamon Press, London, 1961)

high degree of development. The semi-continuous process installations of the GIRDLER Corporation are a distinct improvement on the batch process. In it the oil is continuously steamed in batches.

It is obvious from this description that the whole procedure of refining can be regarded as a purification of crude oils to make them edible. A series of materials are removed (see Table 7) — some of natural origin, some acquired from the apparatus during the production of the crude oil and some formed as secondary oxidation products of oils. Many of these render the crude oils unstable and prevent their usage for stable fats and fat products.

Table 7 Heavy metal content of crude oils and refined oils (in mg/kg)

Type of fat	Cu	Fe
soya oil, crude	0.1-0.5	1-5.6
soya oil, refined	0.004-0.045	0.005-2.5
corn oil, crude	0.5	12.7
corn oil, refined	0.015	0.1

On the other hand some of the nutritionally desirable materials in the fats, e.g. tocopherols and carotenoids, are also removed by refining. Present day processes however keep this loss to a minimum (see Table 8), so that edible fats will still contain a relatively large amount of tocopherols. This is important both from a practical, commercial and a nutritional point of view. Both the vitamin properties of the tocopherols and their favourable anti-oxidant effect on the keepability of the oils are important.

Table 8 Tocopherol content of various vegetable fats before and after refining.

Type of fat	crude	neutralised	bleached	deodourised
coconut oil	3-5	–	–	3
corn oil	119	–	–	95
cotton seed oil	110	105	95	95
ground nut oil	52	–	–	45, 48
soya bean oil	152, 212	–	–	110, 175
palm oil	50, 52	–	–	35, 40

(given in mg total tocopherol/100 g fat) according to J. BALTES.

Storage of refined oils should be at below room temperature and preferably under nitrogen or carbon dioxide in suitable oxide free iron tanks. Refined or partially refined oils used for margarine production are not stored but used immediately.

When refined oils are to be used as edible oils, winterisation (making them stable against low temperatures) must be carried out. This separates out the higher melting point triglycerides in the oil by crystallisation during a slow cooling process. For soya bean oil, rape seed oil and poppy seed oil, normal storage temperature of 15-20°C are adequate. Cotton seed oil, sun flower oil, sesame oil and corn oil should be held at +5°C for two to four days. The solid glycerides which have crystallised out are then removed; the liquid portion is optically brilliant after final filtration.

Ground nut oil, because of its content of long chain saturated fatty acids, cannot be winterised; it forms a gel at 8-10°C.

4.7 HARDENING OF FATS

Whereas originally the term hydrogenation was only used for the addition of hydrogen), 2, interesterification, 3, fractionation of fats, i.e. separation of low hardening now includes all processes which tend to raise the melting point of a glyceride mixture. These include: 1, catalytic hydrogenation, (addition of hydrogen), 2, interestification, 3, fractionation of fats, i.e. separation of low melting point glycerides.

Hardening of fat by hydrogenation

In 1901 NORMANN experimented to change liquid oil into solid fats because fats solid at ordinary temperatures could be more easily used for margarine. His idea was to change the unsaturated fatty acids of the oleic and linoleic acid series present in glycerides into saturated fatty acids with a higher melting point by the addition of hydrogen (hardening).

$$—CH{=}CH— \xrightarrow[\text{Catalyst}]{H_2} —CH_2—CH_2— \quad \text{(exothermic reaction)}$$

The most difficult task in solving this problem was to find suitable catalysts; only then was it possible to work out a series of hardening processes suitable for large scale production.

Hardening of fats by hydrogenation is used on a large scale with liquid vegetable oils, because compared with fats of hard consistency they are available in larger amounts. Hardening of marine oils is an important process in their purification and absolutely necessary; these oils cannot be used in their natural form either as food nor in the soap industry because they do not keep and have a particularly unpleasant smell and taste.

Without going into technical details, the principle on which present methods are based is as follows: the degummed and neutralised vegetable oils or animal fats are warmed and mixed with very finely divided nickel. This is prepared from nickel formate or nickel carbonate by reduction with hydrogen and added in amounts of 0.01-0.2%. Hydrogen is then stirred or injected at a temperature of 160-220°C to give close contact with the oil under pressure (to increase the speed of the reaction). Hydrogenation can be carried out as a batch or continuous process; (see Fig. 11).

Under these conditions some of the unsaturated fatty acids are transformed into saturated fatty acids. Some rearrangement (elaidinisation) of the fatty acids and rearrangement of the position of double bonds (positional isomerism) can also take place, so that the so-called 'Iso-oleic acids' are formed. Their presence is a clear indication of the admixture of hydrogenated fats (solid fats), which produce trans-absorption bands in the IR (965-990/cm) (see also p. 81).

As well as the addition of hydrogen to unsaturated bonds and the isomerisation mentioned above, a small amount of conjugation as well as some interesterification and slight hydrolysis of the triglycerides may occur during commercial hydrogenation.

By varying the conditions of the reaction (catalyst, temperature, pressure etc.) it is possible to carry out the hardening in steps, so that only certain groups, for instance the fatty acids with several double bonds, are partially hydrogenated, whereas the oleic acid and linoleic acid are to a large extent preserved.

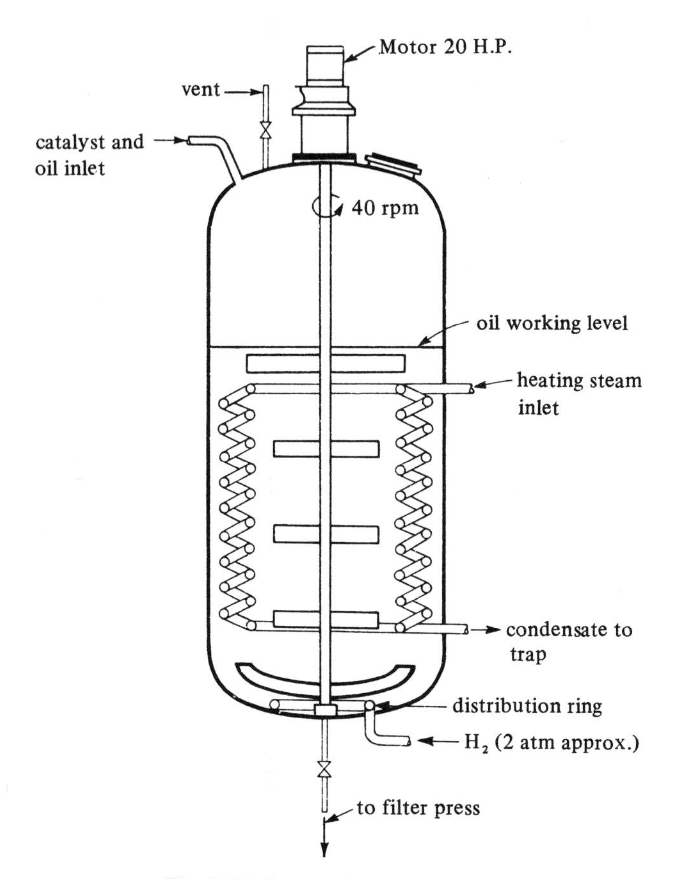

Fig. 11 Hydrogenation apparatus
(from DEVINE and WILLIAMS. The chemistry and technology of edible oils and fats p. 44; Pergamon Press, London, 1961).

In this way hydrogenation can be controlled and softer or harder fats of any desired melting point or consistency obtained. Products corresponding to lard, oleo-margarine, beef or sheep tallow can be prepared. However, absolute selectivity in hydrogenation or complete isomerisation is not possible in fat hardening, (equilibrium reactions).

Hardening of edible fats is always restricted to a partial saturation of double bonds, because complete hydrogenation – with few exceptions – would yield high melting point triglycerides, (saturated fats difficult to digest). Generally edible fats are hydrogenated to a melting point of 32-37°C, which produces malleable fats of an even consistency with a melting point in the physiological range. The constitution of the hydrogenated fat or oil depends on the type and composition of the starting material as well as the (controllable) hydrogenation conditions.

The characteristic taste of hardened fats (hardening flavour), is due to hemi-acetals and their breakdown into aldehydes and alcohols. Refining, particularly deodorisation, reduces this flavour.

The reversion flavour, found in hydrogenated and some refined fats and oils, is caused by the formation of isolinoleic acid (positional isomers), which are oxidise to odoriferous unsaturated carbonyl compounds during storage under the influence of light, trace metals and oxygen attack. Anti-oxidants have no influence on the reversion phenomena.

Hydrogenated fats are usually further refined to remove traces of nickel catalyst hardening flavour, free fatty acids, discolouration etc. This final refining consists of further gentle neutralisation, bleaching and deodorisation.

During **vigorous hardening**, fat-soluble vitamins, (with the exception of vitamin E (tocopherol), and essential fatty acids, all of which are nutritionally valuable, may to a large extent be destroyed. The loss of essential nutrients such as linoleic acid and vitamin A during the hardening of edible fats is made good today by the addition of high grade vegetable oils and vitamin A supplements.

A new improved process for hydrogenation (KAUFMANN) consists in the hydrogenation of a 40% miscella (oil-solvent mixture), which flows continuously past a stationary nickel catalyst. The process works at temperatures below 100°C, so that only insignificant amounts of isooleic acids are formed.

As far as is known at present, edible oils do not acquire any properties during the usual hardening with hydrogen and nickel contacts, which might make them suspect from a nutritional point of view. Even processes which lead to the formation of so-called isooleic acids (see p. 81) during the rearrangement of double bonds or through *cis-trans* isomeration (elaidinisation) can be considered in this light.

Interesterification

Another process for transforming oils into fats, or low melting point into higher melting point fats – without forming fatty acids – depends on the interchange of the fatty acid residues contained in the glyceride molecule. In contrast to alcoholysis and acidolysis this reaction can be called a true interesterification. The ester groups react in the glyceride molecule by exchange of the fatty acid residues. The *inter*esterification may be intermolecular, that is, between different glyceride molecules, or *intra*molecular, that is within the same glyceride molecule.

$$\begin{array}{l} CH_2\text{—O—COR} \\ | \\ CHO\text{—COR}' \\ | \\ CH_2\text{—O—COR}'' \end{array} + \begin{array}{l} CH_2\text{—O—COR}_1 \\ | \\ CHO\text{—COR}_2 \\ | \\ CH_2O\text{—COR}_3 \end{array} \longrightarrow \begin{array}{l} CH_2\text{—O—COR}_2 \\ | \\ CHO\text{—COR}' \\ | \\ CH_2\text{—O—COR}'' \end{array} + \begin{array}{l} CH_2\text{—O—COR}_1 \\ | \\ CHO\text{—COR} \\ | \\ CH_2O\text{—COR}_3 \end{array}$$

intermolecular interesterification

$$\begin{array}{l} CH_2O\text{—COR}' \\ | \\ CHO\text{—COR} \\ | \\ CH_2O\text{—COR}'' \end{array} \longrightarrow \begin{array}{l} CH_2O\text{—COR}'' \\ | \\ CHO\text{—COR}' \\ | \\ CH_2O\text{—OCOR} \end{array}$$

intramolecular interesterification

On the one hand, interesterification can yield homogeneous fats, on the other hand mixtures from which the glycerides containing solid fatty acids can be separated. Both types of interesterification are now widely used commercially, because with them it is possible to change the properties of natural fats without having to use the classical method of hydrolysis and re-esterification.

A basic distinction must be made between random interesterification and directed or controlled interesterification.

In random interesterification the homogeneous, liquid phase is heated for about 2 hours at high temperatures (70-200°C) in the presence of alkali alcoholates or ethoxides as catalysts. The fatty acids distribute themselves over the glycerides on a statistical basis.

Oils used for commercial interesterification in a random fashion contain no water, must be free of fatty acids and hydroperoxides and other autoxidation products. Hence half-refined oils after deacidification or bleaching are used with sodium methoxide or sodium ethoxide in concentrations of 0.5-0.1% (calculated on the amount of fat) as catalysts. After the catalysts have been mixed in by stirring, the fat is heated under vacuum at 70°C in a closed container for one hour. The interesterified contents are then washed with water to remove the catalyst and then bleaching and deodorisation of the interesterified fat follows in the usual way (see p. 100).

In general, the result of this homogeneous phase interesterification is an increase in the number of glyceride types and a wider distribution of the fatty acids. With the formation of eutectic mixtures, the melting point is lowered and generally a softer product is obtained. Interesterification can achieve a wider variety of melting points by means of a relatively simple technical process and from a greater choice of raw materials. Cotton seed oil mixed with coconut oil can be transformed by means of **random interesterification** into a lard-like product, whereas soft fats with a higher melting point can be obtained from the interesterification of cotton seed oil with tallow. These fats have high plasticity and are particularly suitable for production of baking fats and shortenings.

When oils are subjected to **random interesterification**, solid fats (tallow, hydrogenated fats) and low molecular fatty acids (coconut oils, palm kernel oil) can be interesterified to yield products with a better consistency, less dependent on temperature, more resistant to autoxidation and also with a higher water activity. However these products are difficult to store in high humidity, because splitting off of the lower fatty acids can easily produce a soapy flavour, (see p. 110).

In contrast to random interesterification, **directed esterification** results in a transformation of the mixed acid glycerides present in natural fats and oils into glycerides which contain either solid or liquid fatty acids according to what is required. These can easily be separated from the interesterified mixture. Liquid alkali alloys are used as catalysts. This type of interestification is much quicker. The temperatures for directed interesterification are relatively low and must not greatly exceed the expected melting point of the glycerides. Because it is possible to remove one component continuously from the reaction phase, e.g. by the separation of the higher melting point fraction by crystallisation in a cooling process, a fairly high melting point fat can be transformed either into a soft fat (margarine, baking fat) or into harder fats e.g. cocoa butter substitute. If the two components are not separated at the end of directed interesterification, a fat will be obtained which has totally different rheological properties from the original material. It is thought that a change into the more stable crystalline form plays a part here, (see p. 106).

Directed interesterification is widely used for the refining of pork lard, especially in the USA.

Slight interesterification phenomena have been observed as byproducts in certain technical processes; during the neutralisation of acid fats by distilling off the free fatty acids (WECKER process); in the course of fat hydrogenation and during molecular distillation and elaidinisation of fats.

Reduction interesterification is another form of interesterification. Here hydrogenation and interesterification are combined in one process. Because the soft fats obtained by selective hydrogenation readily acquire a **reversion** flavour due to their **isolinoleic** acid content (see p. 113) attempts have been made to interesterify coconut oil and ground nut oil during hardening. This produces products with good plastic properties, in which the flavour does not deteriorate. It has also been possible to interesterify with low fatty acids, such as the coconut oil fatty acids to produce fats which are relatively unaffected by temperature. Products with acetic acid interesterified in them are on the market as aceto-glycerides or as acetins. They are stable to relatively high temperatures but there is a tendency for the acetic acid to split off. They are most suitable as coatings for meat products.

Fractionation processes
Hardening, by means of 'pressing off' low melting point glycerides from fats, is used in manufacturing pressed tallow and oleomargarine. It is also possible to obtain the more saturated glycerides, (i.e. the higher melting point, harder 'fraction') and during 'winterisation' (making stable to low temperatures). During the separation of the higher melting point glycerides fractions from coconut oil, palm kernel oils and palm oil, the more saturated glycerides which have crystallised out during the slow cooling of these oils may be separated by pressing or filtering.

This separation of an oil, e.g. of linseed oil, soya oil or fish oil, into solid or liquid parts is also possible on a commercial scale using mixtures of solvents and oils, for instance from a methanol-oil solution in the Emersol process or from propane oil solution in the solexol process.

4.8 DETERIORATION OF FATS

Chemical changes during the storage of fats can be recognised by the senses and can involve processes generally grouped under the heading of 'deterioration' of fats. Although they are exceedingly complicated, it is possible to distinguish two basic types of change, which occur on storage, and lead to unsuitability of the fats for consumption:

1. purely chemical changes,
2. biochemical and microbiological changes.

Both types of reaction frequently take place at the same time, and depending on the constitution of the fat, trace materials and the external (purely chemical, enzymatic, physical) influences, the following important types of deterioration occur: souring, peroxide formation, aldehyde and ketone formation, appearance of tallow flavour.

Fishy, oily, liver oily, soapy flavours and also the flavour change called 'reversion', are types of deterioration, which again depend on the composition of the fat and its trace materials. They can occur under particular circumstances.

In order to fully understand the type of deterioration, it is necessary to consider the concept of 'fat' as a biologically complete food. Depending on their origin, natural fats consist of saturated and unsaturated glycerides, which can be affected by deterioration, as well as natural trace materials and other reactive substances. These materials, vitamins, phosphatides, lipochromes, and unsaturated hydrocarbons play a vital role in the process of deterioration and its products. They also frequently determine the course taken in deterioration.

Chemical changes in fats

Transformations of a purely chemical nature can be hydrolytic or oxidative.

The main processes of fat deterioration (after K. TAUFEL)

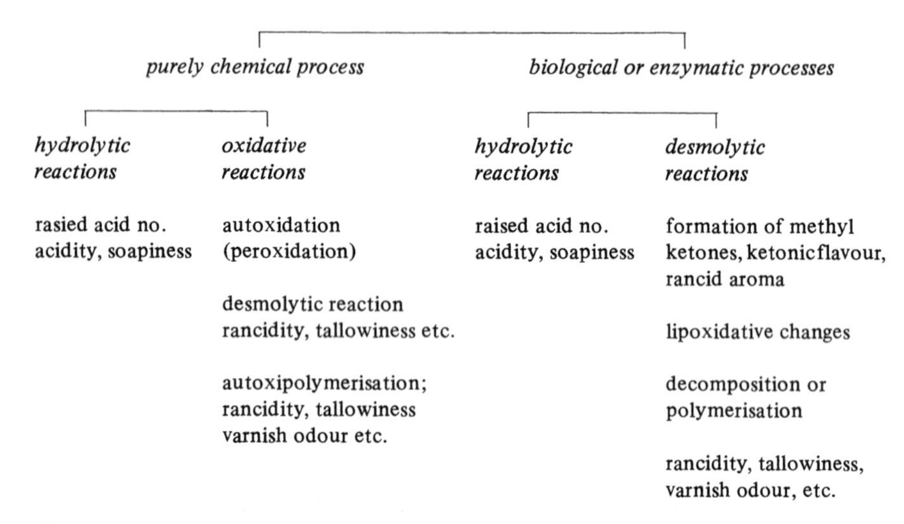

Hydrolytic reactions

Chemically fats are esters, so they are libale to hydrolysis. The final products from the cleavage of glyceride are glycerol and fatty acids. These processes occur during digestion and absorption, during the deposition of fats, in fatty tissue and in foods containing fats at normal or even at relatively low temperatures. They are therefore of special significance in deterioration and preservation during storage of fatty foods and fats.

The occurrence of free fatty acids determines the **acid value** of a fat. Water is necessary for these reactions; heat, light and catalysts encourage this type of deterioration. When the acidity of the fat has risen to a greater or lesser extent, its fitness for consumption will depend on the nature of the fat and on the fatty acids released from it. From glycerides with short chain fatty acids (C_4-C_{10}), for instance in butter, margarine, palm kernel of coconut oil, the release of the smallest amount of these acids (0.1-1 mg/100 g) makes them organoleptically unfit for consumption (soapiness). On the other hand, beef fat with up to 15% of free fatty acids derived from the fat itself is still palatable. Olive oil of the best quality may contain up to 2% free fatty acids. The importance of the degree of hydrolysis depends on the nature of the fat.

Oxidative (desmolytic) reactions

Oxidative changes are accompanied by far reaching decomposition of the glycerides, with the formation of a variety of new, modified, non-fatty materials with important physiological taste and smell.

Autoxidation of the unsaturated fatty acids assumes importance in respect of the nutritional value of fats. The susceptibility of various fats can be very different, due to their varying content of unsaturated fatty acids and to the different content of pro- and antioxidants, free fatty acids in addition to the influence of heat, light and trace metals.

Research into the chemistry of the autoxidation of the olefinic fats and fatty acids is still in its early stages, but oxidation is thought to be based on a free radical chain reaction catalysed by heat, light and other catalysts (chlorophyll, haem, blood pigments, trace metals, enzymes). The reaction is autocatalystic and can take place with increasing speed.

After the induction period, the duration of which depends on the type of fat, its trace materials and external physical influences, (degree of unsaturation, pro- and antioxidants, light and heat), fatty acid radicals are increasingly formed. Their exceptional reactivity with oxygen leads to many kinds of reactions (formation of fat hydroperoxide radicals, fat hydroperoxides, fat peroxides, increase in the size of molecules, and breakdown of molecules) .

The autoxidation of methyloleate exhibits great variety in the formation and decomposition of the primary oxidation products:

Methyloleate

$$CH_3(CH_2)_6{-}CH_2{-}\underset{10}{CH}{=}\underset{9}{CH}{-}CH_2{-}(CH_2)_6COOCH_3$$

$$\downarrow -H \text{ beginning of chain}$$

first stage: formation of four mesomeric C– radicals

$$-\underset{11}{\overset{*}{C}H}-\underset{10}{CH}=\underset{9}{CH}-\underset{8}{CH_2}- \quad\rightleftharpoons\quad -\underset{11}{CH}=\underset{10}{CH}-\underset{9}{\overset{*}{C}H}-\underset{8}{CH_2}-$$

$$-\underset{11}{CH_2}-\underset{10}{CH}=\underset{9}{CH}-\underset{8}{\overset{*}{C}H}- \quad\rightleftharpoons\quad -\underset{11}{CH_2}-\underset{10}{\overset{*}{C}H}-\underset{9}{CH}=\underset{8}{CH}-$$

second stage: formation of peroxide radicals and hydroperoxides and new radicals

(1)
$$-\overset{*}{C}H-CH=CH-CH_2- + O_2 \rightarrow \underset{|\;O-\overset{*}{O}}{CH}-CH=CH-CH_2-$$

(2)
$$-\underset{|\;O-\overset{*}{O}}{\overset{11}{C}H}-CH=CH-CH_2- + -CH_2-CH=CH- \quad \text{(new intact fatty acid molecule)}$$

$$\downarrow -H$$

(3)
$$-\underset{|\;O-OH}{CH}-CH=CH-CH_2- + -\overset{*}{C}H-CH=CH- \quad \text{(new radical as carrier of the chain)}$$

8, 9, 10 or 11 position hydroperoxides may be formed

third stage: the hydroperoxides decompose to carbonyl compounds, acids, alcohols and so on.

(1)

$$CH_3(CH_2)_6{-}\underset{\displaystyle |\atop \displaystyle O{-}OH}{CH}{-}\overset{10}{C}H{=}\overset{9}{C}H{-}(CH_2)_7COOCH_3$$

$$\downarrow$$

$$\underset{\text{aldehydes}}{CH_3(CH_2)_6CHO} + \underset{\text{aldehydes}}{OCH(CH_2)_8COOCH_3}$$

(2)

$$CH_3(CH_2)_7\overset{10}{C}H{=}\overset{9}{C}H{-}\underset{\displaystyle |\atop \displaystyle O{-}OH}{CH}{-}(CH_2)_6COOCH_3 \xrightarrow{-H_2O} \text{Ketone}$$

$$\downarrow$$

$$\underset{\text{aldehydes}}{CH_3(CH_2)_7CH{=}CH{-}CHO} + \underset{\text{alcohols}}{OH(CH_2)_6COOCH_3}$$

In natural fats many different types of reaction are always possible and they may result in intermediate or end products which differ in their sensory effects. In addition materials in the fats (see pro- and antioxidants p. 116) also affect the course of oxidative decomposition, depending on their type and amount.

The main steps in the reactions and their effects on the oxidation of fat can be summarised in the following scheme (according to TAUFEL).

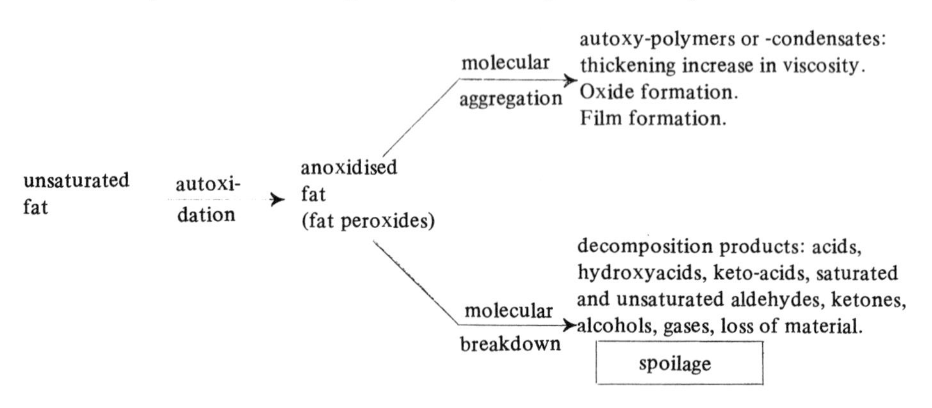

The oxidative process plays a central part – starting from the unsaturated glycerides – in the spoilage of edible fats. The food chemist, the food technologist and the nutritionist are interested in the molecular changes which take place during autoxidation. After the formation of fat peroxides, undetectable to the senses, cleavage gives low molecular materials which then lead to spoilage, detected by smell and taste. This is due to the formation of saturated and unsaturated aldehydes (aldehyde value), acids, (acid value), ketones, (ketone value) and hydroxyacids (tallowiness). All are included in the general term 'rancidity'. The process of decomposition of the hydroperoxides, which are the primary autoxidation products, is frequently catalysed by heat, light, water, oxygen and trace metals.

Determination of the degree of oxidation, for instance by the peroxide number, only gives information about the magnitude of the process in the early stages of autoxidation, because afterwards, decomposition of the hydroperoxides occurs to a greater or lesser extent.

During the storage of refined and hydrogenated edible oils and fats (especially under warm conditions) changes in flavour (hay-like, straw-like, fishy) may occur which are called **reversion**. This reversion flavour can be traced back to the presence of isolinoleic acids (e.g. Δ-9. 10–15, 16-linoleic acid) which have been oxidised during storage, to carbonyl compounds with strong smell and flavour under the influence of light, trace metals and O_2. Compounds such as 2,4-heptadienal, 1.4-octadienal, 2-heptanal, n-butylaldehyde and ethylformate have been identified. Reversion has been observed in linseed and rape seed oil, but particularly in soya bean oil and fish oils. Partially hydrogenated linseed, soya and fish oils are also subject to reversion. Reversion can occur again after deodorisation.

Fat which has become rancid through autoxidation is inedible and can cause changes in the skin and ulcers in animals. Peroxised fats endanger the body's vitamin supply because they destroy the vitamins A, C, E, B_6 and pantothenic acid and also the essential fatty acids in the digestive tract. The formation of 'conjugated' acids also characteristically accompanies autoxidation of fat and has been proved spectroscopically (234 nm). Migration of the methylene interrupted double bonds leads to these conjugated compounds. (See pp. 79 and 93).

Even in the absence of oxygen, fats, saturated and unsaturated fatty acids and glycerides can become ketonic due to the influence of light and elevated temperatures.

Biochemical and microbiological changes in fats

It is necessary to distinguish between hydrolytic and oxidative (desmolytic) processes, catalysed by specific enzyme groups.

Biochemical and microbial fat hydrolysis

Biochemical hydrolytic changes in fats depend on the activity of enzymes present in vegetable and animal tissue as well as on the activity of microbial enzymes. Water is an essential precondition for growth of microbes, so that fats containing water, such as butter, margarine, mayonnaise and fatty tissue (which also contain small amounts of carbohydrates, proteins and minerals as nutrient substrate for microorganisms), are particularly prone to decomposition by microbial attack (in addition to autoxidation). Hydrolytic spoilage frequently predominates in fats or fatty foods when microbes (moulds, yeasts, bacteria) are present.

Fats become sour (as in chemical hydrolysis) when acted on by the hydrolytic enzymes (lipases) from microbes (fungi such as *Penicillium, Aspergillus;* bacteria such as *Serratia marcescens, Pseudomonas aeruginose* etc.). This deterioration described as 'soapiness', occurs when those fats whose glycerides contain fatty acids of medium chain length (e.g. capronic, caprylic, capric acids), are hydrolysed. Coconut oil and palm kernel oils which have deteriorated in this way, and bakery goods made from them, often taste 'soapy'.

The more unsaturated the fatty acids of the triglycerides attackers the more rapid will be the biochemical and microbiological hydrolysis of the fat.

Biochemical and microbiological-oxidative (desmolytic) breakdown
Enzymatic-oxidative decomposition of fat can be of two kinds. The chemical composition of the fatty acids and their glycerides determines the pathway.

According to our present knowledge, the oxidative-enzymatic decomposition of fatty acids in higher plants (oil seeds and grains) is due to the presence of **lipoxygenases** (lipoxidases); (see p. 255).

Lipoxygenases have been isolated from soya beans, vegetables, oil seeds and grain, but not from animal tissue. Lipoxygenases specifically attack polyunsaturated fatty acids; oleic acids and saturated fatty acids are not affected. The attack occurs at the (activated) – methylene ($-CH_2-$) groups in the polyunsaturated acids with the formation of radicals – as in autoxidation. Some conjugated cis-trans-hydroperoxides are formed, whilst in autoxidation (under the same temperature conditions of 0°C) considerable amounts of these hydroperoxides are formed. The mechanism of the lipoxygenase catalysed reaction is not the same as that of autoxidation. Phenolic antioxidants inhibit the lipoxygenase reaction. Lipoxygenases are also active at low temperatures, – and important factor in the storage of deep frozen foods containing fats, (see pp. 294 and 296).

Another form of oxidative-enzymatic decomposition occurs with certain moulds (*Aspergillus, Penicillium, Sclerotina-Monilia*), which attack fatty acids up to C14 in length. When the fatty acids have been set free from the glycerides by hydrolysis, the early stages of decomposition occur similar to β-oxidation, but the β-ketones formed, are found in the form of a CoA compound. This is in contrast to the normal β decomposition of fatty acids where they are converted into the corresponding methyl ketones with cleavage of the CO_2.

$$R-CH_2-\overset{\beta}{C}H_2-\overset{\alpha}{C}H_2-CO-SCoA \longrightarrow R-CH_2-\overset{\beta}{C}O-\overset{\alpha}{C}H_2-CO-SCoA \xrightarrow{H_2O}$$

$$\underset{\text{methylketone}}{R-CH_2-CO-CH_3} + CO_2 + CoASH$$

The formation of the various methyl ketones leads to the so called 'rancidity aroma'.

In cheesemaking the formation of methyl ketone by moulds is deliberately induced and the C_5, C_7, C_9, C_{11} methyl ketones are desired as flavour materials for Roquefort, camembert and other kinds of cheese.

The penetrating smell and flavour of methyl ketones can be recognised in the small concentrations of less than 1μg/g of fat, whereas in edible fats they are only found offensive at 60μg/g. Coconut oil, palm kernel oil, butter fat and pork lard as well as foods containing these fats (e.g. margarine and some bakery goods) are particularly liable to the formation of methyl ketones. The manufacture and storage of these foods must be carried out with special precautions to render the growth of moulds difficult or impossible. (Control of humidity in the food, relative humidity and storage temperatures etc.).

4.9 PRECAUTIONS IN THE STORAGE OF FATS

Fats play an important role in human nutrition. On the one hand they are a rich calorie food for supplying energy (1g fat = 9.3kcal). On the other hand, their content of 'essential fatty acids' (see p. 229) and fat soluble vitamins (see p. 201) provides the body with nutrients necessary for life. The correct storage of fats is therefore concerned with the preservation of the pure fat glycerides and of the raw materials which contain fats. The nutritionally valuable materials or trace materials which accompany fats should also be protected. To achieve this aim and to guard against biochemical and microbial spoilage, the manufacture of foods should be carried out under hygienic conditions. Where this is not possible, the microorganisms should be later suppressed or inhibited.

This can be carried out in the following ways:

1. Use of an elevated temperature, for instance, preparation of lard.
2. Drying of the fats or foods containing fats under vacuum at an elevated temperature (see deodorisation of fats, p. 102) or, as in the case of lard, melted butter and dried full milk, by removal of water or by storing the foods containing fat, for instance bakery goods, a comparatively low relative humidity (see p. 298).
3. Use of suitable sporcidal materials (preservatives), whose safety must have been proved, for instance sorbic acid, p-hydroxybenzoic and ester, benzoic acid (see p. 313).
4. Storage in a suitable gas atmosphere (inert gases: N_2, CO_2); compare storage of refined oils and dried whole milk powder.
5. Storage at suitable low temperatures (cold and frozen storage), bearing in mind that some enzymes in microorganisms, for instance, lipases can remain active though much less so, down to −30°C. Nevertheless, the use of mechanical cooling for the storage of fats and foods which contain fats, (butter, margarine, lard, animal fatty tissue) is widespread and, when carefully used, is one of the best methods for prevention of spoilage of edible fats; compare keeping foodstuffs fresh, p. 291.

Protective measures against purely chemical decomposition are aimed initially against oxidation processes (autoxidation) and the factors which influence them:

1. Prevention or limitation of entry of air (oxygen), suitable packaging, storage in airtight containers, under vacuum or in inert gas atmosphere, for instance storage of refined oils in tanks under CO_2.
2. Exclusion of light, especially the chemically reactive short waves because chain reactions set in train by light (see autoxidation, p. 111), continue in the dark. Exclusion of direct sunlight, use of green or red light in the storage area, use of opaque or coloured (light filtration) or specially impregnated parchment, of coloured cellophanes which absorb ultra violet rays.
3. Use of lower temperatures from −12 to −20°C to slow down any oxidation processes which might have already begun and to guard against further rapid autoxidation during storage (see p. 292).
4. Prevention of access by catalytic, prooxidative materials, such as trace metals (Fe, Cu) during the technological processing of the fat; elimination of prooxidants originating in the fat (chlorophyll, haem blood pigments, lipases, trace metals) by refining whilst preserving the valuable trace materials in the fats (see p. 99 'complete' oils).
5. Use of natural or synthetic **antioxidants** (inhibitors).

Antioxidants react with the fat hydroperoxides which have been formed and prevent the radical chain reaction from continuing further. Antioxidants are consumed during autoxidation and therefore only provide protection for a limited time. An oxidised fat cannot be restored to its original fresh state by the addition of antioxidants. Antioxidants must not be added above a certain level, otherwise a prooxidative effect (inversion effect) would occur. The following conditions must be fulfilled by a material with an antioxidant effect:

An active antioxidant used for fats must be legally permitted and have been proved harmless to health; it must not affect the fat or fatty foods so far as taste, smell or colour is concerned.

The inhibitor must to some extent be fat soluble, so that it can be dispersed and be effective.

Natural antioxidants. Seeds containing fat also contain natural antioxidants. These are found in some crude oils, after pressing and extraction, and protect them from rapid deterioration. Small amounts of such oils (1–10%) – soya bean, rape seed and cotton seed oil have been added to other oils more liable to autoxidation to increase keepability.

Wheat germ oil has a good inhibitory effect and the **tocopherols** (vitamin E, see p. 205) have been shown to be the active antioxidants. α-tocopherol is widely used in industry as a fat stabiliser, sometimes in combination with synergists such as ascorbylpalmitate and citric acid. The good keepability of pork lard, when carefully treated depends mainly on the effect of tocopherol (derived from the seed).

Carotene extracts, which were originally added to the fat (e.g. margarine) for colouring purposes, have an antioxidant effect in fats stored in the dark, but have the opposite, prooxidant, effect in light, so that carotene can be anti- or a prooxidant depending on the conditions. β-carotene and palm oil concentrates rich in β-carotene are frequently used for colouring margarine.

Small amounts of **guaiac gum** and the non-toxic benz-catechin derivate **nordihydroguaiaretic acid** (NDGA) in concentrations from 0.01 upwards, are effective against autoxidation of fats and can be used in lard in certain countries.

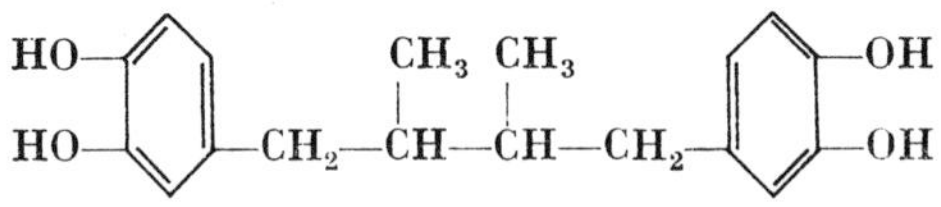

$$(HO)_2C_6H_3{-}CH_2{-}CH(CH_3){-}CH(CH_3){-}CH_2{-}C_6H_3(OH)_2$$

Natural (phenolic) antioxidants are also present in nearly all spices (pepper, cloves, ginger, marjoram, vanilla pod, woodruff etc.) The phenolic substances in wood smoke absorbed by the fatty tissues of the food being smoked have an antioxidant effect (and to some extent an antimicrobial effect).

Synthetic antioxidants. Considerable research has gone into the synthesis of antioxidants. In many cases the phenolic character was taken as the active centre in the natural material. Only a small number of the synthetic effective active antioxidants so far produced, are suitable for use from safety considerations.

The **gallates** = propyl-, octyl- and dodecyl- esters of gallic acid) are the most important antioxidants which have been tested and permitted in various countries. Other phenolic antioxidants which have found practical use are t-butyl hydroxy-anisole (BHA)* and 2,6-ditert butyl-4-methyl-phenol (BHT, ditert-butyl-hydroxy-toluene also called Ionol); these two have proved to be particularly suitable for bakery goods containing fat (carry through-effect).

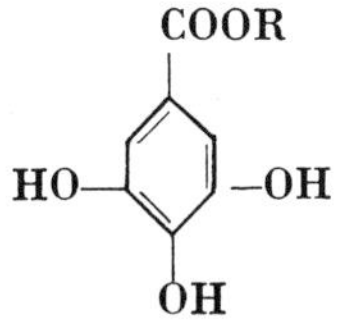

COOR / HO— —OH / OH — Gallate

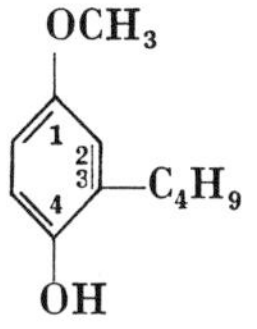

OCH_3 / C_4H_9 / OH — 3-butyl-4-hydroxy-anisole (BHA)

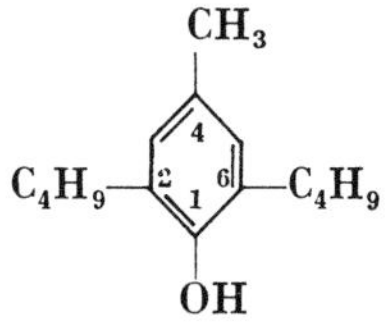

CH_3 / C_4H_9 C_4H_9 / OH — BHT (Ionol)

Thiodipropionic acid and its dodecyl ester have been used to a small extent.

The concentrations of the antioxidants used are kept low: for instance 0.005-0.01% of the gallates are used.

*BHA = butylated hydroxyanisol = a mixture of 2-tert butyl-4-hydroxy-anisole and 3-tert butyl-4-hydroxy-anisole. The commonest form is the 3-isomer (see formula); according to WHO and FAO, not more than 12% of 2-BHA may be present.

The action of many phenolic antioxidants is increased by so-called synergists, such as citric acid, phosphoric acid, amino acids, ascorbic acid, lecithin and so on. Their effectiveness has not yet been fully explained. They act partly as complex formers for the binding of trace metals which have a prooxidant effect (see the use of citric acid in deodorisation). In addition they partly take over the regeneration of the the phenolic antioxidants used up in the course of autoxidation (hydrogen buffer action of synergists). Because of this, in practice, a mixture of antioxidants and synergists is used. Such a mixture is produced and sold in the USA under the name Tenox I and contains BHA, propyl gallate and citric acid as the active agents. In some countries these fat antioxidants and synergists are used for the stabilisation of milk, egg, fish and cereal products.

It must be emphasised that if antioxidants are to be used effectively, it is necessary to consider the nature of the fat to be protected (whether animal or vegetable), its composition (degree of unsaturation) and the condition (fresh, pure, oxidised, free fatty acid content) so that the amounts needed for protection vary considerably. Little is known so far about the fate of antioxidants used up in the protection.

4.10 ACCOMPANYING MATERIALS OF FATS (LIPIDS)

As has been mentioned already, natural fats and oils contain, apart from the glycerides, a number of lipophilic materials of the most diverse chemical make up. Among the most interesting are the **phosphatide** and **lipochrome** groups, the **fat soluble vitamins**, the **sterols** and **hydrocarbons**. The fat soluble vitamins (vitamin A, carotene, vitamin D and E) are of particular importance in nutrition. To preserve these trace materials, the food chemists and food engineers responsible responsible for processing and preparation of fats and fatty foods for storage must use suitable methods, particularly on an industrial scale. They must consider that a number of the nutritionally important materials, because of their chemical make up, are particularly reactive with air (oxygen), humidity, raised temperatures and are sensitive to specific enzymes, metals and chemicals.

Phosphatides (glycero-phosphatides)

These are glycerol esters, and are the most similar to fats of all the lipids. They contain among their characteristic basic building blocks, in addition to glycerol, fatty acids and phosphoric acid, a component containing nitrogen (in the form of an amino alcohol). The cyclic alcohol inositol may be present also. The fatty acids and the phosphoric acid are esterified to the glycerol like an ester, the amino alcohol or the inositol are linked in their turn to the phosphoric acid via an ester link. According to which amino compound, choline or colamine, is present, they are called **lecithins** or **cephalins**. Lysolecithin and lysocephalin contain only one fatty acid residue.

$HO—CH_2—CH_2—N(CH_3)_3]^{\oplus}OH^{\ominus}$ = choline

$HO—CH_2—CH_2—NH_2$ = colamine

$HO—CH_2—CH(HN_2)—COOH$ = serine

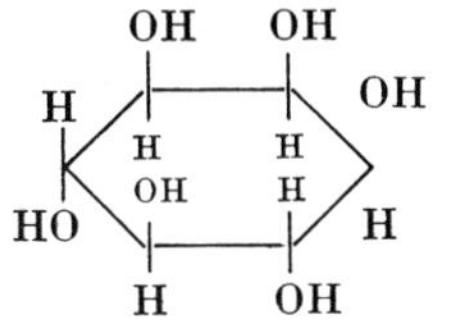

= inositol (myo form)

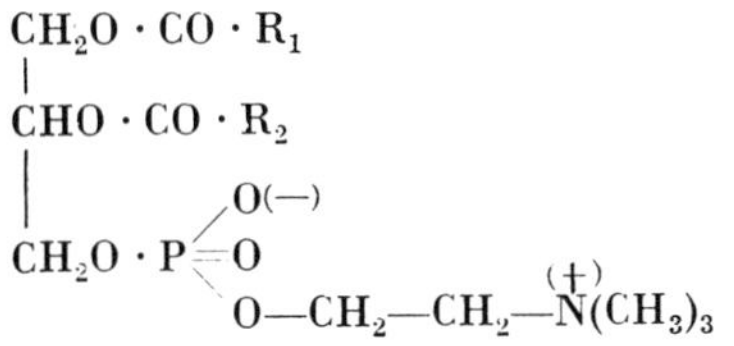

$CH_2O \cdot CO \cdot R_1$ | $CHO \cdot CO \cdot R_2$ | $CH_2O \cdot P(O^{(-)})(=O)—O—CH_2—CH_2—\overset{(+)}{N}(CH_3)_3$

α-lecithin

$CH_2O \cdot CO \cdot R_1$ | $CHO \cdot CO \cdot R_2$ | $CH_2O \cdot P(OH)(=O)—O—CH_2—CH_2—NH_2$

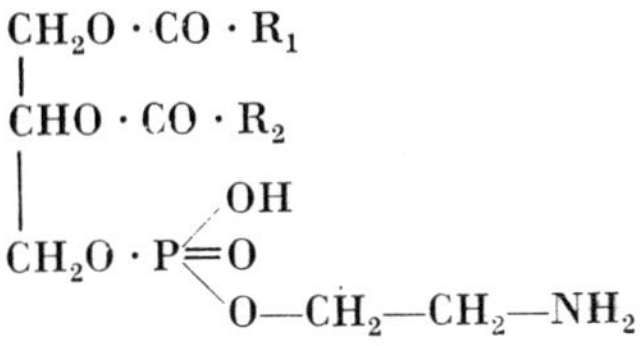

α-cephalin

The strongly basic methyl amino groups of choline in lecithin form an internal salt with the phosphoric acid (with exclusion H_2O) α-cephalin is a weak acid.

The compounds called **serine cephalins** contain the amino acid serine in the place of the amino alcohol colamine (see p. 42). It is thought that the cephalins arise secondarily from them by decarboxylation.

Choline is formed by complete methylisation of the nitrogen atom colamine. Biogenic rearrangements are therefore possible between the individual phosphatides; see p. 121. The linkage to protein in lipoprotein is by serine.

In **phosphoinositols**, which are widely present in vegetable oils, the hexavalent cyclo alcohol inositol is esterified with phosphoric acid.*

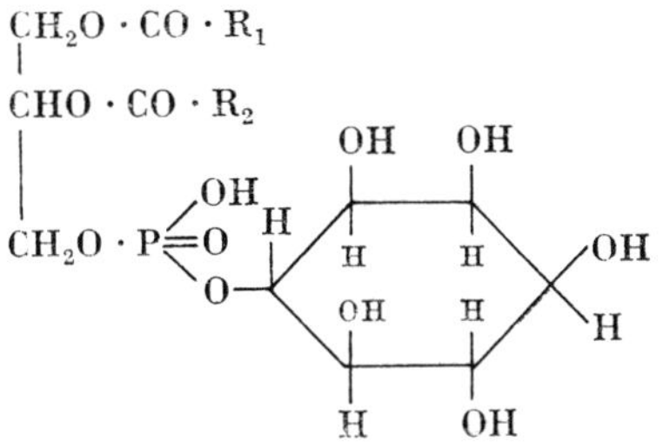

phosphoinositide

*Phytic acid is the hexaphosphoric acid ester of inositol. The K or Ca or Mg salts, potassium phytate, calcium phytate, magnesium phytate, collectively called phytin, are present in the husk of cereals.

The commonest fatty acids which are most frequently found attached to the glycerol as esters are palmitic, stearic, oleic, linoleic, linolenic and arachidonic. No natural lecithin or cephalin contains exclusively saturated fatty acids, one of the unsaturated fatty acids listed above is always included. The analytical separation of the phosphatides is difficult. This is understandable when one considers that the most diverse fatty acid residues occur in the lecithin or cephalin molecule, that these substituents can be present in various arrangements (α and β forms), see p. 76 and also that, due to the presence of an asymetric C atom, optically stereoisomeric (enantiomers) forms are also present. Each fat has its own phosphatide mixture. In the same way that each organ of the body has its own protein.

Lecithin and cephalin are soluble in ether, alcohol and chloroform, but not in acetone or glacial acetic acid. The difference in solubility in alcohol is often used to separate the phosphatides from the fats. The phosphatides are precipitated with metal salts (Cd, Au, Pt). In water they swell easily and are sometimes soluble in colloidal form because of the hydrophilic phosphoric acid and the basic groups, whereas the hydrophobic (water repelling) fatty acid residues in the phosphatide molecule determine the lipophilic (fat soluble) properties. Because the molecule is made up in this way the phosphatides occupy an intermediate position between water and fat soluble substances; they are polar substances and therefore good emulsifiers due to this combination of lipophilic and hydrophilic groups.

Emulsions are built up of two non miscible components (droplets in liquids) – usually water and a lipid (oil, fat, wax, paraffin). One of these components is dispersed in the other. When oil is dispersed in water, it is called an 'oil in water emulsion' (When oil is dispersed in water, it is called an 'oil in water emulsion' (milk, cream, cod liver oil emulsion), the reverse with water dispersed in oil is called a 'water in oil emulsion' (butter is a solidified water in oil emulsion). If two non-miscible liquids are shaken together, an inhomogeneous, cloudy mixture is formed which is unstable and breaks up after a time. (Separation). The total surface area is greater the more finely distributed the dispersed phase is, and the greater the forces at the interface, the so-called surface tension. In the case of an oil in water emulsion, one speaks of the surface tension of the oil in water. This surface tension energy exists in all two phase systems, whether they are liquid-liquid (emulsions), solid-liquid, solid-gas, solid-solid or liquid-gas. The surface tension force depends on the area of the surface. The tendency of the disperse phase to break down is brought about by the force required to reduce its surface to its minimum. When it is desired to stabilise the emulsion for a longer period, emulsifiers are used. Because of their physico-chemical properties, (polarity, adsorption enrichment at the interface) emulsifiers reduce the surface energy of the disperse phase in relation to the dispersing agent and so make it more difficult for the dispersed particles to aggregate together. Emulsifying agents have a certain 'orientation effect' because of their polarity – the molecule contains both lipophilic groups (e.g. aliphatic and aromatic hydrocarbons) and hydrophilic groups (e.g. COOH, OH, NH_2, SO_3H groups). The lipophilic groups, soluble in the oil phase, and the hydrophilic group, soluble in the water phase, align themselves, so that the emulsifier binds the two phases together. In an oil in water emulsion, for instance, the emulsifier (e.g. lecithin or a monoglyceride, see p. 120) enriches the interface of each droplet to such an extent that the hydrocarbon residues of the fatty acids penetrate the droplet and the polar portion of the lecithin or OH groups of the monoglyceride rearrange in the dispersion agent (water) as an electrical dipolar force. Emulsifiers, which reduce the surface tension are surface active or capillary active. The enrichment of such materials at the interface is called adsorption. Adsorption and exchange processes at interface play a large part in biochemistry, industrial chemistry and in food chemistry: chromatographic adsorption, paper chromatography; ion exchangers.

The phosphatides are widespread and of great importance to all cell metabolism because of their polar make up. It has been established that the more important an organ is for life, the greater the phosphatide content in both animals and plants. Bone marrow (6.3–10.8%), brain (3.7–6.0%), liver (1.0–4.9% and heart (1.2–3.4% in animals have a high phosphatide content; egg yolk is also rich in them (8–10%). Phosphatides are found especially in the seeds, roots and underground storage organs of plants. The leguminosae have a particularly high phosphatide content (up to 2.2%). Phosphatides are also found in yeasts, fungi and bacteria. Natural crude oils and fats have a variable phosphatide content: rape seed oil and sesame oil 0.1%, wheat germ oil up to 2%, barley germ oil 3–4%, rye germ oil 3–4%, soya bean oil up to 3.2%;

Phosphatides seldom occur free in nature because of their high surface activity. Usually they are found in association with fats and other lipids or they are present in symplexes more or less strongly associated with proteins (lecithalbumin) or carbohydrates. They are difficult to isolate in a pure state.

For the **commercial production** of phosphatides, for instance lecithins, mainly vegetable oils (e.g. soya oil) which are rich in lecithin are used. The process involves extraction with ethyl alcohol and other solvents. Crude oils can be used after previous treatment with water, steam, or salt solutions (settling of phosphatides) followed by centrifugation and purification similar to the first stages of refining, de-gumming, see p. 99.

Phosphatides are used as **binding agents** in the food industry because of their ability to emulsify. In the production of pasta, phosphatides made from vegetable sources substitute for the binding power of egg lecithin. In the chocolate industry they play a similar role in that by their binding power they give the chocolate a desirable smoothness. The largest quantities of phosphatides are used in the manufacture of margarine, where they assist the emulsification of the water in the fat; in this way the margarine foams and browns in the pan as desired but unwanted spattering is prevented.

Because they contain unsaturated fatty acids and have four ester linkages in the molecule, the phosphatides are chemically and biochemically highly reactive. Phosphatides are physiologically important because they provide a method of transport of fatty acids within the body. This applies not only to the transport of fatty acids in the blood, but also to the transport across the cell walls, which are permeable to phosphatides but not to pure fats (proof by means of radioactive ^{32}P).

When the ester bonds are broken (hydrolysis) various products arise from phosphatides, which are also frequently found associated with them, for instance phosphatidic acids, (after the nitrogenous part has been split off) in leaves and in other parts of plants. Egg pasta contains special enzymatic phosphatide cleavage products which affect the storage and quality of these products. See p. 109. Lysolecithins in cereal starch from cereals is similar. See p. 260.

Decomposition processes of this kind play a direct role in the deterioration of fats, because this particular deterioration called fishiness is related to the cleavage of the nitrogen component (choline, colamine) of the phosphatides accompanying the fat (amine formation). Fishiness also occurs in other products apart from butter and margarine, which contain lecithin, such as milk and egg yolk.

Other lipophilic materials which accompany animal and vegetable fats are the **sphingolipids**, **saccharolipids** and **lipoproteins**, three groups of materials with a complicated chemical structure. Because of their polar properties they occur linked as symplexes with each other and with fats. See also the chapter on symplexes p. 65.

Lipochromes and lipovitamins

Many yellow and red to dark violet pigments, which are found in the animal and plant world, are called lipochromes because of their solubility in fats and fat solvents. They occur in free or esterified form accompanying fats and lipids, for instance in plants as **wax pigments*** (=esters of a hydroxyl group in a carotenoid with a higher fatty acid). These are also loosely bound to protein as as a **chromoprotein** (p. 64).

Carotenoids, the most important group of lipochromes, are closely associated with the colour of natural fats. **Chlorophylls** are another group of lipochromes widely found in nature, although their yellowish colour is sometimes masked by the red and yellow colours of the carotenoids and xanthophylls. The lipochromes which accompany fats are responsible for: the characteristic colour of red palm oil (carotene), of greenish red pumpkin kernel oil, of yellow linseed oil, of olive oil which is often greenish (chlorophyll), and of the greenish animal fats (milk fat, cream, butter, etc.).

*The waxes also belong to the lipid group because of their solubility. They are esters of a monovalent high molecular fatty acid. For instance bees wax contains an ester from myricyl alcohol ($C_{31}H_{63}OH$) and palmitic acid ($C_{15}H_{31}$).

Carotenoids

The carotenoids are basically isoprene derivatives, whose colour is constitutionally due to their polyunsaturated nature (double bond in the conjugated position). They are sometimes pure hydrocarbons, for instance carotene (red pigment of carrot) and lycopene (red pigment of tomato). Sometimes they also have OH groups, as in the case of zanthophyll (=lutein) in egg yolk, milk, butter, zeaxanthin (maize), capsanthin (paprika), cryptoxanthin (maize, paprika, oranges), or COOH groups as in crocetin which is esterified with the disaccharide gentiobiose in crocin, the colouring matter of saffron.

β-carotene, occurs most widely in nature and may serve as an example of the three natural carotenes from which the name of the whole class derives.

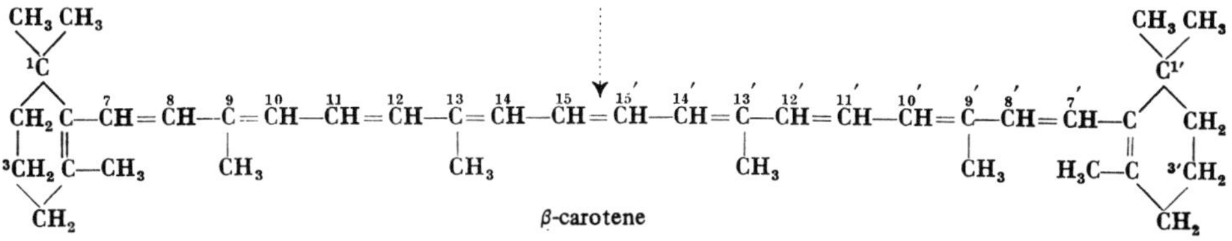

```
CH3 CH3                                                                                              CH3 CH3
  \ /                                                                                                  \ /
  1C                                           ↓                                                       C1'
 /  \    7    8     9  10   11   12   13  14   15   15'  14'  13'  12'  11'  10'  9'  8'   7'         /   \
CH2 C—CH=CH—C=CH—CH=CH—C=CH—CH=CH—CH=C—CH=CH—CH=C—CH=CH—C     CH2
|   ||          |               |                    |               |          ||     |
3CH2 C—CH3     CH3             CH3                  CH3             CH3     H3C—C   3'CH2
  \ /                                                                             \  /
  CH2                                  β-carotene                                  CH2
```

The other two carotenes, α- and γ-carotene differ from β-carotene in the ring at the end of the molecule; in α-carotene the double bond is displaced in the α-ionone ring, whereas in γ-carotene the ionone ring is opened.

All three carotenes are **precursors** of vitamin A (see p. 201), so-called **provitamins**; they yield vitamin A when they are broken down in the body: two molecules of vitamin A are formed from β-carotene and α- and γ-carotene each only yield one molecule. As can be seen, the biological activity depends on the presence of the β-ionone ring in the molecule. (Position cleavage shown by arrow).

Xanthophyll (lutein) is the 3,3′-dihydroxy derivative of α-carotene.

The various carotenoids are widespread in nature, and usually occur together. Chromatographic adsorption analysis is generally used to separate them into individual pigments.

Carotene can be obtained from natural products by any of the following means: by extraction from fresh or dried carrots rich in carotene with oils and fats, from raw palm oil by molecular distillation or by adsorption from the extraction miscella during the refining of oil. Carotene is now produced synthetically on a large scale.

Bixin is a carotenoid (but not a provitamin A0) from the seed of the tropical annatto bush. Annatto is an oily or watery alkaline extract from seeds containing orange-red bixin as the colouring component. This can be used to colour margarine.

Because of the many conjugated double bonds, most carotenoids or carotene pigments are sensitive to oxygen, especially in the presence of light.

They are readily autoxidised, and the decomposition yields epoxides and furanoid oxides. As lipids accompanying edible fats they encourage oxidation (pro-oxidants) in the presence of light: when light is excluded however it has been shown that carotene has an antioxidant effect (see p. 117). After autoxidation the carotenes loose their colour (bleaching). During storage of fats and foods containing caroten light and oxygen should be excluded. Absorption of carotene from the digestive tract is much increased in the presence of edible fats and fatty foods.

Chlorophyll

Chlorophyll is the green pigment in plants, which makes photosynthesis possible. It is usually found in the form of so-called chloroplastic (chromoprotein, see p. 64) in the chloroplats with the yellow pigments carotene and xanthophyll. The pigment itself is a mixture of two porphyrin derivatives containing magnesium, which differ in colour and in chemical composition. Chlorophyll b is yellowish-green while a appears more blue-green. The chlorophylls are esters with phytol and hence are wax pigments (see p. 122). Phytol (a higher terpene alcohol) and methanol are the alcoholic components.

The greenish colour of olive oil, rape seed oil and soya bean oil (from unripe or frost-damaged beans) is due to the presence of chlorophyll. Chlorophyll is not desirable in oils – with the exception of olive oil – because it is difficult to remove during refining and usually has a prooxidant effect in oxidative deterioration of fats.

Both the chlorophylls, while retaining their complex bound magnesium, are split by weak alkalis to give the alcohols (phytol and methanol) and the alkali salts of the respective (acid) chlorophyllins. These are soluble in water in contrast to the fat soluble chlorophylls. Weak acids on the other hand remove

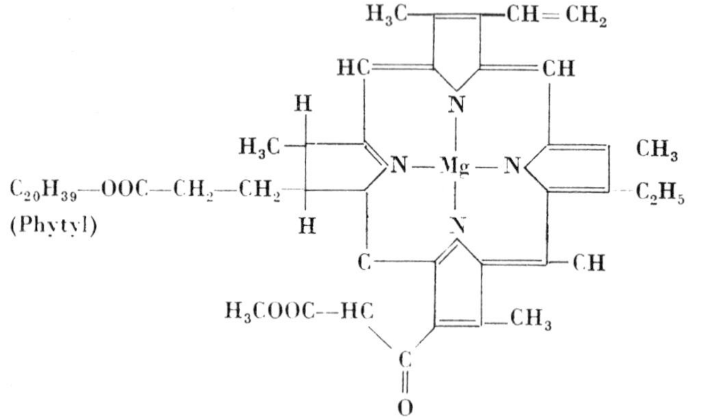

H_3C — $CH{=}CH_2$
HC = CH
H N
H_3C N — Mg — N CH_3
$C_{20}H_{39}$—OOC—CH_2—CH_2 C_2H_5
(Phytyl) H N
C — CH
H_3COOC—HC — CH_3
C
‖
O

Chlorophyll a [1]

*In chlorophyll b the methyl group in the ethyl substituted pyrrole ring is replaced by a formyl group.

the magnesium from the chlorophyll molecule to give the greenish-brown phaeophytin. Although all plant cells have a slightly acid reaction no phaeophytin is formed in them, because the chlorophyll molecule in the plant cells is protected by protein or lipoprotein (symplex formation).

Lipovitamins

Only a summary is given here of fat soluble vitamins which occur in company with fats, because they are treated in detail in the chapter on vitamins:
vitamin A and A_2 and provitamins A (carotenoids)
vitamin D (D_2, D_3) and their provitamins
vitamin E (tocopherols)
vitamin K, although fat soluble, is hardly ever contained in edible fats.

The materials which protect the skin, formerly called vitamin F, are not vitamins as they are only effective in large amounts although essential in nutrition, (see p. 209).

The lipovitamins are particularly sensitive to oxygen. Only vitamin A is sensitive to light; all the lipovitamins are heat stable.

Sterols

The sterols are another class of lipid materials which always accompany phosphatides. Sterols are hydroxyl derivatives of cyclopentano-perhydrophenanthrenes (steranes); they are found free or esterified with fatty acids and are widely distributed in nature.

The **steroids** have a similar chemical structure to the D vitamins – gallic acids, sex hormones and saponins the specific active agent of the suprarenal cortex among others – belong to this group. Sterols can transform into steroids this is seen most clearly by the formation of the D vitamins from specific sterols by ultra violet irradiation (see p. 203), they are therefore of physiological importance as provitamins. They also take part in the formation of the cell membranes and in the processes of detoxication in the organism.

The sterols all have similar chemical and physical properties; they are insoluble in water and readily soluble in lipophilic solvents and in fats. The similarity in the structure of the molecule explains their tendency to form mixed crystals. As secondary alcohols they can form esters with acids, which is useful in identification. Digitonin forms insoluble molecular dompounds with them. Sterol acetates can be distinguished in this way.

The sterols are divided into animal sterol (zoo-sterol), vegetable sterols (phyto-sterol) and fungal sterol (myco-sterols). **Cholesterol** is the most important **zoo-sterol**. It is present in all animal fats in amounts from 0.1-0.4%, in butter 0.18% in larger amounts in liver oil (1%) and in egg yolk (1.3%).

Like other lipids, cholesterol is soluble in ether, chloroform, light petroleum and in fats and oils. Its solubility in cold ethyl acetate enables it to be easily separated from other lipids. Cholesterol melts at 148.3°C.

Cholesterol is formed mainly in the liver and is secreted from the gall bladder, into the duodenum, where it is esterified with free fatty acids by cholesterinase. These fatty acid cholesterol esters can pass through the intestinal wall and are then broken down, an important process in the transporting of fatty acids from food.

Dihydrocholesterol is usually found with cholesterol. Its isomers are present in the form of coprosterol in animal excreta. 7-dehydrocholesterol is found in the skin of animals (as provitamin D_3); a rich source is pig skin.

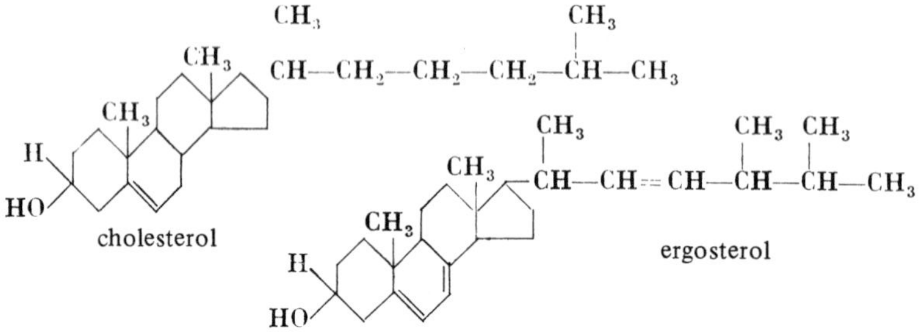

CH_3 (side chain): $CH—CH_2—CH_2—CH_2—CH(CH_3)—CH_3$

cholesterol

CH_3 (side chain): $CH—CH=CH—CH(CH_3)—CH(CH_3)—CH_3$

ergosterol

The following are the **phytosterols**: **stigmasterol** in rape seed oil; **sitosterol** which is found in wheat germ, maize, linseed, cotton seed, sesame and olive oil and in coconut oil. They give the same colour reactions as cholesterol. **Ergosterol** (a mycosterol) is found in yeast, ergot and other fungi in large amounts and also in a few seed oils and roots.

In spite of the close resemblance in their chemical reactions, cholesterol and phytosterols show considerable differences in their pure crystalline form and also in the melting points of their esters. A. BOMER based his classical phytosterol and phytosterol acetate tests which distinguish between vegetable and animal fats and prove the presence of vegetable fats in animal fats, on the melting points of the esters and the precipitation of the sterols by digitonia.

Ergosterol is the provitamin of antirachitic vitamin D_2, but 7-dehydrocholesterol in the provitamin of vitamin D_3. 7-dehydrocholesterol occurs in liver oil and is also formed in human skin on exposure to sunlight. Both D vitamins are also formed from their provitamins by exposure to artificial light containing ultraviolet radiation.

$$\text{7-dehydrocholesterol} \xrightarrow[\text{light}]{\text{UV}} \text{vitamin } D_3$$

$$\text{ergosterol} \xrightarrow[\text{light}]{\text{UV}} \text{vitamin } D_2$$

Cholesterol, when subjected to ultraviolet irradiation, does not give a product active against rickets. It must first be changed into 7-dehydrocholesterol, as occurs in the mucous membranes of the gut by the action of cholesterol dehydrogenase. Vegetable and animal organisms can synthesise the sterols themselves, proved by the radio-active ^{14}C-isotopes.

The starting point for biosynthesis is 'activated acetic acid' (= acetyl-CoA), from which an important key substance, mevalonic acid is formed. After phosphorylation and a number of transformations the intermediate product **isopentenyldisphosphate** (also called 'active isoprene') important for many biosyntheses is formed. This is transformed, via farnesyldiphosphate, squalene and lanosterol, into cholesterol.

Vitamin D_3 is the physiologically active D vitamin, and is almost the only one found in nature as a complete vitamin; it occurs mainly in the liver oils of fishes.

Hydrocarbons

The six fold unsaturated hydrocarbon **squalene**, $C_{30}H_{50}$, is found in large amounts in shark livers (20–85%), in yeast fat (5–16%) and also in vegetable oils (olive oil up to 0.6%, wheat germ oil 0.35%, etc).

Its physiological importance is not yet known for sure. Vitamin K_2 (see p. 208) contains squalene in its side chain. Chemically, squalene is a compound in which the six partly hydrogenated isoprene residues are linked to one another. The starting point for the biosynthesis – as in cholesterol – is 'activated acetic acid', the acetyl-CoA.

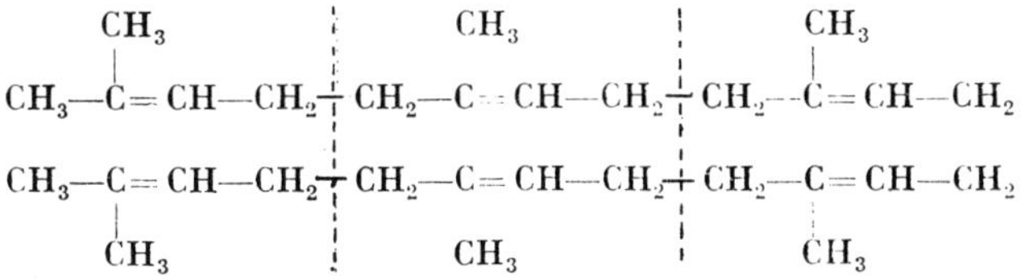

$$\begin{array}{l}
\quad\;\; CH_3 \qquad\qquad\qquad\qquad CH_3 \qquad\qquad\qquad\qquad CH_3 \\
\quad\;\; | \qquad\qquad\qquad\qquad\qquad\quad\;\; | \qquad\qquad\qquad\qquad\qquad\; | \\
CH_3{-}C{=}CH{-}CH_2 \,\vdots\, CH_2{-}C{=}CH{-}CH_2 \,\vdots\, CH_2{-}C{=}CH{-}CH_2 \\
CH_3{-}C{=}CH{-}CH_2 \,\vdots\, CH_2{-}C{=}CH{-}CH_2 \,\vdots\, CH_2{-}C{=}CH{-}CH_2 \\
\quad\;\; | \qquad\qquad\qquad\qquad\qquad\quad\;\; | \qquad\qquad\qquad\qquad\qquad\; | \\
\quad\;\; CH_3 \qquad\qquad\qquad\qquad CH_3 \qquad\qquad\qquad\qquad CH_3
\end{array}$$

squalene

Apart from squalene, other hydrocarbons have been isolated from fats, for instance **gadusene** $C_{18}H_{32}$ from wheat germ, rice germ, soya and fish oils, **hypogene** and **arachidene** $C_{19}H_{38}$ from ground nut oil, and a whole series, the so-called **oleadecenes** from olive oil.

These hydrocarbons are responsible for the peculiar smell and taste of these fats; in particular autoxidation products of these unsaturated hydrocarbons give an unpleasant fishy or disgusting taste and odour.

CHAPTER 5

Carbohydrates

Carbohydrates are a large group of nutrients forming a basic part of our food.

Carbohydrates are formed in plant tissues during the process of photosynthesis and while in nature they may exist as simple sugars (monosaccharides), more frequently they are compound sugars (disaccharides or polysaccharides). Amongst the latter compounds the two most important are, starch, a reserve food material and cellulose, a building or structural material.

Whereas in the animal cell proteins quantitatively make up the greater part of the solids content in plant cells, and therefore in plants used as foods, carbohydrates are the main components. In the animal (including human) organism, carbohydrates are not deposited to any great extent after their ingestion as a food, but after digestion and absorption they function as a preferred source of energy and are partly 'burnt up' in the process. During the metabolic process they are also partly transformed into fats or into some of the basic building blocks of protein compounds.

Plants, and the parts of plants, which are rich in carbohydrates form the basis for the manufacture of starch, for the preparation of fruit products and are the starting point for the paper and fibre industries. They are the starting materials in a large number of fermentation processes and biosyntheses, as in the manufacture of ethyl alcohol, butanol, glycerol, lactic acid, citric acid and so on. Their presence is usually necessary for the biological synthesis of protein and fats (see p. 277).

In addition, carbohydrates are the starting material for the formation of vitamin C in the plant and for commercial vitamin C synthesis.

The word 'carbohydrate' was originally used to indicate that these compounds correspond to the general formula $C_n(H_2O)_n$, that is they were supposed to be hydrates of carbon and to contain the elements oxygen and hydrogen in the same proportions as water. This view is now out of date, because by no means all the known compounds with the composition of $C_n(H_2O)_n$ have the characteristics of carbohydrates (e.g. acetic acid $C_2H_4O_2$, lactic acid $C_3H_6O_3$ etc.). On the other hand a great many compounds are known which, because of their behaviour, must be regarded as carbohydrates, but do not correspond to the formulation quoted above, for instance the desoxy sugars, branched sugars, methylated sugars, amino sugars, oxidation and reduction products of sugars (sugar acids, sugar alcohols) and many glycosides. This also applies to various groups of the so-called polysaccharides (more correctly polysaccharide-like materials) such as pectins, alginates, carrageenans, chitin and similar materials. The word 'carbohydrate' has now become a collective name for a whole class of materials.

In general carbohydrates are optically active compounds which, chemically, are primary oxidation products of polyvalent alcohols, that is polyhydroxy-aldehydes or polyhydroxyketones. They are usually known as saccharides and have the ending 'ose' (e.g. glucose, sucrose). The name sugar is used for the simpler members of the carbohydrate group.

Carbohydrates are usually divided into the following groups:

1. **Monosaccharides**, also called simple sugars which cannot be further decomposed by hydrolysis. Dextrose and fructose belong to this group.

2. **Oligosaccharides** are composed of a small number of monosaccharides which can be set free by hydrolysis. The transition from oligo- to polysaccharides is continuous and any exact division is arbitrary. The normal division lies around ten residue units. Di-, tri- or tetrasaccharides are distinguished according to the number of monosaccharide building blocks present (see pp. 130, 160, 168).

3. **Polysaccharides** are built up from a large number of monosaccharides. For instance starch (amylose) consists of 100–2,000 monosaccharide units (glucose). During hydrolysis the monosaccharides are set free as basic building blocks. The polymeric saccharides are macromolecular compounds and differ in their chemical and physical properties from the monosaccharides and oligosaccharides. The most important materials of this kind in nature are starch, glycogen and cellulose.

5.1 MONOSACCHARIDES

Configuration and classification of sugars

The individual monosacchardies are called trioses, tetroses, pentoses, hexoses, heptoses and so on according to the number of carbon atoms*. According to whether an aldehyde or a keto group is present in the molecule, the monosaccharides are called aldoses (e.g. glucose) or ketoses (e.g. fructose). The ketoses have the systematic ending 'ulose' (e.g. fructose is a hexulose) in addition to their trivial names. The numbering of the C atoms in the molecule is carried out

*The nomenclature follows the I.U.P.A.C. rules (International Union of Pure and Applied Chemistry Rules).

in such a way that the reducing group carries the lowest number, that is 1 in the case of the aldoses. As sugars are optically active compounds, such a reference is necessary. The simplest sugars are the trioses, that is glyceraldehyde and dihydroxyacetone (see Table, p. 132). Both are intermediate products during the biochemical synthesis and decomposition of the hexoses and therefore they are dealt with more fully here.

As far as the stereochemistry of simple sugars is concerned, they contain as many assymetrical C atoms as HCOH groups. Glyceraldehyde,which acts as reference substance has one assymetrical C atom (C_2) as the first member of the series. Dihydroxy acetone is optically inactive. According to whether the position of the OH group in the projection formula is to the right or left of the C chain, the two forms of glyceraldehyde, which are mirror images of each other, will be called D- or L-glyceraldehyde.

All sugars which have the same configuration as the middle C atom of D-glyceraldehyde on the lowest asymmetrical C atom, that is the one with the highest number, belong to the D-series; all others belong to the L-series*.

The table gives a summary of the trioses, tetroses, pentoses, hexoses and hexuloses of the D-series. Those sugars marked with * are found in nature. These formulae are written according to the scheme of EMIL FISCHER; the C chain, which in nature has angles in it, is written as if in the plane of the paper, so that all H and OH groups are drawn right and left above or below the plane or the paper. The H and OH groups cannot then be exchanged without changing the configuration. The formula may only be turned in the plane of the paper.

Optical activity of sugars

The chemical properties of enantiomers are the same because, in solution, the energy value of both isomers is equal as the distances between any attached atom or group of the carbohydrate and the other attached atoms or groups are equal in both forms. They cannot be distinguished from each other either by chemical reaction or by physical properties (melting point, solubility, specific gravity, refractive index etc.). However polarised light is deflected in an opposite but equal amount. For this reason they are called the **'optical isomers'**.

The optical activity, that is the ability of a body to turn the plane of vibration of polarised light, is also used in the food industry for characterisation and identification of sugars. The direction of rotation of a body to the right or to the left is indicated by the prefix ⊕ or ⊖, less frequently by writing *dextro* (= rotating to the right) or *laevo* (= rotating to the left). The direction of rotation is independent of whether the sugar belongs structurally to the D- or L-series. For instance D-fructose rotates to the left (D-⊖-fructose) and vitamin C, L-⊕-ascorbic acid, rotates to the right.

*Glyceraldehyde is also the reference substance for the configuration of other materials (e.g. amino acids, lactic acids etc. see p. 27).

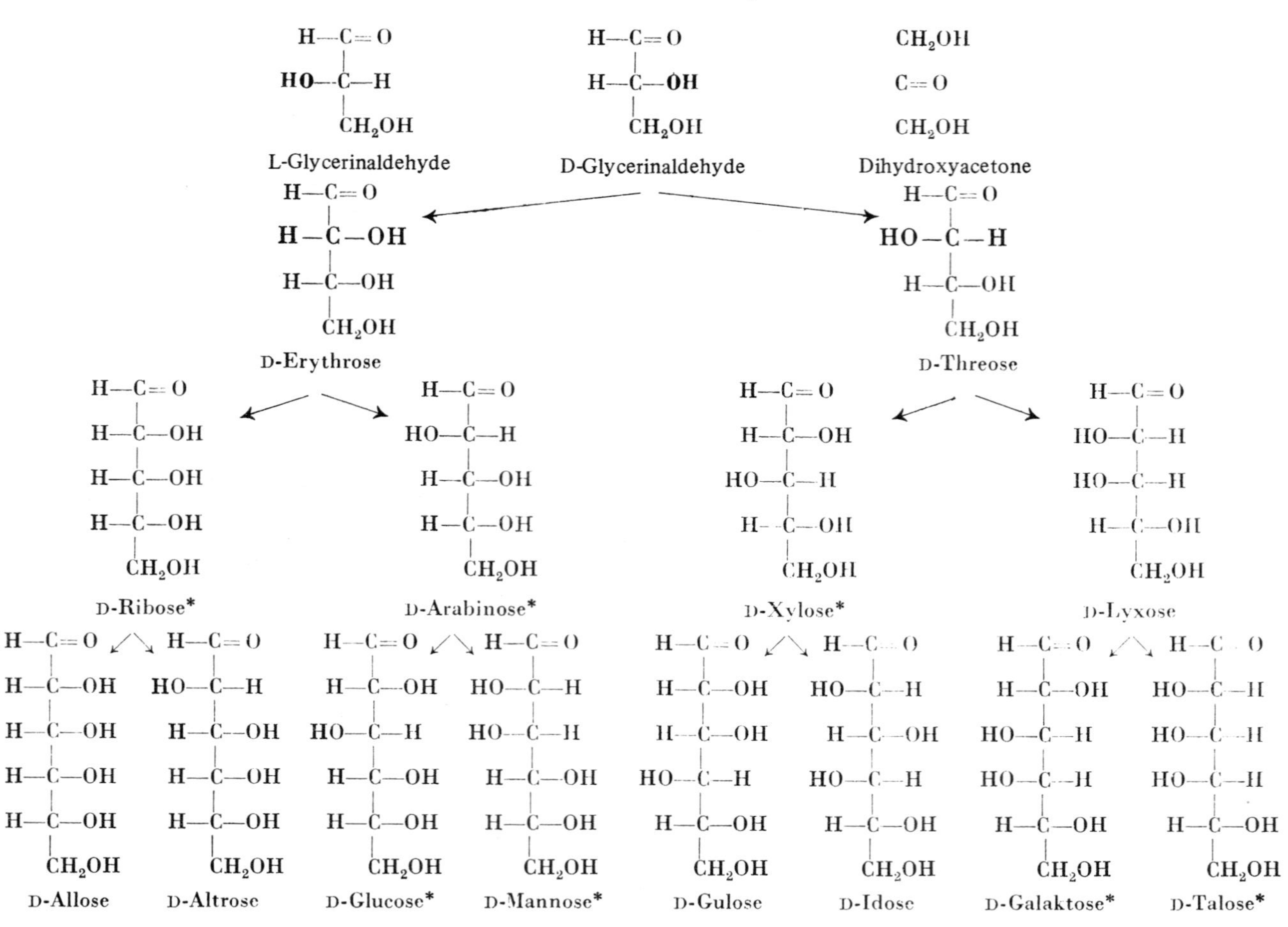
Classification of the D-sugars
L-Glycerinaldehyde
D-Glycerinaldehyde
Dihydroxyacetone
D-Erythrose
D-Threose
D-Ribose*
D-Arabinose*
D-Xylose*
D-Lyxose
D-Allose
D-Altrose
D-Glucose*
D-Mannose*
D-Gulose
D-Idose
D-Galaktose*
D-Talose*

```
    CH2OH              CH2OH             CH2OH             CH2OH
    |                  |                 |                 |
    C=O                C=O               C=O               C=O
    |                  |                 |                 |
 H—C—OH            HO—C—H             H—C—OH           HO—C—H
    |                  |                 |                 |
 H—C—OH             H—C—OH           HO—C—H            HO—C—H
    |                  |                 |                 |
 H—C—OH             H—C—OH            H—C—OH            H—C—OH
    |                  |                 |                 |
    CH2OH              CH2OH             CH2OH             CH2OH
```

D-Psicose* (Allulose, D-Ribohexulose)

D-Fructose* (D-Arabinohexulose)

D-Sorbose (D-Xylohexulose)

D-Tagatose* (D-Lyxohexulose)

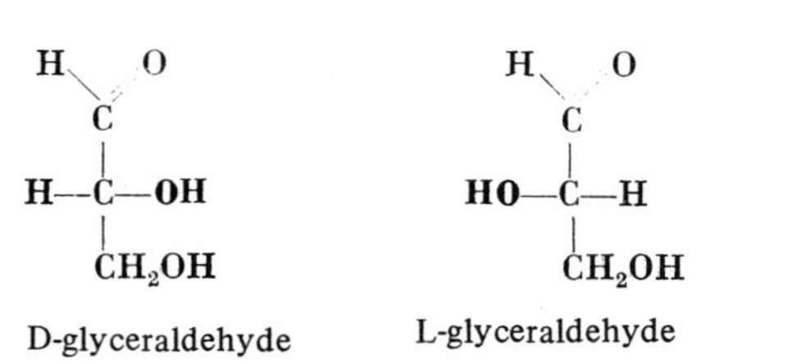

D-glyceraldehyde

L-glyceraldehyde

When equimolar amounts of D- and L- bodies are present in a crystalline mixture or a solution (mixtures, conglomerates), this mixture of antipodes (where no formation of compounds can be observed) is called a racemic mixture and is optically inactive. A mixture of equal parts of the D- and L-form will often crystallize out into a new uniform crystalline form with a different melting point and water of crystallization content. This crystallisate, which is crystallographically uniform from both isomers, is called a **racemate**. The description D-, L- form, (±)-form or inactive form, is used for both mixtures and racemates.

Not only the direction of rotation (+) or (−) but also the amount of rotation as a constant is important from a practical and scientific point of view for identifying a sugar. It is useful to know the 'specific rotation'. The 'specific rotation' of a sugar is the angle of rotation given by a sugar solution with a content of 1 g of active substance per ml with a light path of 1 dm length.

Smaller concentrations are used for measurement and the results are then converted, as the rotation is proportional to the concentration. If c is the concentration of a solution (expressed in grams per 100 ml solution), the angle read off on the polarimeter in degrees and l is the light path in dm, then

$$[a]_D^{20} = \frac{a \cdot 100}{l \cdot c} .$$

The indices D and 20 denote the wavelength of the light used for the observation (usually monochromatic light, corresponding to the wavelength of the sodium D line) and a fixed temperature (generally 20°C).

The specific rotation $[a]$ of some sugars is:

D-glucose	+52.5°	lactose	+52.5°
D-fructose	−92.3°	maltose	+136.8°
sucrose (saccharose)	+66.5°		

The relationships given in the above equation can be used quantitatively in food analysis. When the sugar present is known (even when in a mixture with others) this equation can be used to calculate the concentration of a sugar solution using the specific rotation and the rotation value measured on the polarimeter:*

$$c = \frac{a \cdot 100}{[a] \cdot l} .$$

*When the readings for a cane sugar solution are = +13.41° with a tube length = 2 dm; using the specific rotation = +66.5°, c, as the concentration (content in 100 ml) of the cane sugar solution, is obtained as

$$c = \frac{13{,}41 \cdot 100}{66{,}5 \cdot 2} = 10{,}08\ \mathrm{g/100\ ml}$$

In some sugars we can observe that the rotation value and with it the specific rotation, measured immediately after dissolving, gradually changes. This change in the rotation to a constant final value is called mutarotation or multirotation (see p. 144).

General properties and reactions of monosaccharides

Simple sugars are colourless, solid compounds, sometimes difficult to crystallize, which often form syrupy supersaturated solutions; they have a neutral reaction and, because of their hydrophilic OH groups, they dissolve readily in water but less easily in methyl and ethyl alcohol, pyridine, dimethylformamide and dimethylsulphoxide. They are insoluble in lipophilic solvents such as chloroform, ether, benzene, light petroleum and so on. Many of them taste sweet (see p. 164), others are tasteless or even bitter. Sugars decompose easily on heating because of their tendency to lose water.

The most important chemical properties of simple carbohydrates (monosaccharides) are based on the presence of the **carbonyl group** (aldehyde or keto group) and the alcoholic nature of the hydroxyl groups. The carbonyl group may affect the neighbouring hydroxyl groups, since hydroxyl groups which belong to α-carbonyl groups are readily oxidised (dehydrated). Hence the ketoses (e.g. fructose) are strong reducing agents in contrast to the simple ketones (e.g. acetone).

Various reactions are also used for the analytical characterisation of sugars. For the sake of clarity, the analytical proofs will be discussed under the heading of the reaction on which they are based.

Reactions of the reducing group. Sugars are able to react with the typical carbonyl reagents at their –C=O group: in this way the corresponding cyanhydrins are formed by addition of hydrocyanic acid

$$\text{H—}\underset{|}{\text{C}}\text{=O} + \text{HCN} \longrightarrow \text{H—}\overset{\overset{\text{C}\equiv\text{N}}{|}}{\underset{|}{\text{C}}}\text{—OH}$$

Sugars form oximes with hydroxylamine:

$$\text{H—}\underset{|}{\text{C}}\text{=O} + \text{NH}_2\text{OH} \longrightarrow \text{H—}\underset{|}{\text{C}}\text{=NOH} + \text{H}_2\text{O}$$

Reducing sugars give **hydrazones** with phenylhydrazine and when reacted for longer periods with at least 3 moles of phenylhydrazine the even more important **osazones** are formed:

$$\begin{array}{c}\text{H—C=O}\\|\\\text{H—C—OH}\\|\end{array} \xrightarrow{+\text{H}_2\text{NNHC}_6\text{H}_5} \begin{array}{c}\text{H—C=NNHC}_6\text{H}_5\\|\\\text{H—C—OH}\\|\end{array} \xrightarrow{+2\text{H}_2\text{NNHC}_6\text{H}_5} \begin{array}{c}\text{H—C=NNHC}_6\text{H}_5\\|\\\text{C=NNHC}_6\text{H}_5\\|\end{array}$$

Hydrazone Osazone

The osazones are very important as a help in recognising and distinguishing between individual sugars and in separating mixtures of sugars, because of their different solubilities, their tendency to crystallise, their characteristic crystal forms and melting points. The ketoses frequently react more easily than the aldoses. EMIL FISCHER used the hydrazones and osazones of sugars to clarify the constitution of the various sugars. It must however be pointed out that glucose, fructose and mannose give the same osazone (not hydrazone), because the asymmetrical C atoms, which do not take part in osazone formation, have the same spatial arrangement (configuration). Cane sugar (as a non-reducing sugar) forms no osazone.

The oxidation of the osazones with copper sulphate (HUDSON) gives the very stable **osatriazoles**, which can be more easily purified than the osazones. They have characteristic melting points and their rotation can be more easily measured than those of the osazones, because the osatriazoles are colourless.

$$\begin{array}{l} H-C=NNHC_6H_5 \\ \quad | \\ \quad C=NNHC_6H_5 \\ \quad | \end{array} \xrightarrow{CuSO_4} \begin{array}{l} H-C=N \\ \quad | \qquad \rangle N-C_6H_5 \\ \quad C=N \\ \quad | \end{array} + C_6H_5-NH_2$$

osazone osatriazole

Aldoses react with mercaptans to form mercaptals. This reaction is suitable for separating aldoses and ketoses, as the latter react with much more difficulty as seen in the equation

$$\begin{array}{l} H-C=O \\ \quad | \\ H-C-OH \\ \quad | \end{array} \xrightarrow{+2\,RSH} \begin{array}{l} H-C\begin{array}{l} SR \\ SR \end{array} \\ \quad | \\ H-C-OH \\ \quad | \end{array} + H_2O$$

The reaction of sugars with **amines** is of great importance in food chemistry and food technology: the reducing group reacts under very mild conditions with a large number of primary amines, with amino acids, peptides, proteins, arylamines and so on, to form the so-called **glycosylamines**.

Similar products can also be obtained from secondary amines under somewhat more drastic conditions. In weak acid solution the glycosamines rearrange readily into **1-amino ketoses** (AMADORI rearrangement). In the same way ketoses react with amines to give ketosylamines, which are arranged to the epimers **2-amino aldoses** in the presence of weak acids (HEYNS rearrangement).

These compounds play a part (in addition to the formation of simple Schiffs bases) in the preliminary stages of **non-enzymatic browning**, the so-called MAILLARD reaction, in which high molecular, brown coloured products (melanoidins) containing humic acid are formed from complicated condensations between carbonyl and amino compounds followed by further rearrangements. In general the MAILLARD reaction leads to a reduction in nutritive value of foods. Sugar-protein compounds may be formed which cannot be readily attacked by the digestive enzymes, or can only be split very slowly. Several essential amino acids (e.g. lysine) of proteins are blocked during the browning reaction and are therefore no longer available to the body. In food technology it is often essential to prevent the MAILLARD reaction. It is however desirable and encouraged in some baking processes (browning of bread crust, manufacture of biscuits), in the brewing of beer, in the roasting of coffee, cereals, potatoes and in frying meat and fish, because here the development of flavour and aroma go hand in hand with browning. It is undesirable, on the other hand, in pasteurisation, sterilisation and drying processes with milk, fruit juices and egg products.

Action of alkalis. The action of dilute alkalis on simple sugars is initially to produce isomerisation, where at first there is no destruction of the molecule. The formation of sugar enols as intermediates is important. The OH group changes its position on the C atom (= change of configuration), a process which is called **epimerisation**; then the actual isomerisation aldose $\rightleftharpoons$ ketose takes place; for glucose the transition is to fructose. Glucose and is epimer, mannose as well as ketoses (fructose) result from the action of alkali on aldoses (LOBRY-DE-BRUYN rearrangement), see p. 192.

Similar rearrangements are known to occur in the vegetable and animal organism through enzymatic action when the various sugars may be transformed from one to the other. For instance, in the mammary gland the glucose of the blood is changed into galactose (for the formation of milk sugar (lactose)).

More profound changes occur in the sugar molecule subjected to the action of more **concentrated alkali**, especially at elevated temperatures. Yellow to brown discolouration appears on the surface. Chemically, after the primary enolisation and the resulting breakdown of the sugar chain, the two isomeric trioses, glyceraldehyde and dihydroxyacetone and methylglyoxal are formed with the release of H_2O. At the same time lactic acid, formic acid, formaldehyde, acetin and diacetyl are also produced. Such processes of decomposition also occur during the oxidation of sugar with FEHLING's solution (see p. 141) which will be discussed later. The decomposition products which are formed also have the same self reducing effect as does sugar itself.

Special emphasis must be given to a further important decomposition product found in the alkaline break up of the enol form of the hexoses. The compound **triose reductone**, formerly simply called reductone, possesses a high power of reduction. Chemically triose reductose is similar to ascorbic acid (vitamin C, see p. 226) because of its enediol structure, but it is not effective as an antiscorbic. It has the same reducing effect as FEHLING's solution, iodine solution, silver nitrate solution, TILLMAN's reagent, methylene blue and so on, and frequently leads to difficulties and errors in the chemical analysis of foods.

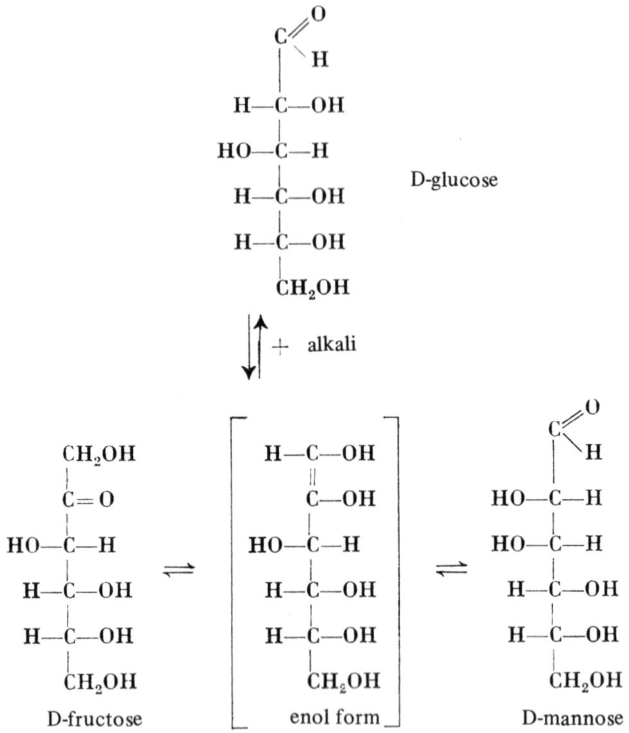

```
        O
      C//
      | \H
   H—C—OH
  HO—C—H
   H—C—OH          D-glucose
   H—C—OH
     CH2OH
```

⇅ + alkali

```
     CH2OH          [   H—C—OH  ]          C//O
      |             [     ||    ]          | \H
      C=O           [     C—OH  ]     HO—C—H
  HO—C—H      ⇌     [ HO—C—H    ]  ⇌  HO—C—H
   H—C—OH           [  H—C—OH   ]      H—C—OH
   H—C—OH           [  H—C—OH   ]      H—C—OH
     CH2OH          [    CH2OH  ]        CH2OH
  D-fructose        [ enol form ]     D-mannose
```

Errors can arise in the titrimetric determination of vitamin C (for instance in many fruit juices, jams and confectionery), because the redox potential of the triose reductone is almost the same as that of L-ascorbutic acid. The formula of the triose reductone as enol-tartronic aldehyde shows clearly its reducing ability.

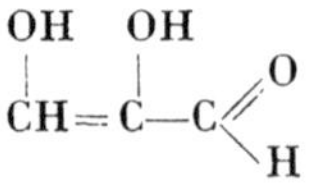

```
OH   OH
|    |     O
CH=C—C//           triose reductone
         \H
```

In solution **triose reductone** exists in tautomeric equilibrium with another substance, α-hydroxy malondialdehyde, which is also very reactive. Each is easily transformed into the other, as we have here a stable conjugated system, which exhibits tautomerism.

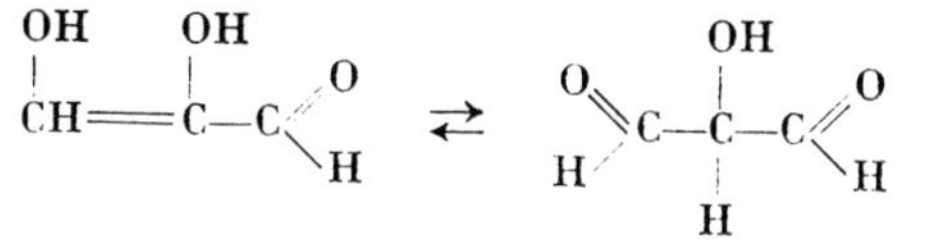

$$\underset{\text{triose reductone}}{\overset{\text{OH}}{\text{CH}}=\overset{\text{OH}}{\text{C}}-\text{C}{\overset{\text{O}}{\underset{\text{H}}{\lessgtr}}}} \rightleftarrows \underset{\alpha\text{-hydroxymalondialdehyde}}{\text{OHC}-\underset{\text{H}}{\overset{\text{OH}}{\text{C}}}-\text{CHO}}$$

Among the reductones are a number of other organic substances, with simple or complicated structure and of an aliphatic, aromatic and heterocyclic nature, all of which are conspicuous for their reactivity.

It is now known that a number of reductones, derived from carbohydrates, are involved as intermediates in the process of non-enzymatic browning (MAILLARD reaction) and the development of aroma in the preparation of many foods (see p. 136). These reductones then combine with amino compounds (humin bodies, melanoidins etc.).

Action of acids. Whereas under cold conditions, dilute acids hardly affect the monosaccharides and concentrated mineral acids only attack the sugar molecule slowly, reactions are in both cases rapid under warm conditions; the effects are stepwise according to whether pentoses or hexoses, aldoses or ketoses are involved. With HCl or H_2SO_4, disaccharides and higher saccharides are formed through *inter*molecular splitting off of water; through *intra*molecular splitting off of water furfural is formed from pentoses, methylfurfural from methylpentose (e.g. rhamnose) and hydroxymethylfurfural (HMF) from hexoses, see also p. 148.

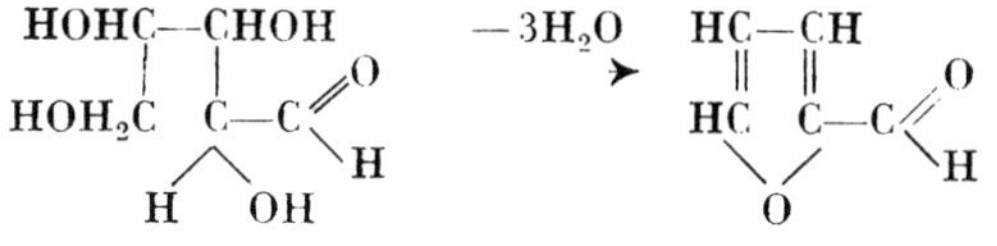

pentose (HOHC—CHOH / HOH$_2$C C—CHO / H OH) $\xrightarrow{-3H_2O}$ furfural (HC—CH / HC C—CHO / O)

hexose (HOHC CHOH / HOH$_2$C—C C—CHO / H OH H OH) $\xrightarrow{-3H_2O}$ hydroxy methylfurfural (HC—CH / HOH$_2$C—C C—CHO / O) $\rightarrow$ $CH_3COCH_2CH_2COOH$ laevulinic acid + HCOOH formic acid

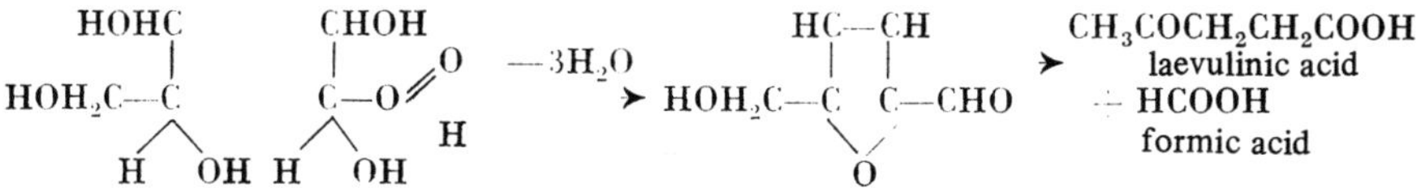

These compounds give typical **colour reactions** with specific reagents, for example phenol derivatives, which permit identification of sugars.

The general test for the presence of carbohydrates (MOLISCH) is positive for all sugars and polymeric saccharides (starch, cellulose). With concentrated H_2SO_4, all carbohydrates are dehydrated to yield furfural or hydroxy methylfurfural which condenses with α-naphthol to give a violet colour. The MOLISCH reaction is not a protein reaction as is frequently wrongly claimed in the literature.

If it is found to be positive, for instance as in mucinins, then this is due to the presence of carbohydrates in the proteins or to the fact that carbohydrates cling stubbornly to proteins (see symplexes p. 65).

SELIWANOFF's reaction makes use of the appearance of the characteristic cherry red colour of the reaction product of resorcinol with hydroxy methylfurfural formed from the hexoses with hydrochloric acid. With ketoses, for example fructose, the colour appears much sooner than with the aldoses, therefore this reaction is suitable for distinguishing between aldoses and ketoses.

This has important applications in Europe where qualitative testing for artificial invert sugar in honey depends on this reaction.

The stability of hydroxy methylfurfural to mineral acids is slight, so that, in a complicated reaction, laevulinic acid may be formed when formic acid is split off (see p. 139).

Pentoses can be quantitatively determined by colorimetric methods, since a dye is formed by condensation of the furfural distilled off with aniline.

Oxidation of monosaccharides. Various oxidation products are formed according to the strength of the oxidising agent. The aldehyde group as well as the final CH_2OH group can be oxidised to carbonyl groups, so that the following oxidation products are formed from a hexose:

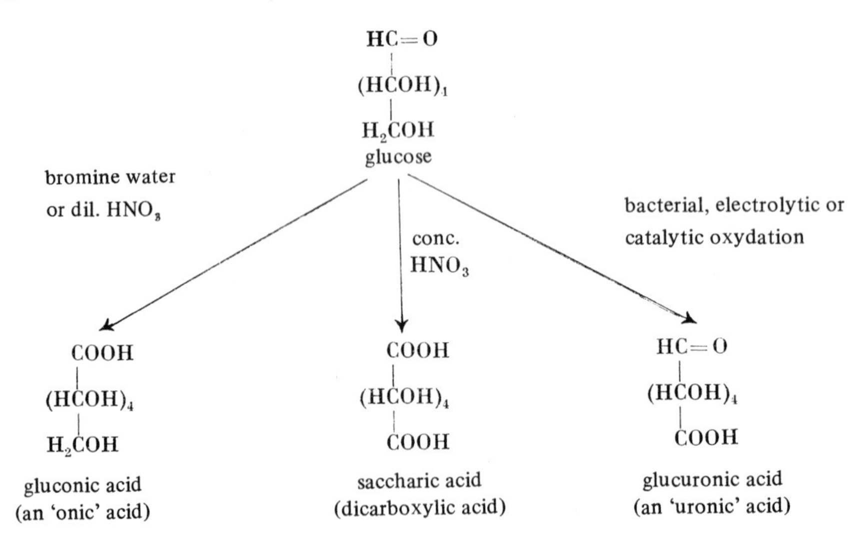

Mucic acid, which is important in the analytical determination of galactose and lactose (milk sugar), is formed in this manner from galactose (see pp. 149 and 150).

The oxidation of the aldoses to uronic acids can be carried out enzymatically by both higher and lower organisms (bacteria); the uronic acids are important as 'coupling agents' for detoxification reactions in the body. Some **biochemical**

oxidation procedures are of considerable technical importance: it is possible to transform by oxidation the hexavalent alcohol D-sorbitol obtained by catalytic reduction of glucose, into L-sorbose bacterium (*Acetobacter xylinum*). This method is important when starting from grape sugar since sorbose is used for the commercial production of vitamin C, see p. 224.

Various oxidation products result from the decomposition of the C chain when strong oxidising agents are used.

Most simple reactions for the presence and determination of sugars are based on oxidation. When oxidising agents are used they are reduced, which explains why the term **reduction test** for sugars is used. These reduction tests are often connected with changes in colour (visual or colorimetric) and hence **qualitative** reactions indicate their presence and **quantitative** methods of determination are based on them.

In these methods the oxidising agents are usually oxides of the heavy metals, copper, bismuth, mercury and silver in alkaline solution. The following are the most important qualitative tests for sugars:

The older TROMMER test and the improved FEHLING test. In these reactions $Cu(OH)_2$ is the oxidising agent; the presence of sugar can be recognised from the brick red copper (I) oxide, Cu_2O which appears. In the TOLLENS test the sugar is oxidised with silver oxide, present in the form of an ammoniacal silver nitrate solution ('silver mirror test').

In the NYLANDER test the presence of sugar is detected by the separation of black, metallic bismuth from basic bismuth nitrate.

FEHLING's test is the most frequently used qualitative test for sugar; it is of importance in the various modifications used in chemical food analysis. In the different methods the quantity of the cuprous (I) oxide formed is determined either by weighing, titration (ALLIHN, BERTRAND, SCHOORL-REGENBOGEN, BRUHNS, LUFF–SCHOORL, POTTERAT-ESCHMANN) or colorimetrically (FOLIN-WU).

When an alkaline solution of heavy metal oxide is used, not only are sugars effective as reducers but in addition the strong alkali leads to secondary, strongly reducing, decomposition products of the sugars (formic acid, glycoic acid, methylglyoxal, reductones etc. see p. 142). For this reason these reactions do not proceed according to a stoichiometric equation (for instance when FEHLING's solution is used as oxidising agent). The amount of Cu_2O obtained is converted by the use of empirical tables to give the amount of sugar present (taking into consideration the kind of sugar and the concentration of sugar and reagent). Materials, for instance proteins, which might affect the reaction must be removed beforehand, as in the determination of lactose in milk.

As well as heavy metal oxides, alkaline iodite solutions (hypoiodite) are also used as oxidising agents for the quantitative determination of sugar. Only aldoses are oxidised to aldonic acids in a reaction which proceeds according to stoichiometric laws, while ketoses and non-reducing disaccharides (e.g. sucrose) are not affected (WILLSTATTER-SCHUDEL method). This is an elegant method for determining both aldoses and ketoses.

Potassium ferricyanide is frequently used as an oxidising agent in the determination of glucose in physiological fluids (e.g. blood) (micromethod of HAGEDORN-JENSEN).

Reduction of monosaccharides. Sodium borohydride, Na amalgam, and electrolytically or catalytically excited hydrogen are used to reduce the carbonyl group (–CHO; > C = 0) to the corresponding alcohol group, forming sugar alcohols. Some of these are sweet tasting, such as the penitols from pentoses, hexitols from hexoses as for instance D-sorbitol from glucose. Two epimeric alcohols are formed from ketoses, because a new asymmetric centre takes the place of the carbonyl group, for example D-sorbitol and D-mannitol are formed from D-fructose.

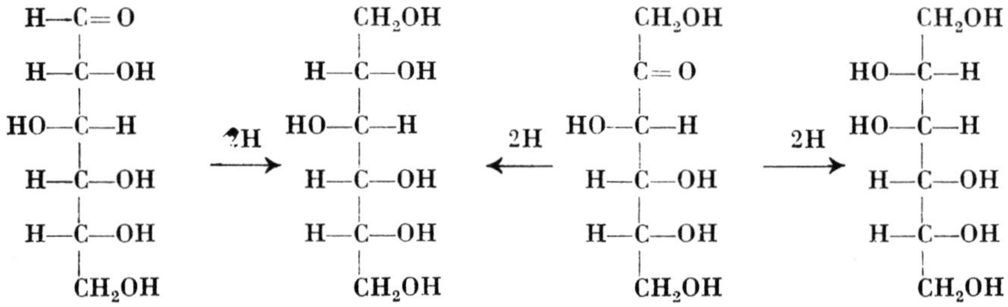

$$
\begin{array}{c}
H{-}C{=}O \\ | \\ H{-}C{-}OH \\ | \\ HO{-}C{-}H \\ | \\ H{-}C{-}OH \\ | \\ H{-}C{-}OH \\ | \\ CH_2OH \\ \text{D-glucose}
\end{array}
\xrightarrow{2H}
\begin{array}{c}
CH_2OH \\ | \\ H{-}C{-}OH \\ | \\ HO{-}C{-}H \\ | \\ H{-}C{-}OH \\ | \\ H{-}C{-}OH \\ | \\ CH_2OH \\ \text{D-sorbitol}
\end{array}
\xleftarrow{2H}
\begin{array}{c}
CH_2OH \\ | \\ C{=}O \\ | \\ HO{-}C{-}H \\ | \\ H{-}C{-}OH \\ | \\ H{-}C{-}OH \\ | \\ CH_2OH \\ \text{D-fructose}
\end{array}
\xrightarrow{2H}
\begin{array}{c}
CH_2OH \\ | \\ HO{-}C{-}H \\ | \\ HO{-}C{-}H \\ | \\ H{-}C{-}OH \\ | \\ H{-}C{-}OH \\ | \\ CH_2OH \\ \text{D-mannitol}
\end{array}
$$

Reactions of the hydroxyl group. Because they are **polyhydroxy** compounds, sugars react with periodic acid or $NaIO_4$, a reagent for proving the presence of α-glycol groupings (MALAPRADE reaction): an aldose is quantitatively decomposed to formic acid and formaldehyde.

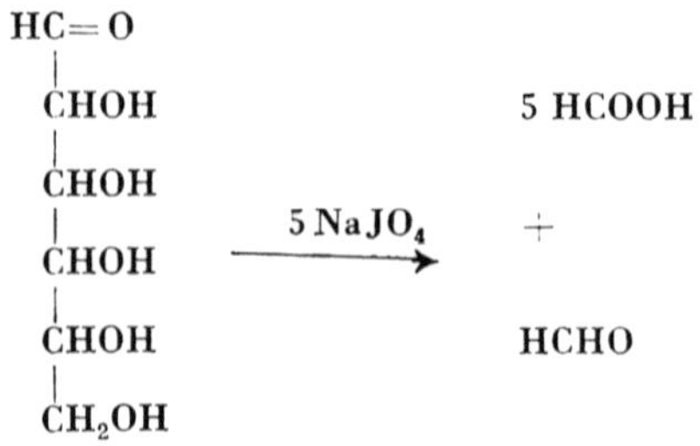

$$
\begin{array}{l}
HC{=}O \\ | \\ CHOH \\ | \\ CHOH \\ | \\ CHOH \\ | \\ CHOH \\ | \\ CH_2OH
\end{array}
\xrightarrow{5\,NaJO_4}
\begin{array}{c}
5\ HCOOH \\ + \\ HCHO
\end{array}
$$

With acid anhydrides or acid chlorides the OH groups are **esterified**. The esters are normally almost insoluble in water, relatively stable to dilute acids, but labile with alkalis. Glucose fatty acid esters act as emulsifiers (ice cream, fat emulsions etc.).

Cyclic acetals are formed with aldehydes and ketones, by the splitting off of water using dilute acids as catalysts. While these are readily split by dilute acids they are stable to dilute alkalis. Tri-(2-chlorbenzal)-sorbitol, which is derived from 2-chlorbenzaldehyde and sorbitol is used to prove the presence of apple wine in wine, see p. 154.

Ether formation from the OH group can occur either *inter-* or *intra-*molecularly. *Intra*molecularly the sugar anhydrides (sorbitans and sorbides) are obtained. With *inter*molecular etherification, methyl ethers as well as benzyl and triphenyl methyl ether have played an important role in the determination of the structure of glycosides, oligo- and polymeric saccharides.

Mutarotation and conformation. In the previous paragraphs various reactions of the monosaccharides showed that free carbonyl groups (aldehyde or keto groups) must be present in the sugar molecule. But on the other hand it is clear from our knowledge and the reactions, that the free carbonyl groups of sugars (e.g. of glucose) are not present in a solid state and they only occur in solution in very small amounts (under 0.1%) in equilibrium with other structurally isomeric forms (α- and β-forms).

With the typical (SCHIFFS) aldehyde reagent, no red colour appears; with hydrogen sulphite no addition compound is formed, and FEHLING's solution only reacts when heated.

The best proof lies in the fact that two D-glucose forms can actually be isolated. α-D-glucose crystallises from water, β-D-glucose from pyridine. The α- and β-forms, the so-called anomeric pairs, form no mixed crystals because they are diasteroisomers, and crystallise from solvents either only as the α-form or only as the β-form. α- and β-D-glucose have different physical properties (melting point, solubility, optical rotation); in aqueous solution they are both present in a dynamic tautomeric equilibrium. Further investigation shows that the carbonyl group (intramolecular) can react with the OH group at Carbon 5 in the same molecule, when a **sugar hemiacetal** (= cycloacetal or lactol form) ring is formed. In pentoses and hexoses the formation of the tautomeric equilibrium is spontaneous and voluntary; the hydroxyl group 5 functions as partner (with ketoses 2,5- and 2,6-ring formation, see p. 147). A new OH group is formed on the orginal carbonyl C atom (with aldoses at the C_1, with ketoses at the C_2): the **acetal or glucosidic hydroxyl**, which in contrast to the other sugar hydroxyl groups has a strong reducing effect, and is particularly reactive. The C atom itself becomes asymmetrical due to the formation of the hemiacetal, so that two other diastereomeric forms, the so-called anomeric (α- and β-forms) with different rotation values are formed.

α- and β-forms are only mirror isomers on the C_1. In the rest of the molecule they are superimposed. They are therefore not optical isomers but diastereo-. isomers.

In the crystalline state only one of the two forms is found at any one time.

In solution the cyclic hemiacetal forms (lactol forms) change into each other through the open carbonyl form:

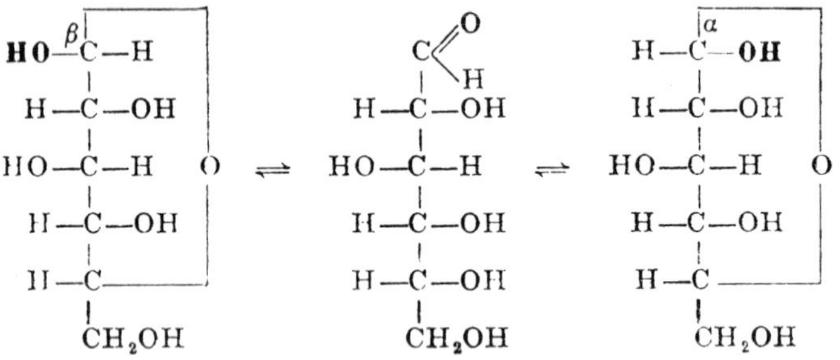

β-D-glucose

open carboyl form (oxoform)

α-D-glucose

When a reducing sugar, for instance glucose, is dissolved in water, some rearrangement occurs to reach equilibrium state; the original optical rotation value also changes to a constant final value. Hence a pure solution of glucose recrystallised from water (α-form) has a specific rotation of +112.2°, which falls to +52.5°. On the other hand the product recrystallised from pyridine (β-form) after it has been dissolved in water, initially has a specific rotation of +18.7°, which then rises to 52.5°. This phenomenon is called **mutarotation**. The equilibrium to the final value frequently proceeds very slowly. However the equilibrium can be catalysed by boiling, or by the addition of acids or bases, so that for instance with a few drops of NH_4OH the equilibrium value is reached immediately. Mutarotation occurs in all reducing sugars.

HAWORTH's ring formula clearly shows the configuration. Compare α and β position of the glycosidic hydroxyl on the C atom 1:

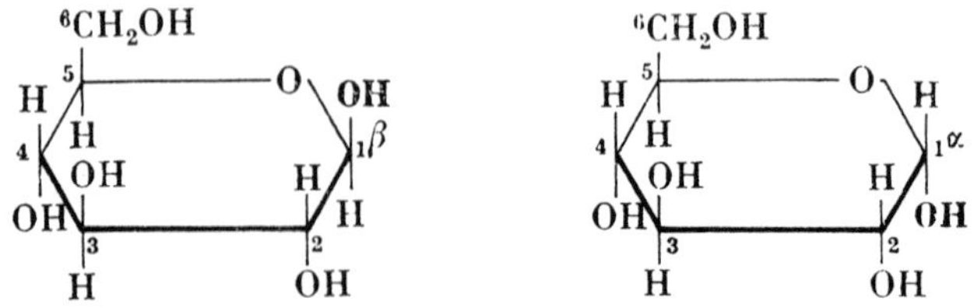

β-D-glucose α-D-glucose

pyranoid form (HAWORTH's ring formula)

In general the cyclic hemiacetals of sugars are in stable 6 membered rings which are relatively strain free of the so-called **pyranose** form. (Derived from the heterocyclic ring systems of pyran and furan). However rings of 5 members may be formed, the so-called furanose form; this form is frequently found in bound ketoses (see sucrose p. 136).

Whereas the tensioned 5 membered ring of the **furanose** form is built up evenly, an angular plan must be assumed for the pyranose form, because the valency angles of the carbohydrate of 109° 28′ do not permit a strainless 6 ring. Therefore similar conditions exist in pyranose ring formation in sugars as exist in cyclohexane, which can be present in two forms, either in the stiff chair or the flexible boat or basin form.

The geometry of the pyranose form of the sugar molecule is only slightly altered by oxygen in the ring, but the asymmetry caused by the oxygen means that two chair forms (C1 and 1C of the 'chair') and 6 boat or basin forms are possible. The pyranoses are usually in the chair form in solution, because this form is more favoured from an energy point of view than the boat form is around 6 kcal more energy rich and is therefore more unstable than the chair form.

If an axis is drawn through the centre of the 6 ring, five valencies of the 5 ring C atoms will lie parallel to the axis, that is they are axial (a), the other five will be horizontal to them, that is equatorial (e). The two possible chair forms C1 and 1C can easily interchange in the model by folding the C atoms of the 1 and 4 ring respectively up or down. In this way one of the basin forms B1 or 1B will be an intermediate.

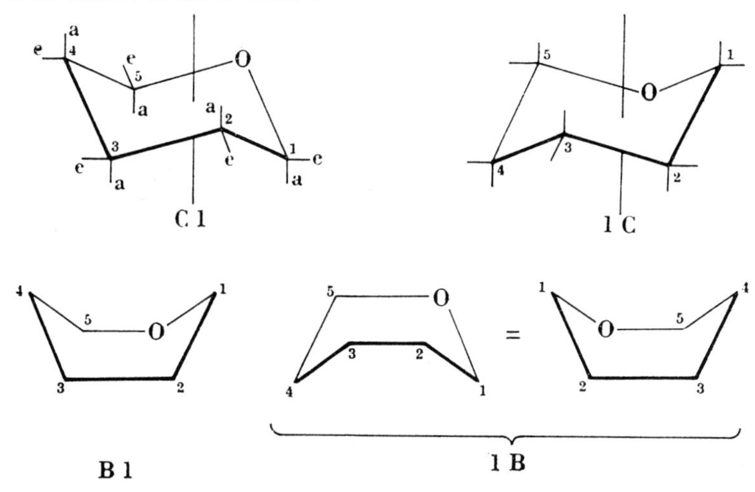

Ligands, or attached atoms which take up an equatorial position in the C1 conformation, are axial when the conformation changes to 1C or vice versa. The most energy favoured, that is most stable conformation for monosaccharides, will always aim towards having the attached atoms filling the greatest possible space. The atoms, with mutually repulsing charges will end up lying as far as possible from each other, that is equatorially. For D-glucose, D-galactose and D-mannose the C1 conformation is therefore the most stable (however sometimes in certain sugars the C1 form and 1C form may be present in equal amounts).

The **conformation formulae** of sugars have heuristic importance, because they reproduce more exactly the spatial organisation of the substituents in the ring and with it the details of the molecular structure, and so give a truer picture of the properties and the reactivity of the materials than all other projection formulae. It is clear, on the basis of the conformation formulae (see p. 143) why in solution the α-form of D-glucose is predominant at the equilibrium (63%). In the C1 conformation of β-D-glucopyranose all the OH groups (and with them all the other groups) on the C atoms 1 and 2, are in an **equatorial** position, whereas in the α-form the glycosidic (acetal) OH groups are arranged **axially**, that is spatially and in an energy unfavoured position to the OH group on the second C atom. Such a molecule naturally tends towards the more stable, equatorial position of the OH group on the C1 atom.

The conformation formulae give a better understanding of the chemical and biochemical reactions of sugars in food chemistry and food technology. More subtle explanations can also be found for the enzyme substrate bonding, substrate specificity and the competitive inhibition of enzymes and other problems in enzyme technology.

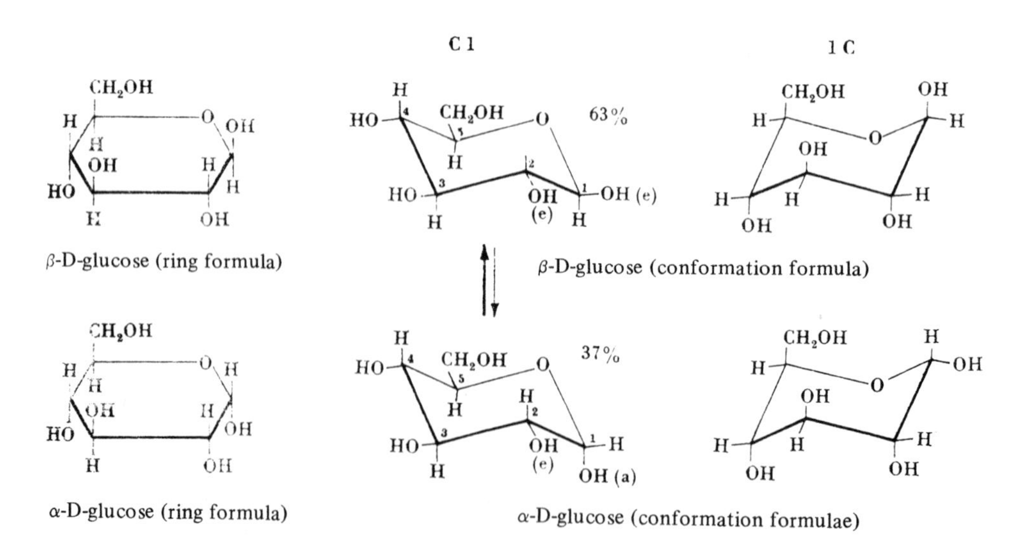

β-D-glucose (ring formula)

β-D-glucose (conformation formula)

α-D-glucose (ring formula)

α-D-glucose (conformation formulae)

Individual monosaccharides

Pentoses

Of the four aldopentoses*, the most important representatives for our purposes are D-xylose, L-arabinose and D-ribose. They are seldom found free in nature, but occur in the vegetable world as polymeric saccharides, the so-called **pentosans** and are therefore consumed in vegetable foods.

D-xylose ($[\alpha]_D^{20} = +19°$), wood sugar, is obtained from the polymeric saccharide xylan (= wood gum), which functions as a supporting tissue in plants and is obtained from oat straw, maize stalks and wood. Xylose tastes sweet (see p. 164. Xylitol is formed by catalytic reduction and is now produced on a commercial scale (see pp. 152 and 156).

L-arabinose ($[\alpha]_D^{20} = +174°$) is found bound in the natural polymeric saccharide araban in cherry gum and in the pith of elderberry, from which it is extracted by acid hydrolysis. Arabinose (in pentosans) occurs in considerable amounts in pectin and pectin materials.

D-ribose occurs as **N-glucosides** (p. 159) of purine and pyrimidine bases, in the building blocks of the nucleic acids, nucleotides and nucleosides which are widely distributed in nature and were discussed earlier.

In nucleotides, in addition to ribose, **2-desoxyribose**, a desoxy sugar is present; here the OH group on the C_2 of ribose is replaced by an H atom, see p. 147. Ribitol, a reduction product of ribose is found in lactoflavin.

In the glycosides of many vegetable pigments (especially flavone materials), **rhamnose**, the most common of the **methypentoses**, can be found; rhamnose is derived from **desoxyhexose**, see p. 153.

D(+)-xylose*	L(+)-arabinose	D(—)-ribose	L(+)-rhamnose
H—C=O	H—C=O	H—C=O	H—C=O
H—C—OH	H—C—OH	H—C—OH	H—C—OH
HO—C—H	HO—C—H	H—C—OH	H—C—OH
H—C—OH	HO—C—H	H—C—OH	HO—C—H
CH_2OH	CH_2OH	CH_2OH	HO—C—H
			CH_3

It can be seen from these formulae that the pentoses as hydroxyaldehydes have the typical properties and reactions of simple sugars, for instance, they possess optical activity, have a reducing effect and form hydrazones and osazones.

*Altogether eight forms are possible: four β-forms and four D-forms.
*D-xylose is usually called L-xylose in the older literature.

Furfural and methylfurfural are formed from pentoses or methylpentoses by distillation with dilute acids (HCl, H_2SO_4) and are used to prove their presence and to distinguish them from the hexoses*.

Suitable reagents (e.g. orcinol, phloroglucinol, aniline, resorcinol etc. produce characteristic colours. The colours are distinguished from one another by their particular absorption spectrum which can be used for estimation of the sugars. The appropriate hydrazone can also be used for identification purposes. Paper chromatography has recently given good results for the separation of sugar mixtures.

Pentoses are not fermented by ordinary yeasts and can be distinguished from hexoses in this way, since hexoses can be separated by fermentation. However certain strains of yeast can develop the ability to ferment pentose (adaptation). In the so-called wood saccharification (BERGIUS, HAGGLUND, SCHOLLER-TORNESCH process (see p. 180) the calcium bisulphite process for preparation of cellulose from beech wood, large amounts of pentoses accumulate.

If suitable inorganic nutrients, for instance salts of calcium, magnesium, nitrogen and phosphorus, are added together with forced aeration, the wild yeast *Torula utilis* can utilise the pentoses quickly while itself multiplying (formation of nutrient yeast). This is called biological protein synthesis. The yeast decomposes the sugars to simple C compounds such as acetic acid, and then uses them to build up the polysaccharides of the cell wall and the amino acids for the proteins (see also p. 54).

Pentoses are also suitable for biological fat synthesis using mycelium fungi (*Endomyces vernalis*) and varieties of Fusarium, (see also p. 98).

Protein and fat synthesis are endothermic requiring a great deal of energy which is compensated for by the enormously increased respiration of the sugar, an exothermic reaction.

Hexoses

The most important and best known sugar types are the hexoses. Most of the oligo- and polysaccharides found in plants and animals are built up from hexoses and many glycosides have hexoses as their constituent sugar. The hexoses are widely distributed in nature in their free form, such as glucose and fructose in sweet fruits.

The hexoses are typical members of the monosaccharides and their characteristic reactions have already been described in detail; they are readily soluble in water, exhibit optical activity and mutarotation, reduce FEHLING's solution and form hydroxymethylfurfural, hydrazones and osazones.

*In contrast, hexoses split off hydroxymethylfurfural, which give other colour reactions and make it possible to distinguish between them, see p. 139.

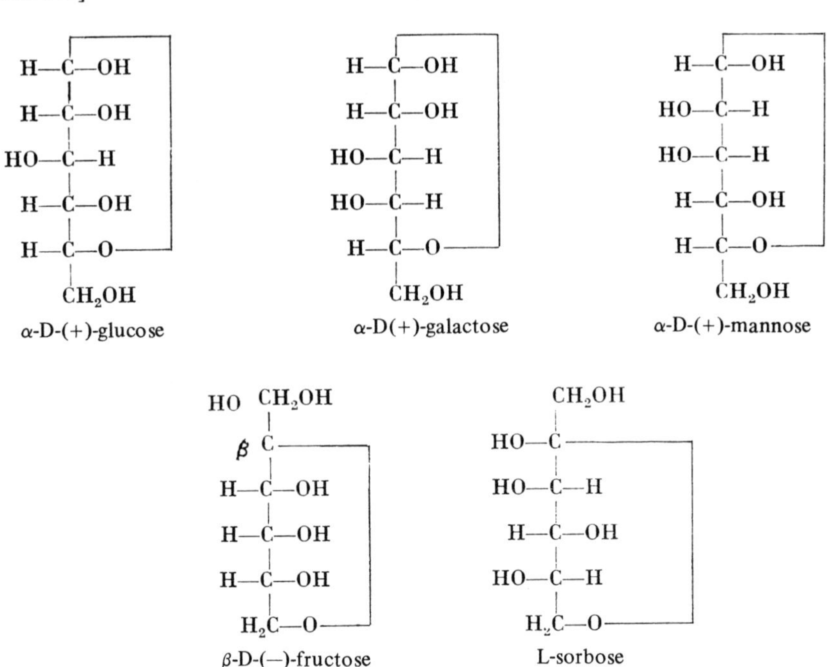

α-D-(+)-glucose α-D(+)-galactose α-D-(+)-mannose

β-D-(—)-fructose L-sorbose

Hexoses are very weak acids (average $K = 10^{-13}$) and they and the saccharides built up from them (saccharose, raffinose etc.) form **saccharates** with bases of calcium earths groups (e.g. calcium or strontium hydroxide). Saccharates are decomposed even by CO_2 as they are salts of exceedingly weak acids. This is used in sugar production (saturation process).

The destruction of hexoses, and also of the reducing mono-, di- and oligo-saccharides in general, by alkali and calcium hydroxides under certain conditions, is used analytically to determine sucrose in a mixture with other sugars; in this case only the disaccharide sucrose remains undamaged (lime process, analysis, see p. 137).

Dry heating of hexoses alone or with a small amount of sodium or ammonium carbonate, yields results in caramelisation which is used for colouring purposes in the spirits industry.

Among the aldoses glucose, galactose and mannose are of practical interest, as are fructose and sorbose among the ketoses.

D-Glucose Glucose is also called **grape sugar** because it is found in grapes and many sweet fruits. With fructose it is the main constituent of honey. Glucose is found in small amounts (0.1%) in blood. Far more important than the occurrence of free glucose is the fact that it is a building block of oligo- and polymeric saccharides such as cane sugar, milk sugar, malt sugar, starch, glycogen and cellulose. It is the most important nutrient carbohydrate. All other carbohydrates (sucrose, lactose, starch, glycogen) are only secondary nutrients, they first have to be broken down in the process of digestion.

Commercial production of glucose is by acid or enzymatic hydrolytic cleavage of potato starch, maize starch as well as by acid hydrolysis of celluloses ('wood saccharification', see p. 180).

Glucose has less sweetening power than cane sugar, see p. 164. It is also widely used in dietetics and medicine.

The usual crystalline D-glucose is α-D-glucose monohydrate (melting point 80° to 86°C). Anhydrous glucose (melting point 146.5°C) crystallises out from alcohol. In the literature anhydrous sugars are frequently (incorrectly) described as **anhydrides**. They should be called **anhydrous**. Because it is dextrorotatory, glucose is also called **dextrose**; the specific rotation is $[\alpha]_D^{20} = +52.5°$. In the determination of grape sugar mutarotation may occur, that is the change in rotation during the establishment of the equilibrium, see p. 143. The name in commerce for glucose (grape sugar) is dextrose.

Glucose is subject to many types of fermentation. Anaerobic cleavage of carbohydrates is fermentation in the narrow sense of the word, but one also talks in a wider sense of aerobic (= oxidative) fermentation.

The important types of anaerobic fermentation as in alcoholic fermentation, lactic acid fermentation in milk, sauerkraut, silage (= souring of green fodder), cheese ripening (kefir, yoghourt, koumiss, propionic acid fermentation in Swiss cheese (formation of holes). Butanol-acetone fermentation is of great commercial importance. The end products are always cleavage products from glucose.

Of the various oxidative (aerobic) fermentations of bacteria and moulds, only acetic acid fermentation, fumaric acid fermentation and citric acid fermentation need be mentioned.

D-galactose (formula p. 149). This hexose occurs mainly as a building block of the oligosaccharides lactose and raffinose and polysaccharides (gum arabic). It can best be isolated from milk sugar, because after hydrolysis, the galactose is easily separated from the glucose. Galactose is only slightly soluble in water and crystallizes readily. The anhydrous form (from alcohol) melts at 165°C (= α-form). The monohydrate crystallizes from water. Galactose tastes much less sweet than sucrose see p. 164. $[\alpha]_D^{20} = +80.5°C$ (check for mutarotation). The presence of galactose is best confirmed by **mucic acid** (melting point 214°C, see p. 168) formation and in addition by conversion of the mucic acid with NH_3 into pyrrol, which in the presence of spruce wood shavings gives a precipitate of pyrrole red. Galactose is not fermented by higher yeasts: it can therefore be distinguished from the other three hexoses – glucose, fructose and mannose.

D-mannose (formula p. 138). D-mannose occurs in nature as a building block of high polymeric polysaccharides, called mannans, in the wood of coniferous trees. In contrast, pentosans predominate in deciduous trees. The kernels of stone fruits, date kernels and carob beans are also rich in mannan. Mannans are present in yeast in the form of gum. The determinantion of yeast gum can be used in the proof and estimation of yeasts and processed yeast products.

D-fructose (formula p. 138). Fruit sugar is sometimes called laevulose, a name derived from the laevorotation of fructose (latin *laevus* = left). The configuration of fructose is that of the D-series (see p. 130).

In its free form it is always accompanied by glucose and sucrose, and is found in large amounts in sweet fruits. Invert sugar, a mixture of equal parts of fruit sugar and glucose, results from hydrolysis (acid, enzyme) of sucrose. This process is called inversion, because invert sugar rotates polarised light in the opposite direction (left) to the original sucrose. In its bound form fructose is a component of several oligosaccharides (sucrose, raffinose, gentianose, melezitose, stachyose) and as a polymeric building block for **fructosans** (see p. 183), whose best known member, inulin, is found as a reserve carbohydrate in many plant tubers such as Jerusalem artichokes, dahlia tubers and in chicory roots. Production of fructose is by careful acid hydrolysis of fructosans, especially inulin.

Fructose as such, exists in the crystalline form in the more stable pyranose structures (2, 6, ring); in the bound form, as in sucrose, it is always in the furanose (2, 5 ring) form. Fructose like glucose exhibits mutarotation in solution but the phenomenon is more complicated because it is a matter of equilibrium between the pyranose and furanose forms, as well as the α and β open carbonyl forms:

Fructose

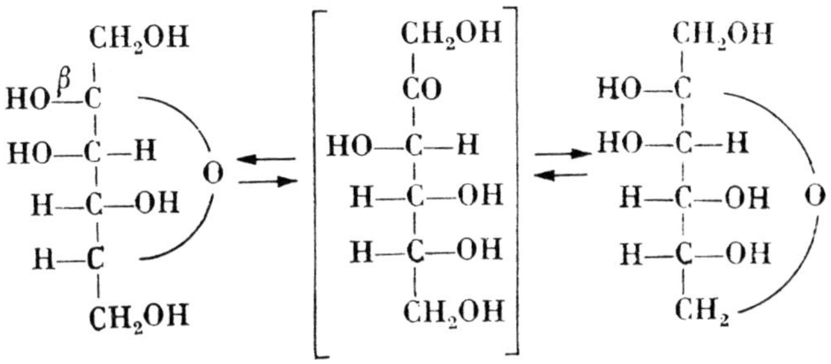

```
     CH2OH                 CH2OH                 CH2OH
    β |                      |                     |
  HO—C ─┐                    CO               HO—C ──┐
     |   \                   |                     |   \
  HO—C—H  O  ⇌        HO—C—H        ⇌    HO—C—H      \
     |   /                   |                     |     O
   H—C—OH               H—C—OH              H—C—OH  |
     |  /                    |                     |    /
   H—C ─┘                H—C—OH              H—C—OH  /
     |                       |                     |  /
    CH2OH                  CH2OH                 CH2 ─┘
```

β-D-fructofuranose — keto form D-fructose — β-D-fructopyranose

α_D^{20} − 21° α_D^{20} − 133° (calculated)

equilibrium $[\alpha]_D^{20}$ − 92°

Fructose is difficult to crystallize; therefore, when honey crystallises, the liquid portion is mainly fructose. Fructose crystallises out of an aqueous solution with half to one molecule of water of crystallisation. Its melting point is 102–104°C. Fructose is hygroscopic and tastes sweeter than all other types of sugar, see p. 140. Yeast ferments it.

When D-fructose is reduced, equal amounts of the two hexavalent sugar alcohols, D-sorbitol and D-mannitol are formed. This is because the CO group on the C_2 atom results in the formation of a new asymmetrical C atom (see p. 142). The osazones of the three monosaccharides glucose, fructose and mannose are the same, m.p. 208°C. SELIWANOFF's reaction is suitable for proving the presence of fructose and to distinguish it from the aldoses (see p. 140).

L-sorbose (structural formula p. 149). This ketohexose – called D-sorbose in the older literature – has so far only been found in certain plant juices. It is probable that in these cases the L-sorbose was formed through bacterial oxidation of D-sorbitol. It can be produced in nutrient media by the sorbose bacterium (*acetobacter xylinum*) or by the Acetobacter suboxydans bacterium growing on the hexavalent alcohol sorbitol (see p. 154). On the other hand sorbose can be transformed into sorbitol by chemical reduction. Sorbose, formerly considered of little commercial importance, has recently become important because, starting from D-glucose, and proceeding via D-sorbitol and its oxidation to L-sorbose, a method has emerged for the large scale production of L-ascorbic acid, vitamin C (see p. 223).

Derivatives of the monosaccharides

Desoxysugars

When one or more OH groups in the sugar molecule are replaced by hydrogen, the compounds are called **desoxysugars**. The figure placed in front of the name indicates the position of the CH_2 group. The desoxysugars which occur most frequently in nature are: 2-desoxy-D-ribose, the sugar portion of the desoxyribonucleic acids, Lβrhamnose (6-desoxy-L-mannose), formula p. 147) and L-fucose (6-desoxy-L-galactose).

```
H—C=O              H—C=O
  |                  |
H—C—H            HO—C—H
  |                  |
H—C—OH            H—C—OH
  |                  |
H—C—OH            H—C—OH
  |                  |
  CH2OH          HO—C—H
                     |
                     CH3
```

2-desoxy-D-ribose L-fucose

L-Rhamnose has been found in glycosides and polysaccharides of many plants and green algae. L-fucose is a component of the oligosaccharides in human milk, of the blood group substances and of many glycoproteins. Fucoidin, a polysaccharide from brown algae, is built up entirely of L-fucose units. Di-desoxyhexoses have been found as components of liposaccharides (pyrogenic substances, that is materials which induce fever) in the cell walls of Salmonella and also in plant glycosides (Digitalis).

Sugars with branched chains

Recently a series of rare, branched sugars have been isolated from widely varying natural materials, such as tannins, antibiotics and plant glycosides. L-streptose (from streptomycin) may serve as an example; it is 5-desoxy-3-C-formyl-L-lyxose.

```
          H—C=O        1
            |
    H  H—C—OH          2
     \      |
  O=C—C—OH             3      L-streptose
            |
       HO—C—H          4
            |
           CH3         5
```

Amino sugars

Monosaccharides, which have one or more hydroxyl groups replaced by the amino group, are called amino sugars. The position of the amino group in the molecule is indicated by a figure in front of the name. The two most important are D-galactosamine (chondrosamine, 2-amino-2-desoxy-D-glucose) and D-galactosamine (chondrosamine, 2-amino-2-desoxy-D-galactose). Both compounds are found in mucopolysacchardes, blood group substances, cartilage and glycoproteins. In nature, the amino sugars are usually found as N-acetyl, or more rarely as N-sulphuryl derivatives. Hydrochlorides of amino sugars are hydrolysed with compounds containing amino sugars are hydrolysed with hydrochloric acid. The MORGAN-ELSON reaction is important for proving the presence of amino sugars.

```
   ┌──────────┐             ┌──────────┐
 H—C—OH     |             H—C—OH     |
   |          |               |          |
 H—C—NH2    |             H—C—NH2    |
   |          |               |          |
HO—C—H      |            HO—C—H      |
   |          |               |          |
 H—C—OH     |            HO—C—H      |
   |          |               |          |
 H—C—O──────┘             H—C—O──────┘
   |                          |
  CH2OH                      CH2OH
```

α-D-glucosamine (chitosamine) α-D-galactosamine (chondrosamine)

Sugar alcohols

Sugar alcohols are polyalcohols; they are products of the reduction of sugars. Important ones in food chemistry are the three hexitols D-sorbitol, D-mannitol, dulcitol and the pentitol xylitol.

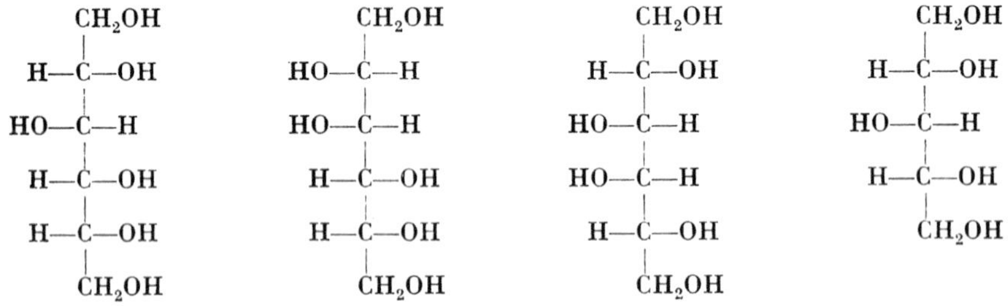

$$\begin{array}{c} CH_2OH \\ | \\ H-C-OH \\ | \\ HO-C-H \\ | \\ H-C-OH \\ | \\ H-C-OH \\ | \\ CH_2OH \end{array}$$

D-sorbitol

$$\begin{array}{c} CH_2OH \\ | \\ HO-C-H \\ | \\ HO-C-H \\ | \\ H-C-OH \\ | \\ H-C-OH \\ | \\ CH_2OH \end{array}$$

D-mannitol

$$\begin{array}{c} CH_2OH \\ | \\ H-C-OH \\ | \\ HO-C-H \\ | \\ HO-C-H \\ | \\ H-C-OH \\ | \\ CH_2OH \end{array}$$

dulcitol

$$\begin{array}{c} CH_2OH \\ | \\ H-C-OH \\ | \\ HO-C-H \\ | \\ H-C-OH \\ | \\ CH_2OH \end{array}$$

xylitol

D-sorbitol. Of all the naturally occurring polyalcohols, D-sorbitol is the most widely found. Fresh fruits frequently contain 5–10% sorbitol. Algae, nut galls and tobacco leaves contain considerable amounts and so do many fruits and berries of the Rosaceae family, such as pears, apples, cherries, plums, peaches and apricots. Grapes, on the other hand contain only very little or almost no sorbitol. This fact makes it possible for instance, to distinguish between wine and cider or other fruit wines. Naturally occurring sorbitol is not sufficient to meet the demand for it, therefore commercial production is particularly important.

Sorbitol can also be obtained by reduction of the naturally occurring hexoses, which possess the same configuration, D-glucose (dextrose), D-fructose and L-sorbose. Nowadays large amounts of sorbitol are produced by catalytic hydrogenation from dextrose, usually using nickel catalyst under high pressure (100–150 atm.) and high temperatures (approx. 150°C). The only other technical method which is of some importance is electrolytic reduction.

Sorbitol is stable in the cold against dilute acids and alkalis and also against atmospheric oxygen. Unlike sugar, it contains no aldehyde groups and does not reduce FEHLING's solution. Nor is sorbitol fermented by yeast; on the contrary, it shows great resistance to bacterial decomposition and can be stored almost indefinitely. Like xylitol, it does not take part in the MAILLARD reaction.

When sorbitol is heated in the presence of acid catalysts, an internal ether is readily formed with the splitting off of water. These compounds are called sorbitans or sorbides, according to whether one or two molecules of water are split off from the sorbitol:

```
    CH2OH                     CH2 ──┐                    CH2 ──┐
     |                         |    |                     |    |
  H—C—OH                    H—C—OH  |                  H—C—OH  |
     |                         |    O                     |    O
 HO—C—H         —H2O       HO—C—H   |     —H2O       ┌── C—H   |
     |         ──────>         |    |    ──────>     |    |    |
  H—C—OH                    H—C ────┘                | H—C ────┘
     |                         |                     O    |
  H—C—OH                    H—C—OH                   | H—C—OH
     |                         |                     |    |
    CH2OH                     CH2OH                  └── CH2

  D-sorbitol              1,4-sorbitan            1,4-3,6-sorbide
```

Isomers may also be formed such as 1,5 2,5- or 3,6-sorbitan.

The so-called SPAN products are simple esters of the sorbitans with fatty acids, such as stearic, palmitic, lauric and oleic acid. After partial esterification with fatty acids and esterification with ethylene oxide, the so-called TWEEN products are obtained from the sorbitans. In the last ten years, these have become particularly important as emulsifiers and solubilisers. Tweens are especially important in the USA in the fat and confectionery industry.

The emulsifier TWEEN 20 is a **polyethylene glycol-sorbitan-laurate**. The corresponding palmitate is TWEEN 40, the stearate TWEEN 60, the oleate TWEEN 80.

Sorbitol (and with it dextrose), is used for the large scale commercial synthesis of L-ascorbic acid (Vitamin C) see p. 223.

Pure sorbitol forms odourless, white crystals with a sweet, cool taste. The cooling effect on the tongue is produced by the negative heat of solution. The melting point of sorbitol is either 92.7° or 97.2°C according to the form of the crystal. The optical activity is slight; the specific rotation is −1.98° (in water).

Sorbitol is readily soluble in water but only slightly soluble in methanol, ethanol and acetic acid. Aqueous solutions of sorbitol are miscible with methanol, ethanol and glycerol under any conditions. Sorbitol is neither volatile nor steam volatile, so that there is no loss when confectionery made with sorbitol is baked or cooked.

Similar to the di- and tri-valent polyalcohols, glycol and glycerol, sorbitol attracts water and holds it, it being hygroscopic. This means it is suitable as a means of retaining softness, as in confectionery. Sorbitol has almost the same calorific value as dextrose and is completely utilised by the organism. For sweetening power see p. 164.

Apart from its use in retaining freshness and softness in confectionery, sorbitol has been used for diabetics for many years. Sorbitol is first changed into fructose in human metabolism by enzymatic oxidation. Fructose is better tolerated by diabetics than dextrose (grape sugar) or sucrose; hence sorbitol is to be preferred to these sugars. In spite of this, diabetics cannot be given uncontrolled amounts of sorbitol without endangering the insulin balance.

D-Mannitol is the main component of manna-ash juice. It is widely distributed in the vegetable kingdom in buds, blossoms, twigs, fruits and roots of the Oleaceae and in fungi and algae. Proust, in 1806, isolated D-mannitol from the juice of the manna of the ash tree (in cambial bark). It is produced on a commercial scale by reduction of D-mannose. Reduction of fructose yields sorbitol as well as mannitol. It is used by diabetics as a substitute for sugar, but care must be exercised as in excess it possesses a laxative action. Mannitol forms complexes with boric acid and this can be used for titrimetric determination.

Dulcitol or melampyritol is found in the sap and skin of various plants. Because of its symmetrical constitution, dulcitol is optically inactive. It is produced technically by the reduction of galactose and from the dulcitol-containing manna.

Xylitol, an optically inactive pentitol with the same degree of sweetness as cane sugar, is produced commercially by catalytic hydrogenation of the cheaper xylose (p. 152).

Xylitol is increasingly used for diabetics as a substitute for sugar. It is as well tolerated, has the same calorific value and is easy to keep crystalline. As it passes directly into the pentose cycle of the metabolism, xylitol, like sorbitol is independent of insulin in controlled amounts and can be fully utilised by the organism with no physiological disadvantages. Xylitol and sorbitol do not take part in the MAILLARD reaction. This fact is important in the manufacture of foods for diabetics, because with xylitol (and sorbitol) there can be no loss of nutrients, as for instance in the blocking of essential amino acids during sterilisation, pasterurisation and drying.

Uronic acids

Uronic acids can be derived from the corresponding aldoses by oxidation of the terminal CH_2OH to a COOH group. Uronic acids are not found free in nature, but have an important function as components of glycosides and polysaccharides such as pectins, alginic acids, plant mucilages and mucopolysaccharides. In nature, D-glycuronic acid, D-galacturonic acid and D-mannuronic acid, D-galacturonic acid and D-mannuronic acids are the most common, L-iduronic acid and L-guluronic acids less so (formulae p. 157).

D-glycuronic acid	D-galacturonic acid	D-mannuronic acid	L-guluronic acid	L-iduronic acid
H—C=O	H—C=O	H—C=O	H—C=O	H—C=O
H—C—OH	H—C—OH	HO—C—H	HO—C—H	H—C—OH
HO—C—H	HO—C—H	HO—C—H	HO—C—H	HO—C—H
H—C—OH	HO—C—H	H—C—OH	H—C—OH	H—C—OH
H—C—OH	H—C—OH	H—C—OH	HO—C—H	HO—C—H
COOH	COOH	COOH	COOH	COOH

Uronic acids have many of the properties of the sugars from which they are derived. They can form α, β-pyranose and γ-furanose forms and reduce FEHLING's solution. The carboxyl group tends to form intramolecular esters (lactones) with the OH groups, most frequently γ-lactones. Uronic acids are usually obtained in the form of their lactones from polysaccharides after hydrolysis. D-glucuronic acid and D-mannuronic acid are obtained as furanosido-γ-lactones. The furanosido-γ-lactone of glucuronic acid is called D-glycurone.

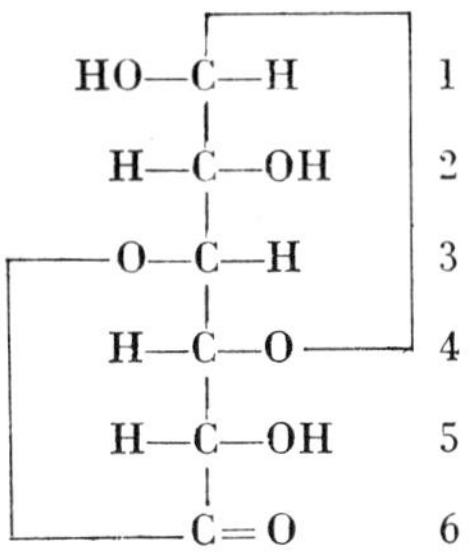

```
              ┌────────────┐
    HO—C—H    │       1
       |      │
     H—C—OH   │       2
       |      │
 ┌──O—C—H    │       3   D-glucurone
 │     |      │
 │   H—C—O ───┘       4
 │     |
 │   H—C—OH           5
 │     |
 └─────C=O            6
```

A number of methods are suitable for determining the uronic acids. CO_2 can be split off with 12–19% HCl at boiling temperature. The mechanism of this reaction is not fully understood. Furfural is formed as well as CO_2. In the presence of strong mineral acids and napthoresorcinol, specific colours are formed, which can be measured colorimetrically. Conversion of the carboxyl group to hydroxamic acid permits determination with Fe^{3+} salts, so long as no other acids are present.

Glycosides

The various hydroxyl groups of the sugar molecule are not all equally reactive. When the cyclic hemiacetal from of a sugar is prepared, a new hydroxyl group is formed by lactolisation of the aldehyde or keto group and these forms are particularly reactive. They are called acetal or glycosidic OH groups. They can combine with the alcoholic or phenolic OH groups of other molecules giving off H_2O to form ether-like compounds (full acetals, lactolides), the so-called glycosides. The conjugated substance or glycoside ether linked to the sugar is called an aglycone (genin). If this conjugated substance is itself a sugar, it is no longer called a glycoside, but a disaccharide (see p. 160); extension of this sugar linkage gives the oligo- and polysaccharides. Glycosides can be present as furanosides or pyranosides according to their ring size, and these may be α- or β-glycosides. Depending on the nature of the sugar, the glycosides are called glucosides (when derived from glucose), fructosides, arabinosides, galactosides, rutinosides (p. 167) and so on.

Methyl-D-glucoside in the α- and β-form is given here as an example of a glycoside:

methyl-α-D-glucoside (as usually depicted)

(according to HAWORTH) methyl-β-glucoside

The configuration of the primary hydroxyl is stable in the glycosides; for this reason no mutarotation develops, the FEHLING's solution is not reduced. The glycosides are stable compounds and frequently highly soluble in water. They are stable to alkalis as they are full acetals. Cleavage occurs with mineral acids and enzymes (glycosidic hydrolyses, see p. 260); enzymatic cleavage depends on the kind of sugar and above all on the configuration (α- or β-form) of the glycosidic linkage.

Many plants which are used as spices or stimulants contain considerable amounts of primary glycosides, but they only break down into the characteristic, desired components on processing (for instance, drying of the vanilla pod, fermentation of tea, preparation of mustard).

vanillin-β-D-glucoside $\xrightarrow[H_2O]{\text{Enzyme}}$ Vanillin + Glucose

In the vanilla pod vanillin occurs as a glycoside (gluco-vanillin) before drying.

The pleasant aroma of black tea, an ethereal oil – contained in green tea as an odourless glycoside – is only freed during fermentation by a β-glucosidase.

Natural glycosides are classified according to the aglycone (genin) into the following five groups.

Alcohol glycosides
Phenol and enol glycosides
Nitrogen or N-glycosides
Mustard oil glycosides
Cyanic acid glycosides

Whereas the **alcohol glycosides** of simple alcohols, for instance methyl glycoside, do not occur in nature, in Digitalis and Strophantus glycosides, as well as in the saponins, we find important natural alcohol glycosides, where the aglycones all belong to the steroids.

Members of the **phenolic glycoside** group are arbutin (hydroquinone glucoside) in bearberry (*Arctostaphylos uva-ursi*) leaves, vanillin (formula p. 158), coniferin, the anthocyanins and flavone glycosides; gentiopicrin the bitter principle of gentian root, is one of the closely related enol glycosides.

N-glycosides are compounds which one can assume have been formed from sugars and primary or secondary amines. Nucleosides, nucleotides, and nucleic acids, which are equally important in the animal and vegetable kingdom, also belong to this group (see p. 215).

The commonest sugar component in these compounds is D-ribose or 2-desoxy-D-ribose. Uridine 5′diphosphate-glucose (UDPG) may serve as an example; it is important for transporting energy and in changing one kind of sugar into another (uridine = uracil + D-ribose). The glucose which has been activated in the UDPG can be transported to other molecules with the OH groups. In this way glycosides and disaccharides can be synthesised.

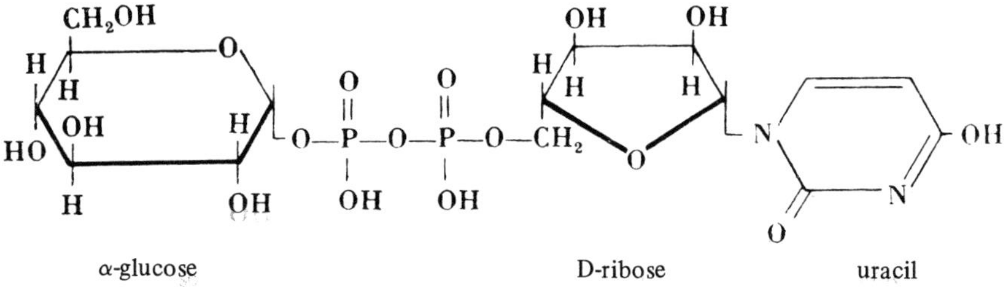

α-glucose D-ribose uracil

uridine-5′-diphosphate-glucose (UDPG)

Mustard oil glycosides are derivatives of thio sugars. Here the sugar is linked to the aglycon through the sulphur and not via an ethereal oxygen as in sinigrin.

White mustard contains the glycoside sinalbin, black mustard and horse-radish contain sinigrin; both are enzymatically split by myrosinase (p. 262). Here the sharp-smelling mustard oils are formed hydrolytically through a LOSSEN rearrangement. For example

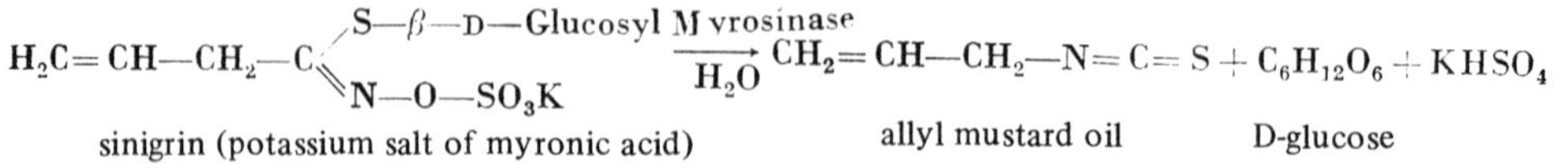

$$H_2C{=}CH{-}CH_2{-}C\begin{smallmatrix}\nearrow S{-}\beta{-}\text{D}{-}\text{Glucosyl}\\ \searrow N{-}O{-}SO_3K\end{smallmatrix} \xrightarrow[H_2O]{\text{Myrosinase}} CH_2{=}CH{-}CH_2{-}N{=}C{=}S + C_6H_{12}O_6 + KHSO_4$$

sinigrin (potassium salt of myronic acid) allyl mustard oil D-glucose

Amygdalin, a glycoside yielding **hydrocyanic acid**, is found in bitter almonds. The enzyme emulsin (β-glucosidase) which is found in the same nut splits amygdalin into benzalcyanhydrin (= almond acid nitrile) and gentiobiose in the presence of water. When bitter almond water is obtained by steam distillation of the de-oiled bitter almond mass, the almond acid nitrile is partly split into HCN and benzaldehyde (0.1% HCN in bitter almond water).

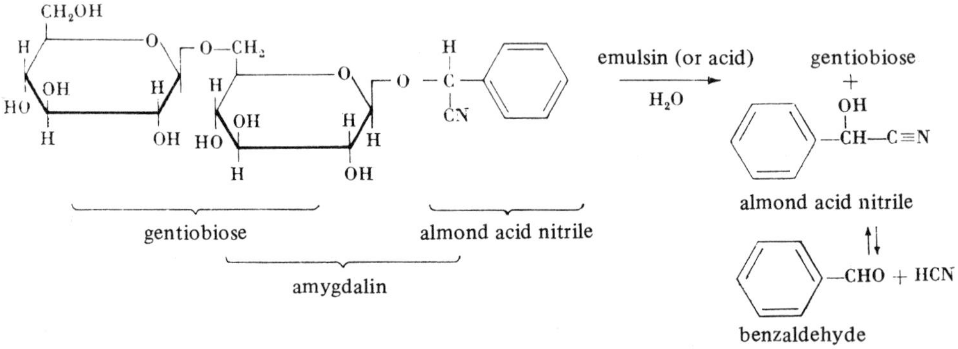

5.2 OLIGOSACCHARIDES

Disaccharides

Compounds in which two monosaccharide molecules are linked together with the loss of one molecule of water are called disaccharides. Naturally occurring disaccharides almost all contain hexoses as one component. Disaccharides belong among the oligosaccharide group, composed of several monosaccharides (2–10). They are split hydrolytically into their components by acids and enzymes.

Two different types of disaccharides are known:

1. Saccharose (sucrose) or **trehalose** type (for formula see sucrose p. 163), in which both monosaccharides are bound through an ether linkage of the reducing group*.

*Both sugars are present as full acetals (see p. 143), and are therefore stable against alkalis and the calcium metals. See the analytical determination of cane sugar in mixtures with reducing sugars after the destruction of the latter with $Ca(OH)_2$ (p. 164).

These disaccharides do not reduce FEHLING's solution; osazone and oxime formation does not occur, hydrocyanic acid forms no addition compounds. Because of the structure mutarotation does not occur and no change from the α- to the β-form, or vice versa, can take place. There are also disaccharides where the reducing groups are blocked. **Sucrose** or cane sugar, is the most important example of this type (see p. 162).

2. **Maltose type** (formula, see maltose p. 166). Here the reducing acetal OH group of one molecule is linked to the (non-reducing) alcoholic hydroxyl of the second sugar molecule in an ethereal manner. Hence the reducing group of one sugar molecule remains free: FEHLING's solution as the monosaccharides. Osazone and oxime formation is possible. Members of this group exhibit mutarotation, because the presence of a free lactonised aldehyde group allows the formation of α- and β-forms which may reach on equilibrium. Maltose, lactose and cellobiose belong to this type.

Disaccharides may also be formed as the so-called **reversion** products. WOHL defines reversion as the opposite of inversion, and so means re-synthesis. Here monosaccharides are heated for longer periods with acids (formation of an anhydride between the monosaccharide). Such slightly reducing anhydrides have also been obtained from fructose in a crystalline form. Starch subjected to acid hydrolysis gives small amounts of gentiobiose and isomaltose, and these are also called reversion products.

In the food industry, in the processing of starch or sucrose by acid hydrolysis and in cleavage products and products of decomposition reversion products may be formed by resynthesis. This occurs in warm starches, jams and processed fruit syrups. They may be formed from the same or different monomers and may be lost in the determination by the usual analytical processes, and so lead to an apparent deficit in sugar. Artificial honey prepared by these methods may contain up to 6% of these reversion products. Laevan is also a reversion product (see pp. 165 and 183).

Much work has been done in recent years on the transformations in the monosaccharides, leading to re-synthesis (reversion); grape sugar has been much studied, as the hydrolysis of the glucose polymers (starch, cellulose) is industrially important.

Reversion is a reaction catalysed by hydrogen ions occurring in all monosaccharides, whether they are aldoses or ketoses. However, the conditions under which it takes place are different for each aldose and ketose. The monomeric sugars are the starting point for the synthesis. A whole series of very different polymeric saccharides can be formed – varying with differing conditions. Analogous processes occur during the acid hydrolysis of oligo- and polysaccharides and reversion occurs secondarily when a sufficiently high concentration of monosaccharides is reached by cleavage.

TAUFEL gives the following schematic diagram for the processes of hydrolysis and cleavage which occur alongside each other:

$$\text{polysaccharide} \xrightarrow{\text{I}} \text{monosaccharide} \underset{}{\overset{\text{II}}{\rightleftarrows}} \text{reversion saccharide}$$

(intermediate products) ↓ III IV
desmolytic decomposition (furan derivatives, laevulinic acid, HCOOH, (browning processes)

If one investigates the factors temperature, time, acidity (pH) and sugar concentration, which influence the effect of acids on mono- and polysaccharides, it becomes clear that, other circumstances being equal, reaction I tends to occur at low sugar concentration, without necessarily leading to reaction II. The latter only manifests itself in addition to reaction I, at higher concentrations of the saccharide. At the same time, depending on conditions, desmolytic decomposition occurs in varying amounts according to reaction IV and leads to cleavage products, caramelisation, browning etc. The retrograde process, the splitting up of the reversion products, can be induced by suitable conditions particularly dilution.

Sucrose, trehalose and lactose are found free in nature. The others are usually found bound in glycosides (for instance gentiobiose in amygdalin) and as building blocks of the higher polymeric carbohydrates (such as maltose in starch, cellobiose in cellulose), from which they can be set free by hydrolysis.

Non-reducing disaccharides

Sucrose or saccharose

Cane or beet sugar, was discovered in 1747 by MARGGRAF in the roots of the beet and later found to be identical to that in cane sugar. Sucrose is widely distributed in the fruits and sap of many plants, especially in sugar cane and in the roots of the sugar beet; modern cultivated sugar beet contain 16–20% of their weight in sugar. Other plants or parts of plants, which are conspicuous for their sugar content (sucrose) are the stems of certain Graminosae, maize (8–12%), sweet millet (10–18%), also the sap of palms (3–6%), birches (2–4%), sweet oaks (3%), and the roots and tubers, carrot (3%), sweet potato (*Ipomea batatas*) (0.5–2.5%) and potato (*Solanum tuberosum*) (0.08–1.6%).

Sucrose is important as a high energy foodstuff: 1 g sucrose yields 4.1 kcal.

Sucrose is a disaccharide of the trehalose type; it is a 1(α)-glucosido-(1 → 5)-2β-fructoside-(2 → 5) or α-D-glucopyranosyl-β-D-fructofuranoside.

One molecule each of glucose and fructose are involved in the synthesis of the sucrose molecule with splitting off of H_2O to form the glycoside: glucose and fructose = sucrose + H_2O or $2C_6H_{12}O_6 = C_{12}H_{22}O_{11} + H_2O$. Those reducing (acetal, glycosidic OH) groups which were formerly free, and the keto groups which were capable of reaction, are now no longer present. For this reason cane sugar exhibits no mutarotation, has no ability to reduce FEHLING's solution, is stable, as an acetal, to dilute alkalis and lime solutions (see p. 164) and cannot form an osazone.

Sucrose forms monoclinic crystals with many faces, and large prisms if crystallised slowly as in candy sugar. Its melting point is 160°C when it is heated carefully after previous recrystallisation from alcohol at 170–180°C. It dissolves, at 15°C in a third of its weight of water, to give a clear and colourless syrup. It is only slightly soluble in ethyl alcohol. Sugar is not fermented directly, only when it has been split up into its components by the yeast enzyme invertase (β-D-fructofuranoside-fructohydrolase, see p. 262).

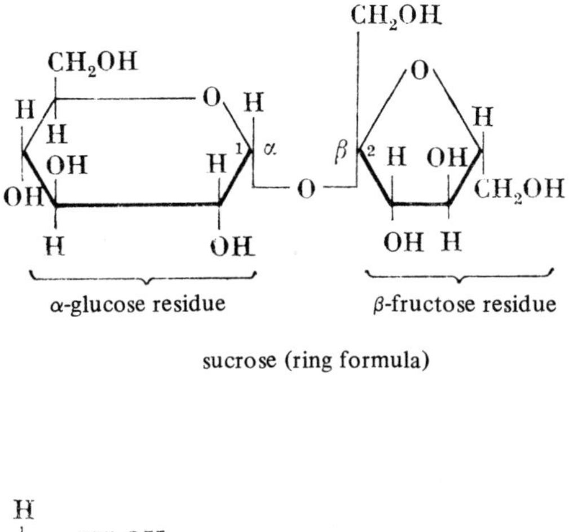

α-glucose residue β-fructose residue

sucrose (ring formula)

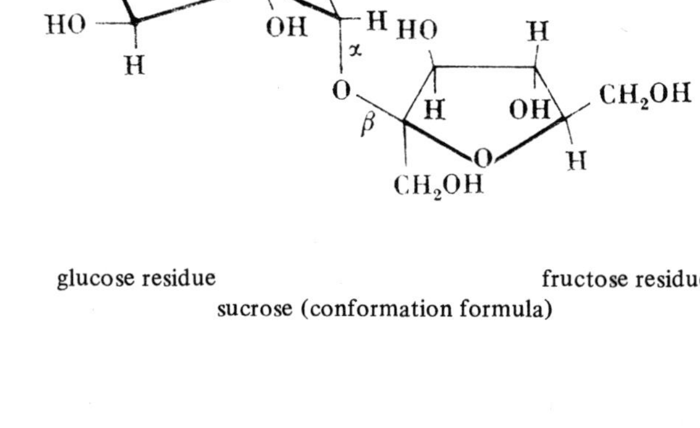

glucose residue fructose residue

sucrose (conformation formula)

Sucrose has a clean, sweet taste. Its relative sweetness compared with other sugars is as follows: (sucrose as the standard)

fructose	115	maltose	46
invert sugar	65	galactose	63
sucrose	100	α-lactose	39
glucose	69	(sorbitol)	54
xylose	67	(xylitol)	100

Sucrose is a weak acid ($K \sim 10^{-13}$) which, because of its stability forms addition compounds with alkali and alkaline earth hydroxides. These are **saccharates** but only the alkali saccharates are readily soluble; the alkaline earth saccharates are only slightly soluble in water. Saccharates are easily decomposed by CO_2 to yield sugar and carbonates. This process is important in the manufacture of sugar and in de-sugaring molasses with calcium hydroxide (STEFFENS process).

Like most glycosidic materials, sucrose is split by dilute acids (even when cold): this acid hydrolysis – which can also occur by means of enzymes, for instance invertase of yeast, decomposes the originally dextrorotary sucrose ($[\alpha]_D^{20} = -20.5°$) made up of equal parts of D-glucose and D-fructose:–

$$C_{12}H_{22}O_{11} + H_2O = 2C_6H_{12}O_6$$

The laevorotation of the fructose is greater than the dextrorotation of the D-glucose, so that the final product, so-called **invert sugar**, is laevorotatory. Hydrolysis of sucrose is called **inversion** because of this change in the direction of rotation.

The furanose ring structure of sucrose in fruits is the explanation for the ready hydrolysis with dilute acids (H^+ catalysis). All **furanose** ring structures show this sensitivity to acids, in contrast to the pyranose structures of lactose and maltose. Whereas **pyranosides** usually require about two hours of boiling with dilute acids for hydrolysis, furanosides, such as cane sugar, hydrolyse about 100 times faster. This is important in the food industry for sugar fermentation from starch, the manufacture of lactose and artificial honey.

The determination of sucrose can be carried out by physical, chemical and biochemical methods (polarimetry, refraction, paper chromatography).

Sucrose can be readily determined quantitatively by measuring its optical rotation before and after inversion. Sucrose can be measured chemically (reductometrically for instance with FEHLING's solution) after hydrolysis with acids or enzymes. The stability of sucrose alkaline earth hydroxides is made use of in the lime process; all sugars except sucrose are destroyed by CaO. A new elegant method for the determination of sucrose in complicated mixtures depends on chromatographic separation of the sugars in the mixture, followed by chemical proof of the sucrose and other accompanying methods, it is also possible to use biochemical methods as in the fermentation with various yeasts.

Cane sugar keeps well in a dry state but *Bacillus subtilis*, present at all stages in commercial production of sugar, decomposes sucrose solution into glucose and fructose; laevan, a fructosan, is then formed secondarily from fructose. Laevan is now considered to be an enzymatic reversion product. Moulds (Penicillium and Aspergillus) also decompose cane sugar with formation of invert sugar. In higher concentrations, sugar solutions have an inhibitory effect on microbes. Sugar is used on a large scale in canned fruit production. Certain osmotolerant yeasts, however, do decompose highly concentrated sugar solution and fruit juice concentrates. Sucrose fatty acid residues are used as emulsifiers in the food industry.

Trehalose (mycose)
The disaccharide, built up of two molecules of glucose, is found in various fungi (Boletus edulis, Aspergillus sp., fly agaric (*Ammanita muscaria*), ergot (*Claviceps purpurea*)) and also in Trehala manna. The two α-glycosidic OH groups of the glucose molecule are bound together through the ether function, so that FEHLING's solution does not reduce trehalose. For this reason osazone formation is impossible.

Three diastereomers are possible from trehalose, according to the type of glycosidic linkage of the two reducing groups: α, α—, α, β— and β, β-bonds. All three forms are known.

Reducing disaccharides
*Maltose or malt sugar**
The sugar $C_{12}H_{22}O_{11}$, is found in nature sometimes in free form at places where there is active breakdown of starch (malt, potato sprouts, leaves). Commercially, maltose is obtained from starch by enzymatic action (fungus, bacteria or malt amylases). The sweet mashes are the starting point for the manufacture of spirits, beer and brandies. Although maltose can be fermented directly by yeast, part of it is split off into glucose by the maltase present in the malt, and this is the basis for fermentation by zymase (whole yeast enzyme).

$$\underset{\text{starch}}{(C_6H_{10}O_5)n} \xrightarrow[]{\text{(diastase) amylase}} \underset{\text{maltose}}{C_{12}H_{22}O_{11}} \xrightarrow{\text{maltase}} \underset{\text{glucose}}{C_6H_{12}O_6}$$

Maltose has all the properties of a reducing sugar, a solution in water exhibits mutarotation. Dilute acids split the disaccharide yielding only glucose. Phenylhydrazine readily forms hydrazones and osazones (m.p. 206°C). Specific rotation

*This compound derives its name from the fact that it is formed from starch by the diastic enzyme of malt.

$[\alpha]_D^{20} = +130°$ (final value; allow for mutarotation). Maltose is a 4-(α-D-glucopyrano-sido)-α-D-glucopyranose, with the following formula

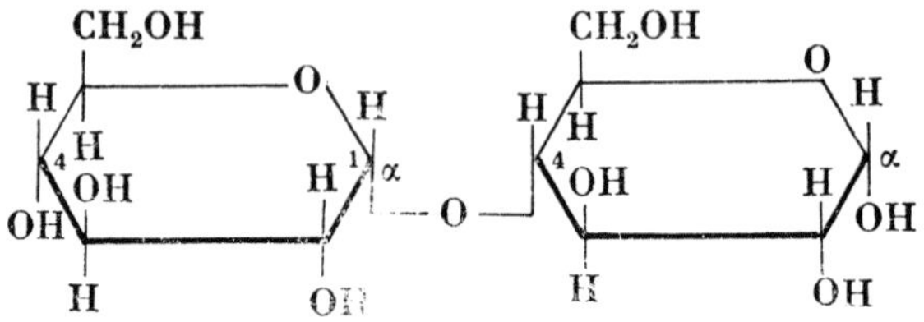

maltose

Maltose, a reducing disaccharide, can crystallize in the α-configuration as well as in the β-formation. Generally the β-form of maltose is obtained from water, as a monohydrate (β-maltose), m.p. 102–103°C.

Isomaltose, celliose and gentioniose

Isomaltose, an isomer of maltose 6-(α-D- glucopyranoside)-β-D-glucopyranose, has been isolated from the partial hydrolysates of dextrans (see p. 183) and from maize starch after hydrolysis and enzymatic decomposition.

During careful acid decomposition of cellulose, a stereoisomeric form of maltose occurs, called **cellobiose**. It reduces FEHLING's solution in the same way as the disaccharide maltose. In contrast there is a β-glycosidic linkage of the basic building blocks. Cellobiose is therefore hydrolytically split to glucose by β-glucosidases (cellobiases) for instance by emulsin. Cellobiose is a white powder; it is soluble in water, the specific rotation is $[\alpha]_D^{20} = +35.2°$, m.p. 225°C.

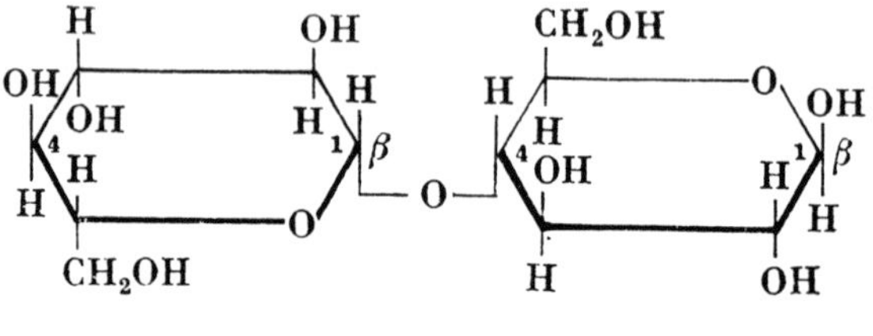

cellobiose

The bitter tasting disaccharide **gentiobiose** also consists of two molecules of glucose and the glycosidic hydroxyl of one is linked with the primary OH group on the C_6 of the other glucose molecule m.p. 190–193°C; $[\alpha]_D^{20} = +9.8°$. Gentiobiose is not fermented by yeast. In nature it is only found in a bound form in glycosides, for instance in amygdalin and gentiopicrin. It can best be isolated from the glycosides. Gentiobiose is also found in the trisaccharide gentianose (1 molecule fructose, 2 molecules glucose; see p. 170). Gentiobiose is also among the hydrolysis products of grain starch. In the yellow saffron colour crocin it is esterified with crocetin (see p. 123). As it occurs in small amounts when glucose is treated with hydrochloric acid, it is obvious that gentiobiose must be considered as a reversion product (pp. 139, 161).

Lactose (milk sugar, sand sugar)

Lactose is found free in the milk of mammals and in human milk (cow's milk 4.5–5.5%, human milk 5.5–8.0%). Recently lactose has been found for the first time in vegetables in the form of a glycoside. Lactose is produced commercially from the whey left after the manufacture of cheese. Rennet whey is used (dry mass 6–7% of which 4.5–5% is milk sugar and 0.8–1% soluble nitrogenous material bcause it contains about 0.7% more lactose than acid whey. Milk sugar consists of glucose and galactose which can be obtained by hydrolysis with mineral acids.

Lactose reduces FEHLING's solution because it is of the 'maltose type' (see p. 165), it forms an osazone and exhibits mutarotation. During isolation lactose crystallises out below 93.5°C as an α-lactose monohydrate; this is the usual commercial lactose and is only slightly sweet (see p. 164), its solubility in water is slight compared with other sugars (1:6). This is the reason why lactose often feels sandy on the tongue. In pure, or even 50% alcohol, lactose is almost insoluble. Its behaviour in 50% alcohol is used analytically in the separation of lactose from foodstuffs. Lactose is a 1(β)-galactosido-(1 → 5)-4-glucose-(1 → 5) or 4-(β-galactosido)-glucose.

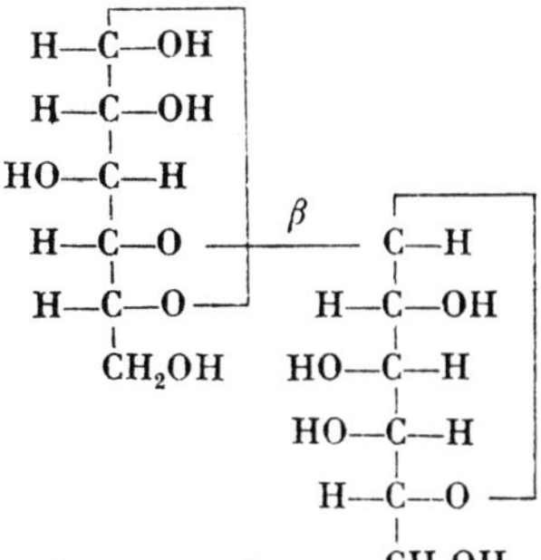

α-glucose portion

β-galactoside portion

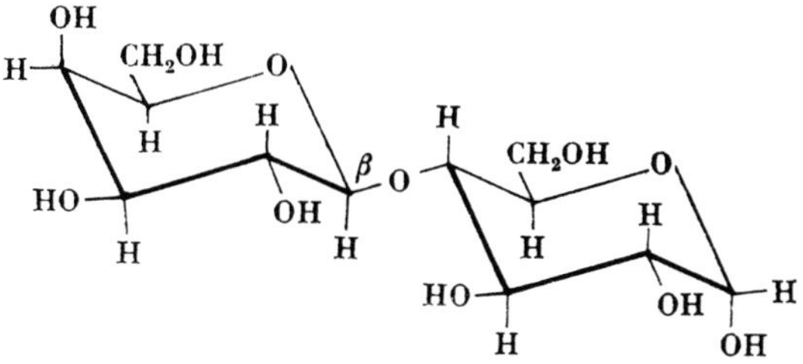

β-galactoside portion

α-glucose portion

α-lactose

When crystallisation from water is carried out above 93.5°C, anhydrous β-lactose is obtained, which is soluble in water and tastes sweeter than the α-form. The equilibrium value of the specific rotation is $[\alpha]_D^{20} = +55.3°$. Melting point of $C_{12}H_{22}O_{11} \cdot H_2O$ is 203°C when heated quickly.

Lactose is not fermented by ordinary yeasts (analytically important), but special yeasts, such as kefir or koumiss enzyme, can cause alcoholic fermentation. Many of these possibilities are used in the preparation of kefir and koumiss. The enzymes in kefir and koumiss cause alcoholic fermentation in the milk (3% alcohol in the product from mare's milk). Lactic acid bacteria ferment lactose to lactic acid (souring of milk). The presence of lactose encourages the growth of Bifidus flora in infants.

The usual methods for quantitative determination of lactose are polarimetry and reduction, but the qualitative proof is by the **mucic acid reaction.** In this the galactose portion of the lactose disaccharide is converted into mucic acid COOH . $(CHOH)_4.COOH$ with concentrated HNO_3; mucic acid (soluble in cold water 1:3000) optically inactive because of its symmetrical structure, and which can easily be recognised by its characteristic crystalline form, m.p. 214°C (with decomposition) and finally by the pyrrol red reaction (spruce shaving reaction) after steaming with NH_3. Many plant mucilages contain galactose among their building blocks and these also give a positive mucic acid reaction.

Melibiose

This disaccharide is present in plant exudates. It is also a component of raffinose. Its building blocks glucose and galctose are set free by hydrolysis. The properties of melibiose are the same as its isomer lactose, apart from the greater dextro-rotation. $[\alpha]_D^{20} = + 126.5°$, m.p. 85°C (dihydrate), FEHLING's solution is reduced.

Trisaccharides

There are three important natural disaccharides: raffinose, gentianose and melezitose. Previously the trifructosan, called trifructose anhydride was also included among the trisaccharides. Recent research, however, has made it clear that it is a polysaccharide with 10 molecules of fructose (see p. 133). The trisaccharides contain three monosaccharide molecules, which may be the same or different from each other. They can be broken down by acids and certain enzymes, into the three monosaccharides, or sometimes into one monosaccharide and one disaccharide (see formation of gentiobiose from gentianose, p. 166).

Raffinose

Raffinose, sometimes wrongly called raffinotriose, is a well known trisaccharide found in many plants, for instance in soya beans and cotton seed (8%). It is also present in sugar beet (but not in sugar cane) and remains in the syrup (1.4–4.0%) and is found as an undesirable impurity in commercial beet sugar.

The trisaccharide is broken down into galactose, glucose and fructose after complete hydrolysis by acids or bottom fermenting yeasts. The linkage bond is on the 3-position of the glycoside (trehalose type). For this reason raffinose cannot reduce FEHLING's solution nor can it form an osazone.

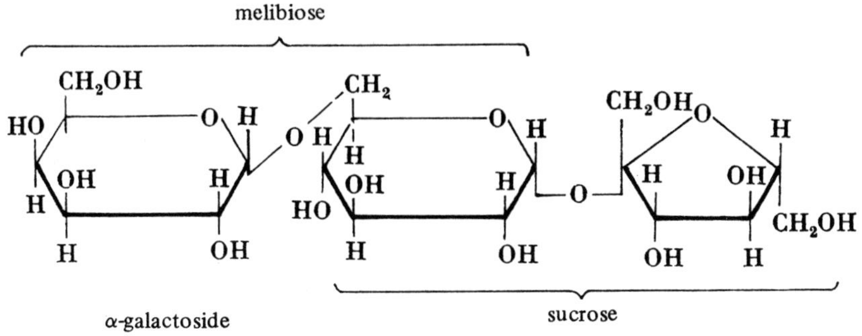

melibiose

α-galactoside sucrose

raffinose

The polarisation value of raffinose is considerably greater than that of sucrose, which leads to confusion in calculating the sucrose content in crude sugar consequently raffinose is often called 'plus sugar' in industry. Raffinose does not taste sweet.

When raffinose is crystallised with 5 mol water of crystallisation ($C_{18}H_{32}O_{16} \cdot 5H_2O$) it has a specific rotation of $[\alpha]_D^{20} = +150°$ (independent of temperature, concentration and alcohol content, however is greatly lowered by lead acetate).

Melezitose

This trisaccharide ($C_{18}H_{32}O_{16}$) found in many kinds of manna, in the 'honey dew' of lime trees (up to 40%), on pine, larch and fir trees and frequently found in honey from firs (especially in dry years when insufficient nectar is formed). It breaks down into glucose and turanose (an isomer of sucrose) under mild hydrolysis. This indicates its composition as such

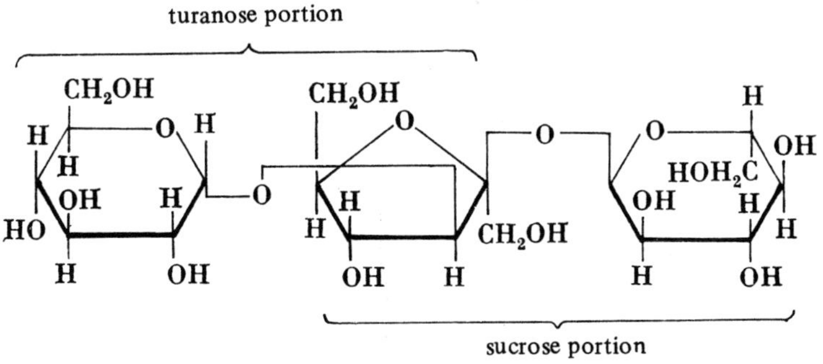

turanose portion

sucrose portion

melezitose

Melezitose does not reduce; it exists as a mono- and a dihydrate, the latter is the stable form $[\alpha]_D^{20} = 88.5°$, which corresponds to a rotation of + 91.7° for the anhydrous form. m.p. 155°C (dihydrate).

Gentianose

Gentianose is found in the roots of gentians. Complete hydrolysis splits it into two molecules of glucose and one molecule fructose, whereas dilute acids only split it partially into fructose and gentiobiose. The glycosidic ether bond between fructose and glucose hydrolyses more readily than the glycosidic bond between the two glucose residues, see furanoside and pyranoside splitting p. 157 Emulsin and other enzyme extracts splits it into glucose and sucrose. This indicates the constitution of this trisaccharide.

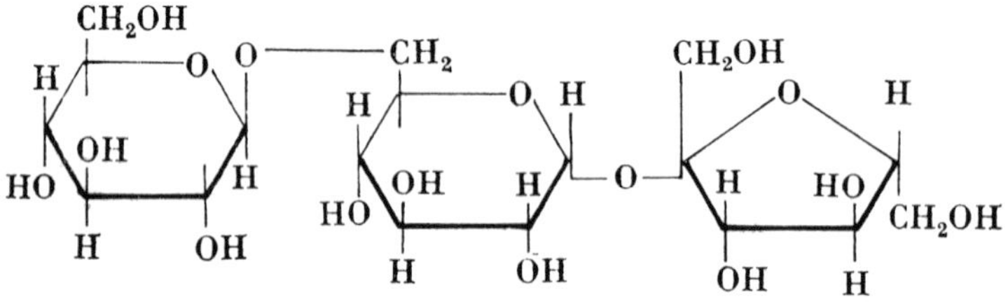

gentianose

Gentianose has almost no sweet taste and is not changed by FEHLING's solution (the reducing groups are missing). m.p. = 209°C; $[\alpha]_D^{20} = +33.4°$

Higher oligosaccharides

Non-reducing oligosaccharides

In addition to the trisaccharides, a number of tetra-, penta-, hexa- and higher saccharides have been found in plant juices, which are also derived from sucrose and do not reduce FEHLING's solution. The tetrasaccharides **stachyose**, **lychnose** and **isolychnose** belong to this group, which can be regarded as the basic members of a homologous series. With the addition of one or more galactose molecules and according to the nature of the bond, the homologous penta- and hexasaccharides are formed, for example the pentasaccharide verbascose which derives from stachyose.

Most of the reducing plant oligosaccharides are obtained by partial hydrolysis of polysaccharides and subsequent chromatographic separation of the cleavage products. Oligosaccharides with six or more sugar units have been separated and isolated in this way from partial hydrolysis of starch, cellulose, xylans and dextrans.

Oligosaccharides of milk

The milk of mammals, especially human milk, contains (apart from the main disaccharide, lactose) a number of other reducing alogosaccharides. A few of them can be prepared in a pure state and their structure clarified. Those tri-, tetra-, penta-, and hexasaccharides so far identified in human milk are all derived from fructose, and glucose is always the reducing sugar residue. They contain

L-fructose N-acetyl-D-glucosamine and N-acetylneuraminic acid in addition to the glucose and galactose. The formula for lacto-N-tetraose will serve as an example of a saccharide isolated from human milk:

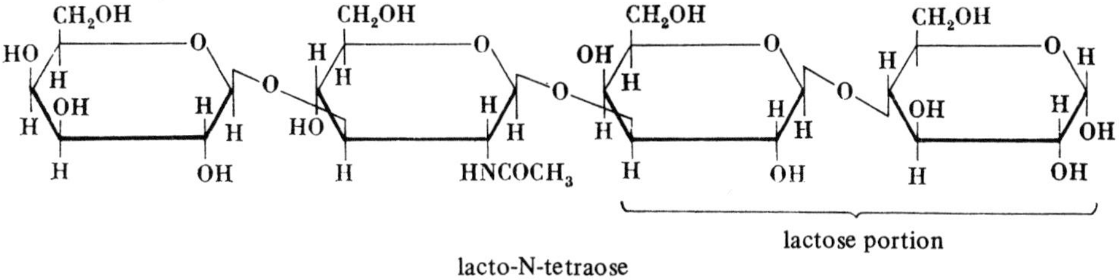

lactose portion

lacto-N-tetraose

Some of these oligosaccharides exhibit blood group activity. In addition they are necessary for the growth of certain strains of *Bifidobacterium bifidum*. A trisaccharide neuraminosyl-lactose (N-acetyl-lactaminic acid-lactose) has been found in the colostrum from cow's and in human milk

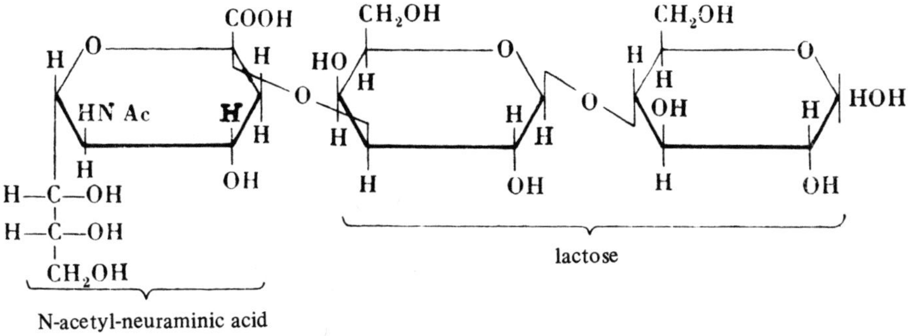

lactose

N-acetyl-neuraminic acid

N-acetyl-neuraminosyl-lactose

The N-acetyl-neuraminic acid (sialic acid), which is bound to lactose, is widely distributed in mature and is found in both the free and the bound form in many glycoproteins from blood, mucus and so on. N-acetyl-neuraminic acid, in addition to the general reactions of sugar, also has a characteristic colour reaction with reagents such as EHRLICH's reagent (p-dimethylaminobenzaldehyde/HCl), BIAL's reagent (orcinol/HCl). This can be used for quantitative determination.

Other oligosaccharides, some of them higher, have been found, but their structure is not yet known. The total amount of all neutral oligosaccharides in human milk, excluding lactose is about 3 g/l.

Chemical methods of separation

As well as reactions to prove the presence of sugars discussed above, most of which cover a whole group of sugars, chromatographic methods are of the greatest value in sugar chemistry. They make it possible to identify and isolate even traces of individual sugars from a complex mixture of various mono- and oligosaccharides, amino acids and protein.

Paper chromatography and **thin layer chromatography** on silica gel, kieselguhr, cellulose powder and magnesium silicate can be used in analysis. Recently gas chromatography of sugars in the form of volatile derivatives such as trimethysilylethers, acetyl or methyl compounds has become increasingly important for their determination.

The **proof** for sugar on the chromatograms is carried out by spraying the finished chromatogram with a suitable reagent such as aniline phthalate, benzidine periodate, (caution – the use of benzidine is now not permitted in many countries, as it can be carcinogenic), ammoniacal $AgNO_3$, triphenyltetrazolium chloride and so on. Various coloured spots appear, according to the reagent, and show the position of the individual sugars on the chromatogram. Autoradiochromatography is used for radioactive labelled compounds.

Column chromatographic and **electrophoretic methods** are available for preparative separation and isolation of sugars. Cellulose powder, kieselguhr (celite) silica gel, ion exchangers, molecular sieves and so on can be used as column packing. For elution various solvents such as water, ethanol, n-propanol, n-butanol, pyridine, ethylacetate, glacial acetic acid and formic acid may be used. These have to be selected to suit the compounds involved.

Chromatography can also be used for **quantitative determination** of sugars. One method is to cut out the coloured spots obtained, extract the colour and measure its intensity photometrically. The exact amount can then be calculated from a calibration curve which has been worked out from known amounts. It is advisable to run the calibration materials in the same chromatogram as the mixture being tested, so that identical conditions are observed. In gas chromatography, 'internal standard' is used which is added in a known amount to the sample.

When evaluating a chromatogram, it is necessary to realise how far the results can be influenced by the effects of **temperature** and accompanying **impurities**. The proof of a particular sugar should never be based on chromatographic data alone. It is always better to isolate, where possible, a crystalline derivative which is characteristic for the sugar in question.

5.3 POLYMERIC SACCHARIDES OR POLYSACCHARIDES

By repetition of the glycosidic linkage of two monosaccharide molecules it is possible to obtain materials of relatively large molecular size via the oligosaccharides (p. 160). They are called polymeric saccharides because they are built up of many monosaccharides. The systematic name for the compound is formed by replacing the ending -ose by the ending -an on the name of the monosaccharide from which the polymeric saccharide is built up; for instance glutan is the name for polymeric saccharides built up of D- or L-glucose; pentosans are those built up from pentoses only; in general the name glycan is used for all polymeric saccharides (from glycose, the generic name for sugar).

Polymeric saccharides are very different from the basic building blocks from which they are made up. They are apparently macroscopically and microscopically amorphous and have only been shown to be microcrystalline when examined under X-rays. They do not taste sweet and are often insoluble or barely soluble in water. When they are soluble, they only form colloidal solutions which diffuse with difficulty. The molecules of the solvent can no longer overcome the molecular attraction (van der Waal forces, dipolar forces, secondary valency, molecular cohesion, residual valency), which is conditioned by the intermolecular forces which hold together the branched and unbranched chains of macromolecules (unlike smaller molecules). This is because the greater the chain length the larger is the intermolecular force.

Colloidal solutions of polymeric saccharides (e.g. starch) are optically inactive, but do not reduce FEHLING's solution and do not form osazones with phenylhydrazine nor saccharates with metallic oxides.

Polymeric saccharides are broken down to their basic building blocks by acid or enzyme hydrolysis, thus giving intermediate steps in the formation of the oligosaccharides. Polymeric saccharides can also be decomposed by oxidation agents and by alkali. The reactions are mostly of a complex nature and do not lead to the monosaccharide building blocks but to derivatives or decomposition products of many kinds.

Table 9 Composition of polymeric saccharides according to their functions (the letters a, b, c, d, e, in brackets give the order of the chemical classification)

Origin	Structural Material	Reserve Material	Other functions
Plant kingdom			
Higher plants	cellulose (a_1)	starch (a)	plant gums (c)
and	hemicelluloses (b)	amylose (a)	
micro-organisms	pectins (d)	amylopectin (a_2)	plant mucilages (c)
	alginic acids (d)	inulin (a_1)	tragacanth
	carrageenan (a_1)	grain fructans (a_2)	gum arabic
	agar-agar (a_1)	laevans (a_2)	
	mannans (a_1)	mannans	
		guaran (b)	carubin (b)
		laminaran (a_2)	
		glycogen (a_2)	

Table 9–*continued*

Origin	Structural Material	Reserve Material	Other functions
Animal kingdom			
Vertebrates	chondroitin sulphate (e)	glycogen (a_2)	heparin (e) hyaluronic acid (e) blood group substances (e) galactans (a)
Invertebrates	chitin (a_1) cellulose (a_1)	glycogen (a_2) galactans (a)	galactans (a)

Classification of polysaccharides

Polysaccharides can be classified in various ways: according to their origin and function (see Table 9), or according to their chemical composition. The following table gives a chemical classification.

(a) **Homoglycans**, that is polysaccharides which are built up of all the same sugar building blocks. There are two sub-groups, the straight chain (a_1) and the branched chain (a_2); amylose is an a_1, amylopectin an a_2.

(b) **Heteroglycans** which are built up of two or more kinds of neutral monosaccharide.

(c) **Heteroglycans** which contain uronic acid as well as sugars.

(d) **Homoglycans** which consist only of uronic acids, that is the glycuronans (polyuronides).

(e) **Heteroglycans** which contain amino sugars as well as uronic acid (mucopolymeric saccharides).

This classification is incomplete since the structure of many polysaccharides is not yet fully known.

Homoglycans from sugar group (a) (Table 9)

Glucans

The following groups of homoglycans from sugars will be discussed here from a food chemistry and physiological aspect.

Glucans are built up from glucose, fructosans from fructose and a few other polysaccharides, such as mannans, galactans and chitin.

Glucans contain only glucose as a basic building block. They include polysaccharides such as starch, glycogen and cellulose which are important in food chemistry.

Starch. **Starch** is the most important final product of photosynthesis, and is therefore found as a high molecular, osmotically inactive reserve material in most plants – except some lower plants. In contrast to cellulose (p. 180) glucose deposited in the form of starch can always be brought back into the metabolism of the cell with the help of diastic enzymes (amylases). In this way it can be used as a source of energy (during breakdown) and also for the synthesis of other materials (fats).

Synthesis and breakdown of starch and glycogen is carried out enzymatically by the phosphorylases which convert the inorganic phosphate into 'energy rich' organically bound phosphate, which in turn sets free energy for further synthesis.

Grain is especially rich in starch, wheat 60–70%, rice 70–80%, maize 65–75%, potatoes 17–24%, rhizones and tubers of tropical plants (e.g. tapioca starch) and the pith of the sago palm (sago).

The starch granules in plants differ from each other in size (0.002–0.15 mm) and in shape (round, oval etc.) and are usually built up from numerous layers round a nucleus. This allows the various kinds of starch to be distinguished from one another microscopically.

Starch is optically active, with specific rotation about + 190° and is a polysaccharide built up entirely of glucose, although the glucose molecule may be combined into larger component units. Thus **amylose**, one of the components, is soluble in hot water and is made up from various long unbranched chains in which 100–2000 glucose molecules are linked to one another by α-1,4 bonds. The second component is **amylopectin** which only swells in hot water and is **branched** (see formula). The main chain (= α-1,4 linkage of the basic molecule) has side chains of various lengths (with approx. 10–20 glucose residues) attached to it by means of α-1.6 bonds. The ratio of amylose and amylopectin in the mixture depends on the origin of the starch.

It is generally assumed today that the linear chains form a spiral, whilst the existence of branches gives a globular and very dense conformation (spherical colloid). The term **spherocolloid** or **globular molecule** is used when the macromolecule is very branched or spatially intertwined (starch, glycogen, albumin, globulin, haemoglobin).

Starch has a crystalline structure. When considering its properties in foods, retrogradation into its component parts is important.

The two polymeric starch fractions (amylose and amylopectin) can be separated from each other, using for example their different behaviour with water at 70–80°C (amylose dissolves out). To distinguish them qualitatively their behaviour with iodine is used. Amylose, which has a spiral form in its secondary structure, gives an intense blue with iodine. This is due to the deposition of the iodine atoms linked in chains in the spiral canal-like structures of the starch molecule. Amylopectin gives blue, violet to red shades with low colour intensity. Ordinary starch solutions give a blue to violet colour with iodine.

Starch dissolves completely in certain salt solutions, in dilute alkali and in some organic media (formamide, formic acid).

MAYRHOFER's quantitative determination of starch is based on the solubility of starch in dilute alkalies.

It is difficult to determine the molecular weight of starch, because solutions of it are subject to aging and easily succumb to decomposition. There are considerable differences between the molecular weights obtained by physical and chemical methods (e.g. end group determination). Chemical determination normally gives the lower weights, as the side chains are already attached by their end groups. Maize amylose for instance has a molecular weight of between 40,000 and 340,000 corresponding to the degree of polymerisation, that is the number of basic units in the macromolecule (250 to 2,100). Starch and other polymeric carbohydrates are polydispersed in that they are mixtures of residues of different molecular weights. Amylopectin in maize starch has an average molecular weight of 45,000.

Native starch does not swell noticeably in cold water, on the other hand, when it is warmed (to about 50°C) there is strong swelling with gel formation. The gel formed has high viscosity, even at low concentrations (1-4%), but is strongly dependent on the electrolyte content of the starch used. Amylopectin is responsible for the formation of the gel which results in a stiff paste when cold.

Native starch is always accompanied by various non carbohydrate components. The phosphoric acid content is greatest in potato starch, followed by wheat and other grain starches. It should be emphasised that phosphoric acid may be present in starch in various bonding forms.

Phosphoric acid in potato starch is mainly bound to the amylopectin as an ester. It may also be present in the form of phosphatides, which are absorbed on the starch and can be removed from it by extraction.

With some kinds of starch, such as potato starch, the non-esterified OH groups of phosphoric acid, are neutralised by cations (Ca^{2+}, Mg^{2+}, K^+, Na^+). The relative amount of these cations determines certain properties of the starch, such as flavour, swelling, solubility, viscosity of the gel, electrical conductance, pH, reactivity with alkaline solutions and so on. In native potato starch it is possible during production to exchange the potassium cation for calcium by means of a permutit type exchange. Potassium starch gives a highly viscous gel in contrast to calcium starch.

Fatty acids (palmitic acid, oleic acid, linoleic and linolenic acids) have been separated out especially from maize starch. These are now considered to be bound to the polymeric starch as esters. Synthetic starch esters are important in the food industry.

Silicic acid is present in small amounts. There is no doubt that these non-carbohydrate components are integral parts of the starch molecule; and they account for the variety of the chemical and physical properties which the different starches exhibit. Starch granules usually contain, in addition to the amylose and amylopectin, H_2O 12-20%, nitrogen 0.004-0.08%, fatty acids 0.04-0.83% (as fat), phosphoric acid 0.0005-0.3% as P_2O_5, small amounts of silicic acid, ash content according to the type of starch 0.2-1%.

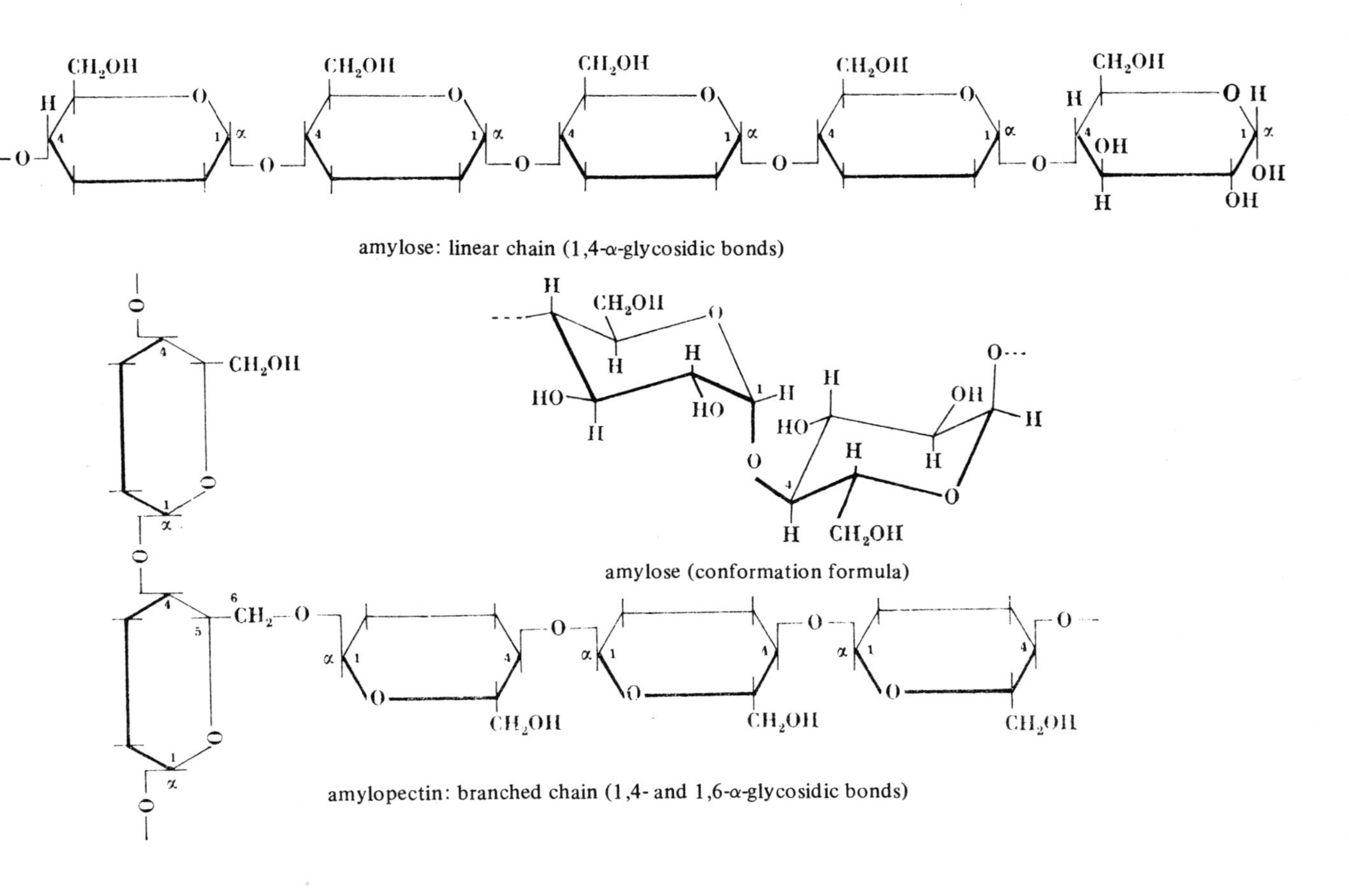

amylose: linear chain (1,4-α-glycosidic bonds)

amylose (conformation formula)

amylopectin: branched chain (1,4- and 1,6-α-glycosidic bonds)

Amylose and amylopectin, together making up starch, are broken down into D-glucose (grape sugar) by boiling with dilute acids (hydrolysis):

$$n(C_6H_{10}O_5) + n(H_2O) = n(C_6H_{12}O_6)$$

Hydrolysis of starch is carried out commercially during the manufacture of sugar products from starch – starch syrup, starch sugar and dextrose.

Starch is also decomposed to maltose and glucose by hydrolytic enzymes. Such enzymes are to be found in the digestive tract, in saliva, pancreas and so on. The so-called amylases (diastases) are particularly important in sugar formation from starch, in spirits distillation, in baking and in brewing (see also enzymes p. 237). The α- and β-amylases decompose starch via different intermediate products and attack the starch molecule differently.

α-amylase splits internal bonds of the starch molecule, and not those at the ends of the chains. The breakdown goes as far as maltose (with some glucose formation).

β-amylase attacks the starch from the non-reducing end of the chain and splits off one maltose after another. Amylose is completely broken down into maltose, whereas with amylopectin (because of the branching of the molecule) about 40% remains behind as 'limit dextrins'.

In the manufacture of beer and brandy, grain starch is converted via malt diastase (amylase) into soluble sugars (maltose), but this can only be fermented into alcohol and CO_2 after it is broken down to glucose by yeast.

Decomposition products of starch (dextrins):
The decomposition of the polymeric saccharides, either by dry roasting, heating with water under pressure or with dilute acids or by treating with enzymes (addition of malt, diastase) results in the formation of materials which, though they still have the empirical formula $(C_6H_{10}O_5)$, have a low molecular weight (between the mono- or disaccharides and starch). Similar decomposition products from the flour starch are formed in baking, particularly in bread crust and are important for a good flavour.

Dextrins is the general name given to these decomposition products of starch and they form an unbroken series of compounds (between starch and the oligosaccharides) with decreasing molecular size. In contrast to starch however, they are all soluble in water but they behave differently towards iodine solution, according to the degree of breakdown.

SCHARDINGER dextrins are a special group of crystalline dextrins which are formed by *Bacillus macerans* degradation of starch. They are also called **cyclodextrins** because of their cyclic structure. They contain six to eight α-1,4 glucose units in a ring formation and are therefore sugar anhydrides, glucosans. The three members of this group with homologous rings found so far, are called α-, β- and γ-dextrin. They do not reduce FEHLING's solution, but are split quantitatively by acid to give glucose. Cyclodextrins form so-called inclusion compounds, as with iodine.

Industrial soluble starch (amylodextrin) is made by milling starch for a prolonged period (= partial mechanical decomposition of starch) or by treatment with dilute hydrochloric acid (= partial chemical decomposition of starch chains, especially amylopectin) to give a product which disperses more readily in water. This removes the gel formation properties, yielding a clear, thin solution. Soluble starch turns blue with iodine and behaves just like starch apart from its solubility in water.

Erythrodextrin is coloured red by iodine solution and begins to reduce FEHLING's solution due to further decomposition.

Maltodextrin or **achrodextrin** is no longer coloured by iodine. Its reducing power is greater than erythrodextrin, but less than the end products maltose and glucose.

Dextrins are not fermented by yeast, are insoluble in alcohol (important analytically) and are strongly dextrorotatory.

The commercial product '**dextrin**', which is produced on a large scale for technical purposes, is not a uniform compound, but is usually a mixture of maltodextrin with some erythrodextrin and also contains some glucose or maltose according to the method used for preparation. Potato starch is usually the starting material and to decompose it, it is first roasted at 200°C (roast gum, roast starch, starch gum) or treated with a malt infusion at 65-70°C or with dilute acids (e.g. oxalic acid) until the iodine reaction does not occur. The excess acid is removed with $CaCO_3$, the dilute aqueous solution is precipitated with alcohol and the precipitate dried with alcohol. Various types of dextrin have a different dextrorotation. Commercial dextrin is usually given values of $[\alpha]_D^{20} = 135°$, regardless of temperature.

Glycogen. **Glycogen** is the reserve carbohydrate of the animal organism, and is mainly stored in the liver (little in muscle). It is also found in fungi, yeasts and bacteria and occasionally also in the higher plants. Glycogen is the storage material formed from part of the animal's intake of carbohydrate. The glycogen store, on breakdown, yields glucose for energy.

Meat contains only traces of glycogen (0.15-0.18%). Horse flesh is an exception with 0.9%; this was formerly used as a proof for this meat. However, larger amounts are present in the liver, 2.88-8.14%.

Although glucose is the basic building block, the morphological structure of glycogen is similar to starch, but it possesses more branched chains. Some of its chemical properties are the same as those of starch. It is a white, odourless and tasteless powder, but dissolves fairly easily in water without forming a gel. Iodine gives a red colour. It reduces FEHLING's solution to a small extent. The specific rotation is about + 190 to 200°. It is broken down into dextrin, maltose and glucose by hydrolysis with dilute acids. Enzymes split glycogen and starch into maltose (and glucose).

Cellulose. **Cellulose** is a polymeric saccharide ($C_6H_{10}O_5$) like starch, but exhibits considerable differences in its physical properties and chemical structure. It forms the main structural component of cell walls in plants and is associated there with other carbohydrates, such as hemicelluloses, pectin materials, lignin and so on. Cellulose is found in an almost pure state in cotton and in the pith of the elder. On the other hand, coniferous and deciduous trees which are used for the commercial production of cellulose only, have a true cellulose content of only 40–60%.

In wood pulp production, the hemicelluloses and lignins, which accompany the cellulose (see polysaccharides, p. 172) are solubilised by hot sulphurous acid solutions (which form soluble lignin sulphonates) and are removed in the spent liquor; the insoluble cellulose remains behind as a pulp. For each ton of pulp, 8–10 cu metres of spent sulphite liquor is produced, containing 50% of the weight of the original wood. The huge amounts of spent sulphite liquor, which may contaminate rivers badly, can be utilised after removal of the sulphurous acid by fermenting the 2–4% of fermentable sugar contained in it, into alcohol sulphite (sulphite spirits). The spent sulphite liquor can be utilised for nutritional food yeasts for human or animals (by the WALDHOF process, the J. G. SCHOLLER process or the biosynthesis process (**biological protein synthesis**). In these the yeast produces protein. Similarly it can be used for fat (biological fat synthesis). Vanillin is obtained on a commercial scale from the lignin in the sulphite spent liquors by nitrobenzene oxidation.

The chemical composition of cellulose has been mainly clarified by X-ray investigations. The parallel assemblage of the long molecular chains (micelles) into fibres is the explanation for the mechanical properties, particularly the great resistance of cellulose to tearing. The fibrous structure and the resistance of cellulose to the enzymes of the human digestive tract indicate the characteristic structure of cellulose as an unbranched, continuous chain of β-glycosidically linked glucose residues. (In contrast to starch, in which the α-glycosidic links give branched and unbranched chains.

Cellulose is totally insoluble in water, it is practically stable to dilute acids and alkalis; FEHLING's solution is barely reduced.

Cellulose is not attacked by human digestive juices.

The cellulose macromolecule only begins to decompose when treated with concentrated acids (60–70% H_2SO_4 or strong and highly concentrated 41% HCl) or with weaker acids at higher temperatures. The breakdown goes via the disaccharide cellobiose to glucose. This reaction has led to the production of glucose directly from wood (wood saccharification). The hydrolysis of cellulose is usually acheived by one of two of the following methods. 1. With concentrated HCl at normal pressure and temperature (20°C) by the BERGIUS-HAGLUND process or 2. With dilute H_2SO_4 (0.2–0.6%) with heat (170°C) and under pressure (8 atmospheres) by leaching out in diffusion cells, in the SCHOLLER-TORNESCH process. But in this case other sugars are also formed from the

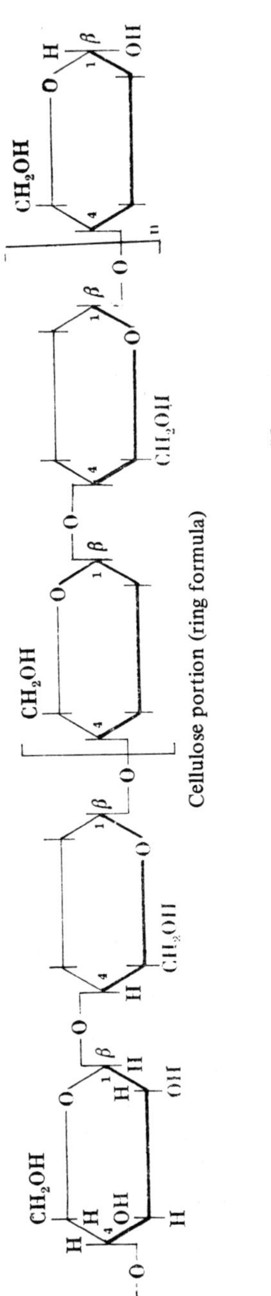

Cellulose portion (ring formula)

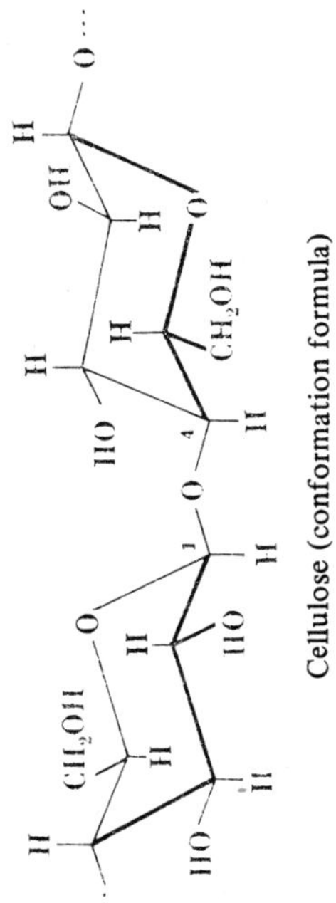

Cellulose (conformation formula)

accompanying polymeric saccharides (hemicelluloses, pentosans) so that the hydrolysates contain sugars which can be processed into dry sugar (for feed), alcohol (by fermentation) or into dried yeast ('Nutritional yeast'), see biological protein synthesis p. 180).

Sugar formation from coniferous wood yields monohexoses while deciduous wood yields more pentosans; therefore the former are preferred for alcohol production and the latter as substrate for the growth of yeasts (see p. 180).

Vegetable parchment is produced by treating sheets of cellulose with strong sulphuric acid; then, after the acid has been washed out, the sheet is dried.

Cellulose is one of the most basic industrial raw materials and is used in many branches of industry: paper and packaging industry, textiles, synthetics, explosives, film industry (acetate film), pharmaceutical and food industries (packaging), fermentation (C_2H_5OH), production of animal foodstuffs (yeast protein and fat), charcoal burning (carbonisation), wood distillation, fuels, chemicals, paint industry, adsorpants and clarifying agents.

In the glucose residues in the cellulose there are always a number of free alcoholic hydroxyl groups. These give the various characteristic reactions of the hydroxyl groups: cellulose forms esters with acids, nitric acid residues (nitrocelluloses) are made into filters, threads, foils, gun cotton, collodion, celluloid and so on. Acetic acid residues of cellulose (acetylcelluloses) are manufactured in considerable quantities. Softeners help to produce soft, non-flammable materials. The acetic acid residues of cellulose are used in the production of acetate silk. Other esters of cellulose, so-called 'viscoses' are cellulose xanthogenates. Acids are used to regenerate the cellulose, a process which is used in the manufacture of viscose silk and cellophane. These and other cellulose derivatives are used in the food industry as packaging materials (packaging films) for protective coverings and for lacquers to coat metallic foil against corrosion, (e.g. bottle tops).

The free hydroxyl groups of cellulose can also form ether links, for example to produce **methyl celluloses**. With only approximately half of the hydroxyl groups in methyl cellulose in the ether form they have desirable solubility in water or gelling properties. Fully methylated derivatives on the other hand are insoluble in water. The use of cellulose ethers such as cellulose methyl ether ($R\text{-}O\text{-}CH_3$) and cellulose glycolic acid ether $R\text{-}O\text{-}CH_2COOH(Na)$ (R = cellulose residue; this can be described chemically as carboxymethyl cellulose and sold as a water soluble Na salt), cellulose ethyl ether (= ethyl cellulose), benzyl cellulose (= cellulose benzyl ether), as water soluble gelling and binding agents, protective colloids or glues, are becoming increasingly important. These cellulose ethers are not fermentable and therefore keep well and are stable to light, atmospheric oxygen and alkalies.

Other glucans. The polymeric saccharide dextran is frequently found as an undesirable by product in the sugar industry and adversely affects the process of raw sugar manufacture. Dextran is a highly branched polymeric saccharide from α-D-glucose with 1 → 4- and 1 → 6-bonds. Nowadays it is obtained by biosynthesis from raw sugar solution by bacterial fermentation and is used in cosmetics and as a substitute for blood plasma. *Leuconostor mesenteroides* is used for the production of dextran from raw sucrose.

Laminaran. Laminaran is isolated from brown seaweed. It is a branched glucan with β-D-(1 → 3) bonds in the chain and β-D-(1 → 6) bonds in the branches.

A number of other glucans have been isolated from many red seaweeds, yeasts and so on. So far they are of little importance in nutrition.

Fructans (fructosans)

Fructans are made up of fructose. The inulins, grain fructan and laevan are of interest to food chemists.

Inulin group. Inulin is found as a reserve carbohydrate in the roots of many plants, especially the Compositae (artichoke, dahlia, chicory). It is extracted from these plants and used for diabetics. Inulin, a white powder, dissolves in hot water without forming a gel; it gives no colour reaction with iodine. The molecule is made up of about 35 fructose units in which fructofuranose residues are linked by β-D-(2 → 1) bonds, so that hydrolysis also yields small amounts of glucose. It is therefore assumed that glucose occurs at the ends of the chains similar to sucrose. This would also explain why the inulins do not have any reducing action. Inulin is broken down to fructose by the plant enzyme inulase (inulin-1-fructo-hydrolase).

Cereal fructans. Fructans, different in each kind of cereal, are found in grain. The name 'trifructosan' (trifructose-anhydride) is now no longer used, because modern research has shown that it consists of 10 fructose residues and is therefore a decafructosan. Graminin is the decafructosan of rye. Because the fructosan in individual cereals differs, it can be used to prove the presence of rye in wheat flour and so on.

Laevans. Laevan is also a polymeric saccharide built up from fructose. It is produced by bacterial action on those oligosaccharides that have a fructose at the end of the chain (raffinose, saccharose). Laevan may be formed by the action of *Bacillus subtilis* during the extraction of sugar. The laevans from various bacteria contain small amounts of glucose like inulin. It is therefore thought that it is synthesised by the transfer of β-fructofuranose residues to a sucrose molecule (for instance) with lengthening of the chain at the β-D-(2 → 6) bond and branching at the β-D-(2 → 1) bond. Because of the glucose content, laevans and inulins cannot be classified in the homoglycan group.

Other homoglycans

Other important homoglycans contain as their basic building blocks: mannose in the mannans, galactose sulphate in the galactons and acetyl glucosamine in chitin.

Chitin. Chitin is found as a structural component in many lower animals (invertebrates, especially arthropods). It is also found in lower plants (yeasts, fungi). D-glucosamine was first isolated from the chitin of lobster shells by LEDERHOSE in 1876. Chitin is a polymeric saccharide, built up entirely of N-acetyl-D-glucosamine units with β-D-(1 → 4) linkages.

Mannans. Mannans have been found in many plants. Mannan from the red seaweed or algae Porphyra umbilicalis is used in the Far East as a food (Nori). The mannose residues are linked with β-D-(1 → 4) bonds. Yeast gums contain mannans.

Galactans. Macromolecular materials, called alginic colloids, are nowadays produced on an industrial scale from red and brown seaweeds. Under the names **Carrageenan** and **agar-agar**, they have many uses. (Alginates which are poly-uronides are also alginic colloids, see p. 192).

Carrageenan or Irish Moss extract is obtained from dried red seaweed or algae by hot water extraction. Cellulose-like materials and pigments are removed from the hot water extract by filtration with kieselguhr (diatomaceous earth). After the filtrate has been concentrated under vacuum, the colloid is precipitated out and dried. There are two chemical types, χ- and λ-carrageenan. χ-carrageenan contains D-galactose-4-sulphate alternating with 3,6-anhydro-D-galactose in a straight chain with α-D-(1 → 3) and β-D-(1 → 4) bonds. λ-carrageenan is also linear and consists of a series of galactose-4-sulphate units, which are linked by α-D-(1 → 3) bonds. In nature, carrageenan is a mixture of calcium, magnesium, potassium and sodium salts.

When they are obtained in the form described above – by hot water extraction involving no chemical changes – they are called carrageenans. Carrageenans are insoluble; in cold water a gel is formed, in hot water they become slimy and set like jelly when cooled. The water content of such a gel can be very high (over 90%). Carrageenans are suitable as protective colloids for dispersions and for emulsions as gel formers and thickeners. They are used in ice cream, chocolate milk drinks, sweets and as thickeners and emulsifying agents; they are also used in drugs and in cosmetics. Alginic colloids are considered physiologically safe, as are pectins, because they are naturally occurring materials.

Another alginic colloid **agar-agar** has for many years been obtained from certain red seaweed. The same method is used as for carrageenan, hot water extraction, cleansing, flocculant precipitation and drying. Agar-agar is a mixture of two galactans, agarose and agaropectin. Agarose is a linear galactan made up of alternate D-galactose and 3,6-anhydro-L-galactose building blocks which are aligned alongside each other by β-D-(1 → 4) and α-L-(1 → 3) bonds. Part of the D-galactose residue is esterified with sulphuric acid in the 6-position. Agaropectin has a similar structure to agarose, but has the majority of the sulphate groups of the native polymeric saccharide. Chemically it is interesting to note the presence of D- to L-galactose within a single molecule. The symmetrical structure of the sugar makes the transformation of D- to L-galactose particularly simple because only the reducing group has to be transferred from C_1 to C_6. Agar-agar is soluble in hot water, insoluble in cold water. It – like the carrageenans – readily forms a gel; even 1% agar-agar solution forms a firm gel. As a replacement for gelatine it is used as a gelling agent and thickener in jams, milk, ice cream and sauces, both industrially and in the home.

Heteroglycans from sugars group (b)

In recent years a large number of heteropolymeric saccharides have been found in plants, particularly grasses, woods and cereal grains. Some of them are important for food chemistry and the manufacturing industry.

Glucomannoglycans. Polymeric saccharides built up from glucose and mannose have been found in many plants. An example is the glucomannan from the tubers of *Conophallus konjaku* (Araceae) used in the Far East as a food. It contains D-glucose and D-mannose in the proportion of 2:3 with (1 → 4) bonds in the chain and (1 → 3) bonds in the branches. Varieties of Amorphophallus contain similar glucomannans.

Galactomannoglycans. A galactomannoglycan (**carubin**) is obtained on a commercial scale from the seeds of the Carob or Locust bean tree (*Ceratonia siliqua*). It is used in the manufacture of paper and textiles and also as a thickener for foods in a number of countries.

Guar flour, or **guaran**, also called guar or guar gum (from *Cyamopsis tetragonalobus*), also used in the food industry, belongs to the galactomannan group. This polymeric saccharide is built up of galactose and mannose where the backbone consists of 1.4-β-glycosidically linked mannose units, whereas the galactose in the side branches is bound 1.6-α-glycosidically. Guar flour is obtained from the endosperm of the seeds of the guar bean (cyamopsis tetragonalobus) containing D-mannose β-(1 → 4), D-galactose in (1 → 6) branches.

Hemicelluloses. Polymeric saccharides which contain D-xylose are found almost exclusively in land plants and frequently in those parts of the plant which are later lignified. Crude hemicellulose is obtained from the spent sulphite liquors in the manufacture of cellulose. After suitable pre-treatment (hydrolysis) it is converted into protein or fat by suitable yeasts. The crude hemicellulose can be separated into a neutral (hemicellulose A) and an acid fraction (hemicellulose B). Hemicellulose B is found particularly in hardwoods. Both hemicelluloses have a basic xylan chain with β-D-(1 → 4) bonds. In hemicellulose A the chain has a large number of short side chains made up of arabinose residues. D-glucose, D-galactose and D-mannose are also present. The arabinoxylans from wheat, barley and rye flour are typical of this group.

Hemicellulose B has mainly 4-O-methyl-D-glucuronic acid residues in place of arabinose and this determines its acid nature.

Homo- and heteroglycans which contain uronic acids, groups (c, d, e)

This covers the most complicated area in carbohydrate chemistry. Nature here exhibits the greatest variety both in the type of bonds and branching and in the number of carbohydrate building blocks combined in one moleculae. Examples of the following groups will be discussed:

Heteroglycans (c) which contain uronic acids as well as sugar such as plant gums and plant mucilages;

Homoglycans (d) which only consist of uronic acids such as pectins and alginates;

Heteroglycans (e) which contain amino sugars as well as uronic acids such as mucopolysaccharides.

Plant gums and plant mucilages

Plant gums and plant mucilages are obtained from plants or trees by breaking the rind or bark and are of a slimy or rubbery consistency. They usually contain one uronic D-glucuronic acid or D-galacturonic acid and three or more neutral monosaccharides such as: D-galactose, D-mannose, L-arabinose, D-xylose, L-rhamnose, D-glucose, L-galactose, D-fructose and L-fucose. Gum arabic from *Acacia senegal,* is the best known of them containing D-glucuronic acid, L-arabinose, L-rhamnose and D-galactose (1:3:1:3) which are linked to one another in a complicated double branched manner. Tragacanth, an exudate from the stem of varieties of Astragalus is also one of the heteropolymeric saccharides. It contains as building blocks, L-arabinose, D-xylose, L-fucose, D-galactose and D-galacturonic acid partly esterified with methanol. These gums are used as glues and gums are used as glues and thickeners and are also used as thickeners for foods (fruit jellies and salad creams), as stabilisers for sweet products and as binding agents in ice cream.

Pectins

This area, important in food chemistry and technology and in industry has been worked on extensively during this century by F. EHRLICH (discovery of galacturonic acid as the basic building block of pectin), by F. A. HENGLEIN and his co-workers (proof of the macromolecular nature of pectin), by PALLMANN AND DEUEL, KERTESZ and others. The pectins (pectos = the stiff one) are now considered to consist of polymeric homologous series of galacturonic acid, in which the carboxyl groups are partly esterified with methyl alcohol. **Pectic substances** must be distinguished from pectins. Pectic substances are mixtures of pectins and accompanying substances, such as polyoses, for instance araban and galactan.

Pectins always occur with cellulose and form a major part of the cell structure and framework of the plant. They are found as an independent layer of the matrix of cell walls and in the so-called central lamella. The morphological and physiological importance of pectins lies in the fact that they act as supporting and strengthening materials in the cellular structure by encrustation. Due to their ability to swell and their colloidal nature, they also play a decisive role in the water ratio in the plant.

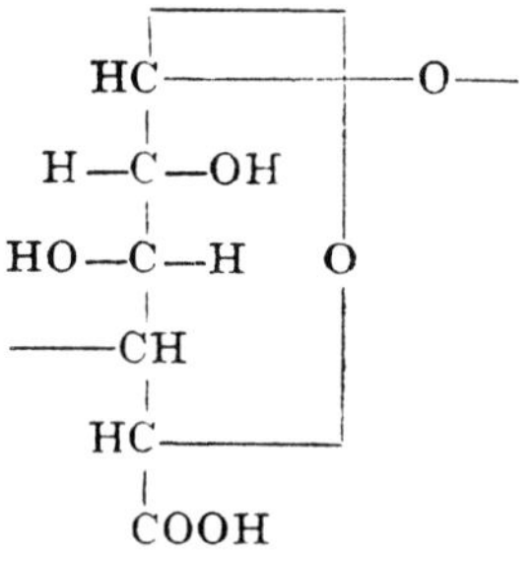

α-galacturonic acid residue of pectin and pectic acid

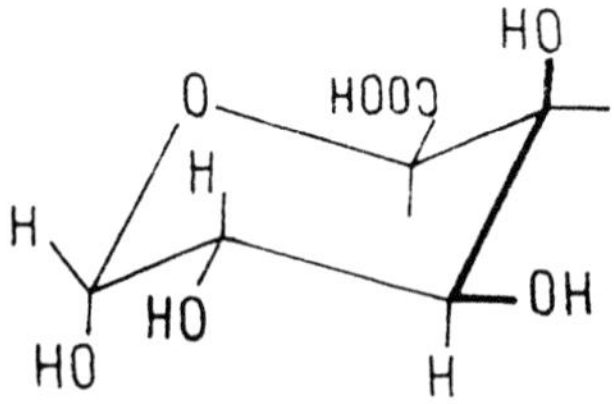

HC_1 conformation

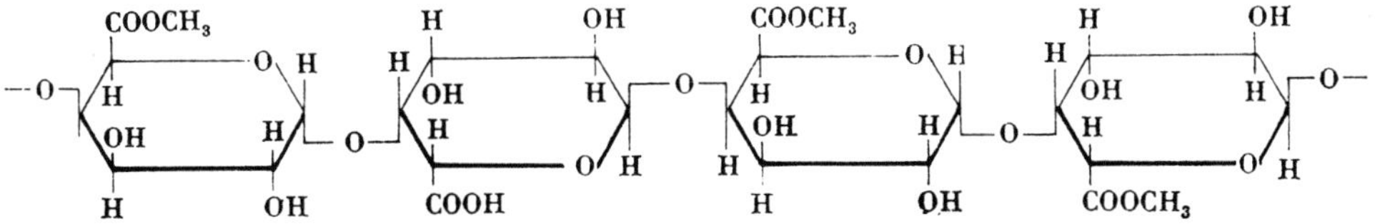

pectin

The pectins, as polyelectrolytes, have a particularly high ability to combine with water due to the presence of (-OH and -COOH) hydrophilic groups. As a result of these properties the pectins are widely distributed in the plant world. Usually they are found mainly in fleshy fruits and roots, in leaves and in the green part of stems; there are relatively small amounts in the lignified part of plants. The flesh of many edible fruits and the green leaves contains 10–15% pectin calculated as dry weight. Sugar beet slices (with about 20% pectin) and dried apple residues (as a waste product from non-alcoholic cider and apple wine manufacture) are used for the industrial production of pectin. Orange or lemon waste products, especially after juice extraction, can also be utilised.

Native pectin found included in the pectic substances of plants is called **protopectin**. Its structure has not yet been fully elucidated. At present it is thought to be as follows: the free COOH groups of the pectin molecule, with the COOH containing cellulose, the COOH containing polyoses (e.g. araban, xylan) and the protein materials, are linked together in a network bound together by polyvalent metal (Ca^{++}) ionic bonds. Phosphoric acid can also behave similarly and act with ester links between pectin molecules and cellulose or sugar. Acid, alkaline or enzymatic hydrolysis of protopectin yields a water soluble pectin.

The protopectin of lemon skins contains many methoxyl ($-OCH_3$) groups and little calcium. It is therefore hydrophilic and swells readily (turgo-regulation in the tissue). Carrot and flax protopectin, with its high calcium content serves mainly to keep the plant tissue rigid (cementing, building up, supporting and strengthening material).

Pectin has other accompanying substances as well as galacturonic acid, such as galactose and arabinose, which are difficult to remove. For this reason it is easy to obtain pectin from the protopectin by treatment with acid. Therefore it would appear that pectin may be linked, perhaps by glycosidic bonds, to other components of the cell wall. The proportions of methoxyl and Ca content in protopectins varies according to their function in the plant. This has a strong influence on the properties of the protopectins; its behaviour with acids (solubility, speed of solution) affects the methods used in commercial production.

In the **manufacture** of pectins, the protopectin is first treated with dilute acids, which leads on the one hand to the breaking of the calcium ion bonds in the pectin chains (= intermolecular reaction) and on the other hand to hydrolytic cleavages within the galacturonic acid chain (= intramacromolecular reaction, shortening of chains). The extraction must be carried out in such a way that soluble pectins with high gelling ability are obtained from the various raw materials (apple residues, beet slices, citrus fruit waste materials) when treated with suitable acids – HCl, H_2SO_4, lactic acid, SO_2 solutions (bleaching effect) – at suitable concentrations under mild temperature and extraction conditions. The higher the acid concentration – 1-3% HCl, 0.2-6% lactic acid or citric acid commonly used – the shorter the time allowed for extraction. This means that the inevitable cleavage of the chains remains at an acceptable level. In present day commercial practice, counter current extractors and continuous process screw extruders are used – similar to those used in modern oil extraction. The extract obtained contains pectins and also the usual accompanying materials (cloudy material) 'polyoses' (araban), starch, protein and its cleavage products and fragments of the cell wall. The clarification of the extract is achieved by filtration combined with continuous centrifuges, followed by the use of bleaching earths (adsorption effect) or particular enzymes to remove the starch and the proteins which render the extract cloudy. The filtered clear juice is concentrated by surface evaporation or multi-stage vacuum evaporation to the desired dry weight concentration (usually 8-10% in commercial pectin extracts), or else dried to pectin powder by spray drying with or without the addition of sugar.*

Pectins from beet slices do not gel as they contain acetyl groups.

*The pectins are clarified using aluminium salt precipitation. The powder can be extracted with acidified alcohol to reduce the ash content. Recently, in the USA, ion exchangers (resin exchangers) have been used to remove the Ca^{++} and Mg^{++} ions in the production and cleansing of citrus pectins.

Table 10. Pectin chemistry nomenclature

Description	Chemical characteristics
pectic acid	Almost completely de-esterified pectins, i.e. chain of galacturonic acid residues; acid form is insoluble in water; soluble after partial neutralisation; also forms gel with Ca.
pectates	Neutral or acid salts from polygalacturonic acids (pectic acid); contain only very few methoxyl groups.
pectin	Partially or fully methylated polygalacturonic acids. There are therefore different pectins according to the methoxyl content and degree of polymerisation.
pectinates	Salts of high or low esterified pectins, they therefore contain methoxyl groups.
pectic substances	Mixtures of pectins with accompanying substances such as polyoses, e.g. araban, galactan, starch, protein.
protopectin	Insoluble in water, the native plant pectin; basically a network of pectin chains linked by metal ions (Ca, Mg) through the non-esterified COOH groups. H_3PO_4 occurs to a small extent.
pectin derivatives	Pectins with special bound groups as acetyl pectin in sugar beet and pears.
pectase (pectin-esterase)	Causes hydrolysis of pectin methyl ester.
pectinase (polygalacturonido-glycanohydrolase)	Causes splitting of the α-1,4-glycosidic bonds of the pectin chain.

Quality evaluation of pectins – an important factor in the food industry – is carried out in various ways. Chemical determination of the galacturonic acid or methoxyl content are inadequate. Physical methods are now used: determination of viscosity, elasticity and resistance to fracture of sugar gels using pectinometers and similar apparatus. These experiments show that the gelling ability depends largely on the molecular size and degree of esterification of the pectin. However it must be emphasised that the quality evaluation of pectins for practical use as gelling agents can only be achieved by gelling tests under specified conditions.

Gelling. The **gelling process** and gelling conditions for pectin vary considerably according to whether the pectin is highly esterified (9–12% methoxyl; esterification 55-75%) or slightly esterified (up to 7% methoxyl, esterification 15-44%), and whether the pectin is of low or high molecular weight. For this reason there are various theories on gelling. The molecular weight of pectins varies between a few thousand and several hundred thousand.

In **highly esterified pectins**, the degree of esterification, the molecular weight, the amount of sugar used and the pH of the solution are important. Firm gels results from the use of one to five grams pectin per kilogram with the sugar content over 50%. Best gelling occurs between pH 3.0 and 3.4 in quick gelling pectins, and for the slower gelling ones between pH 2.8 and 3.2. The presence of Ca^{2+} ions in highly esterified pectins does not directly lead to the gel formation (in contrast to the slightly esterified pectins), but they influence the way in which the acid-sugar-gel is formed. **Acid-sugar-gels** of the highly esterified pectins are considered to be **subsidiary valency gels**, that is the structure and bonding together of the individual pectin chain molecules can be seen as due to formation of **hydrogen bonds** from subsidiary valancy forces (van der WAAL forces, dipolar effect). A high sugar content results in a high abstraction of water, that is strong bonding and orientation of the '**free**' water in the solution. The hydrated envelope of the pectin on the sugar molecule (hydration effect), and the presence of acid leads to reversal in the pectin molecule of the dissociation of the carboxyl groups which are still free. This makes the original solution unstable; the pectin

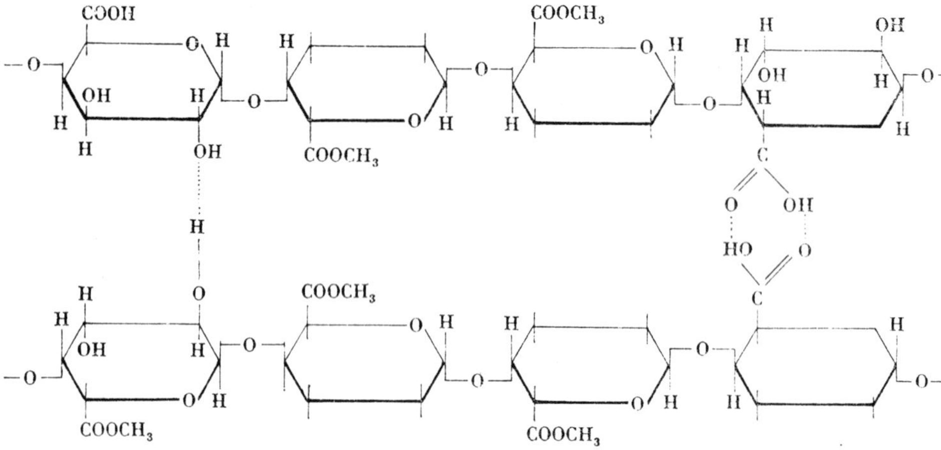

subsidiary valency gel of pectin

molecules form a network in a thermoreversible gel structure utilising the H-bonds.

Pectin gels from **partially esterified** pectins are formed with sugar + acid or even with Ca ions alone, or in the presence of other polyvalent cations, for instance Al ions. For the latter sugar is not necessary and the pH must not be as low as in the case of the subsidiary valency gels, because the salt formation is important and not the reversal of the dissociation point in the pectin molecule.

A jelly can only be formed at pH 3.5–6.5. When calcium pectates are formed, so-called heteropolar, thermo-irreversible **ion-(main) valency gels*** occur.

In this case the metal ions (Ca^{2+}) form the network bonds between the pectin chains by heteropolar main valency bonds (ion bonds), resulting in the formation of a three dimensional gel structure.

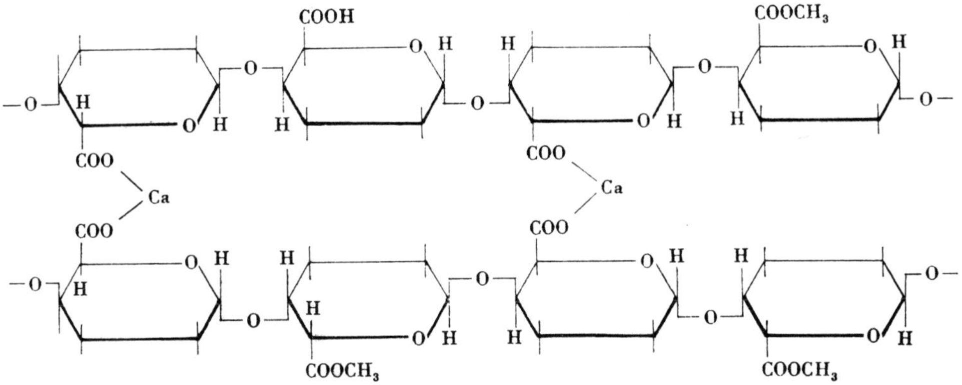

Main valency gel of pectin.

The **use** of pectins as gelling agents, emulsifiers and stabilisers in foodstuffs is considered safe from a physiological point of view. **Highly esterified pectins** are used mainly in the preparation of jams, marmalades and jellies and also for emulsions, coatings, whipped cream coating, ice cream, fruit pastes, fish conserves, sauces, mayonnaises, cheese spreads and cocoa drinks. They are also used to stabilise the sediment in fruit juices, for jellied coatings on fruit tarts and to retain moisture in baked goods with a long shelf life.

Partially esterified pectins and pectic acid are used in preparing sugar-free or low sugar gels, for instance vegetable jellies, fillings for meat pies, also milk puddings, fish in jelly, meats in aspic or as a covering for candied fruits.

Pectins are chemically very different from the neutral polymeric saccharides because they are built up of linear chains of α-1,4-linked galacturonic acid. The glycosidic bonds on the chain, for instance, are very stable to acids.

In contrast the neutral accompanying polymeric saccharides are readily split by dilute acids and can be eliminated in this way.

The non-esterified **pectic acids** behave in the same way as the neutral polymeric saccharides when treated with alkali.

*The formation of homopolar main valency gels results when the macromelecules form a network and are linked together by homopolar bonds. H. DEUEL made such gels from the reaction of the secondary OH groups of pectin with formaldehyde. They can be used as cation exchangers.

After rearrangement of the aldehyde group on the reducing end to a 2-keto group (LOBRY DE BRUYN), the glycosidic bond on the C_4 is split by a **β-elimination**, which leads to a step-wise decomposition from the reducing end. The galacturonic acid at the end of the chain is split off mainly in the form of a **saccharic acid**.

Pectins in which the glycosidic bond is rendered so unstable (presumably by the adjacent ester groups) that β-elimination occurs inside the chain, behave quite differently. Such pectins are therefore very labile to alkaline leaching and most stable at slightly acid pH (3–4).

Table 11. Enzymes which split pectin and pectic acid (after NEUKOM)

	endo-enzymes		exo-enzymes	
enzymes which split pectin	1. polymethylgalacturonases; endo-PMG	2. pectin trans-eliminases; endo-PTE	3. exo-PMG	4. exo-PTE
enzymes which split pectic acids	5. polygalacturonases; endo-PG	6. pectic acid transeliminases; endo-PSTE	7. exo-PG	8. exo-PSTE

PMG = polymethylgalacturonases
PG = polygalacturonases
PTE = pectin transeliminases
PSTE = pectic acid transeliminases

Enzymatic decomposition of pectins takes place by pectinases (Polygalacturonidp-glycanohydrolases), which are widespread in microorganisms, but found only infrequently and in small amounts in plants. The pectinases can be divided into two groups, those which affect the pectic acids and those which affect the pectins. These can be further subdivided into enzymes which attack the polymeric saccharide chain from the reducing or non-reducing end (exo-enzymes) and those which split the chain from the inside (endo-enzymes).

In addition there is another group of enzymes which affect the polymeric saccharide in the same way as alkali (pectin-trans-eliminases) and yield unsaturated decomposition products. An enzyme has been isolated from bacteria in the soil, which decomposes pectin down to 4-desoxy-5-keto-galacturonic acid.

Enzymatic and **non-enzymatic** decomposition of pectin is important in the food industry: clarification of fruit juices by addition of pectolytic enzymes renders filtration easier and increases the yield of juice. On the other hand it is often important to inactivate the hydrolytic pectinases at the correct stage, as for instance in manufacturing so-called 'naturally cloudy' citrus fruit juice, tomato juice, tomato purees, or in canned or bottled fruits, so as to prevent sedimentation of the cloudy matter or to prevent the fruit becoming soft and breaking up.

Alginates

Alginic acid, is a polyuronide like pectin. It is built up in the form of a chain from L-guluronic acid and D-mannuronic acid building blocks, linked by 1,4-β-glycosidic bonds. The molecular weight varies considerably (12,000 to 200,000). The alginates are the salts of alginic acid. Alginates are obtained in large quantities from seaweed (brown algae) in soluble form using extraction with a solution of soda or other alkalis. These solutions of alginates are freed from protein and pigments by filtration, chemically bleached and decomposed with hydrochloric acid to obtain alginic acid. Alginic acid is insoluble in water but can absorb large amounts of water and consequently swells. It is usually converted with soda into soluble alginate which gives a white powder when dried. The water soluble alkaline ammonia and Mg alginates (in contrast to agar-agar, gelatine and pectin) form pastes rather than firm gels. Gels can only be formed from alginates by acid precipitation or by the addition of calcium salts. Alginate gels are stable to heat. Alginates which are sensitive to acids can be used between pH 4 and 10.

Alginic acid and its derivatives are used in the food industry as thickeners and emulsifiers for chocolate drinks, ice cream, custard, sauces and mayonnaises. It is used as a gelling agent in jellies, jams and puddings, and in coatings for the preservation of meats, cheeses and frozen fish. It is also used to stabilise and clarify fruit juices (prevention of settling), for clarification and improving wines and sweet juices, and for improving the foaming of beer.

Muco-polysaccharides

This group includes polymeric saccharides from animal tissue which can be isolated free of proteins without noticeable decomposition. They all contain a uronic acid, which alternates with an amino sugar in the chain. In some cases the acidity of the polymeric saccharide is increased by the presence of sulphate groups.

Hyaluronic acid. Hyaluronic acid has been found in a great variety of tissue, synovial fluids and in skin and umbilical cord. It is an amorphous white, water soluble powder, $[\alpha]_D^{20} = -70°$ to $-80°$. Dilute aqueous solutions have a high viscosity, dependent on pH and strength of ions as in the alginates and pectins. Molecular weights have been found between 5.10^4 to 8.10^6 according to method of preparation and origin of the material. Hyaluronic acid is made up of D-glucuronic acid and N-acetyl-D-glucosamine, linked alternatively in the chain. The glucosamine is attached to the nearest sugar by β-D-(1 → 4) bonds and the glucuronic acid by β-D-(1 → 3) bonds.

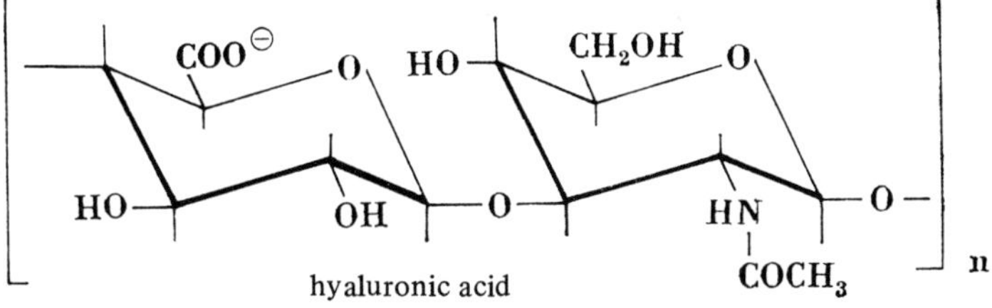

hyaluronic acid

Chondroitinsulphuric acid. **Chondroitin-4-sulphate** is a component of cartilage and can make up to 40% of the dry weight. It is soluble in water and, as a neutral salt has a rotation of $[\alpha]_D^{20} = -28°$ to $-33°$. The molecular weight lies between 40,000 and 50,000. Like hyaluronic acid, chondroitin-4-sulphate is a polyelectrolyte, but because of its extra sulphate group, it migrates faster in the electrical field than the former. This enables it to be easily separated by electrophoresis. Complete hydrolysis yields equal amounts of acetic acid, sulphuric acid and D-galactosamine. Glucuronic acid is almost completely destroyed by the conditions of hydrolysis (see p. 157), but its presence can be demonstrated in another way. Chondroitin-6-sulphate had also been found in cartilage and sinews.

Dermatan sulphate. Dermatan sulphate has been found with other muco-polymeric saccharides in connective tissue, the amount increasing with age. In young animals a small amount is present. It is difficult to separate from the other muco-polymeric saccharides, as its properties are hardly distinguishable. Its structure is similar to chondroitin-4-sulphate, but D-glucuronic acid is replaced by the very rare L-iduronic acid (p. 156).

Heparin. Heparin, which prevents coagulation of the blood, is found in liver and muscle and to a small extent also in the heart, kidneys, thymus, spleen and blood. It is made up of equal amounts of D-glucuronic acid and D-glucosamine, which are linked through α-D-(1 → 4) bonds. This polymeric saccharide is strongly sulphated.

CHAPTER 6

Minerals and Trace Elements

To form living protoplasm, the animal and vegetable organism needs **minerals** as well as carbohydrates, proteins, fats, vitamins and water. Like the purely organic materials, minerals help to create the correct physical and chemical conditions for the functioning of cells and tissues; the regulation of the osmotic pressure and the degree of water absorption by proteins. They control the development of the electric potential at the interface, activation or inhibition of enzyme sytems and functioning of the buffer systems.

Minerals are also responsible for the slightly alkaline reaction of the blood and tissue fluids; they take part in growth and blood formation and are involved in many ways in the functioning and synthesis of hormones, vitamins and enzymes. They are especially important as building blocks for bone, teeth and certain tissues. They are subject to constant change and loss like the rest of the body's building blocks, and the body has to make good these losses.

Sodium and chloride ions are present in food mainly in the form of salt. They make food taste better; regulate the osmotic pressure of the body fluids, form acid in the mucous membranes of the stomach (activation of pepsin and enzymes of the salivary glands of the throat) and keep digestive processes normal. Potassium has a similar function.

Calcium is especially important as it only occurs in a few foods in sufficient amounts. Calcium deficiency is the most frequent mineral deficiency in nutrition. Calcium ions and phosphoric acid in combination with vitamin D must be available in the right proportions for the formation of bone. Not all the calcium in food is adsorbed, and example is the calcium phosphate in cereals*. The type of bonding is important. Absorption is also dependent on the protein content of the diet. Easily soluble calcium salts, such as chloride, lactate and gluconate are absorbed as well as the less soluble calcium sulphate, calcium phosphate and

*The greater part of the phytin (= Ca Mg salts of the inositol hexaphosphoric acid ester, see p. 119) is found in the bran after milling of cereal grains. As the rate of extraction of bran is increased, so the amount of phytin in bread decreases and the absorption of calcium is decreased due to the formation of insoluble and undissociated calcium phosphate.

calcium carbonate. A normal diet must therefore contain an adequate supply of calcium, which is why many countries enrich certain foods (bread, flour) with calcium in the form of calcium carbonate.

The metabolic requirements for phosphoric acid (with calcium and vitamin D), is not only important for the formation of bone, but the phosphates and polyphosphates are essential in the intermediate metabolism and energy processes of all living cells, (see for instance ATP, p. 62).

The following average figures are at present accepted as the daily requirement of minerals for a normal adult:

sodium 5 g	calcium 0.8–1.5 g	phosphoric acid 4 g
potassium 4 g	magnesium 0.4 g	iron 15 mg

The so-called **trace elements**, which are present in the body in very small amounts, have important functions and therefore the body must receive adequate supplies. Serious symptoms result from a deficiency of these elements. Iron and copper are important elements in some enzyme systems: iron in catalase, peroxidase, the cytochromes and cytochromoxidases of the respiratory system; copper in tyrosinases, polyphenolases, ascorbase, and uricase. Iodine is needed for the formation of the hormone thyroxin in the thyroid gland. Manganese is the co-enzyme of arginase and alkaline phosphatase. Zinc is a component of the pancreatic hormone, insulin and of certain enzymes of yeast such as carbonic anhydrase, carboxy polypeptidase and alcohol dehydrase. Vitamin B_{12} contains cobalt needed for red blood cell formation. Molybdenum is a constituent of xanthinoxidase and also necessary for the fixation of nitrogen in the air by soil bacteria (Azotobacter etc.), and is therefore essential for life. *Aspergillus niger* needs gallium to grow. A lack of boron leads, in the higher plants, to severe disturbance of growth, for instance, to rotting of the core of sugar beet. Selenium, which is toxic in large amounts is now considered to be of physiological importance as a trace element. Zinc is probably essential in trace amounts.

Aluminium, nickel, chromium, titanium and uranium are still considered to be of doubtful importance. It is not known what part fluorine plays, but presumably it helps protect the teeth. Fluorine inhibits caries, but caries is not a deficiency disease of fluorine, but is due to bacterial attack.

Lead and mercury are toxic elements, as are also arsenic, antimony, thallium and the rare earths.

CHAPTER 7

Vitamins

Vitamins are 'active agents' like enzymes and hormones. They are **'biocatalysts'** and play a decisive role in the organism by regulating reactions within the body. Nowadays it is realised more and more clearly that enzymes, vitamins and hormones have a close material and functional relationship as many vitamins are constituents of enzymes (co-enzymes).

In comparison with the main nutrients (proteins, fats and carbohydrates) the body only needs very small quantities of vitamins; their energy contribution is insignificant compared with other energy sources.

Table 12. Biochemical Action of Vitamins (after K. LANG)

Fat soluble vitamins:	
Vitamin A (Retinol)	Synthesis of mucopolysaccharides and corticoids
Vitamin A aldehyde (Retinin)	Perception of light, a building block of optical pigments in the retina
Vitamin D	Resorption of Ca^{2+} from the gut. Inhibition of the parathyroid gland
Tocopherols	Antioxidant effect. Electron transport in the respiratory cycle
Vitamin K	Biosynthesis of prothrombin. Electron transport in the respiratory cycle.
Water soluble vitamins:	
Thiamin (Vitamin B_1)	In the form of thiamin pyrophosphate, prosthetic group of the enzymes involved in the break down of α-ketonic acids (e.g. of carboxylase) and of transketolase
Riboflavin (Vitamin B_2)	In the form of flavin mononucleotide (FMN) and flavin adenine dinucleotide (FAD), prosthetic group of the flavin enzymes

Table 12. Biochemical Action of Vitamins (after K. LANG)–*continued*

Water soluble vitamins–*continued*

Niacin	In the form of diphosphopyridin nucleotide (DPN, NAD) or triphosphopyridin dinucleotide (TPN, NADP), prosthetic group of the dehydrogenases (pyridine enzymes)
B_6 group	In the form of pyridoxal-5-phosphate, prosthetic group in the amino acid metabolism (transaminases)
Pantothenic acid	In the form of co-enzyme A, takes part in transacylisation
Biotin	In the form of $CO_2 \sim$ biotin compounds, takes part in carboxylation reactions.
Folic acid (Pteroylglutamic acid)	In the form of tetrahydrofolic acid derivatives, co-enzyme for the transfer of C_1 residues.
Vitamin B_{12}	In the form of cobamide co-enzymes takes part in various reactions
Ascorbic acid	Cofactor in the homogentisic acid oxidase system and hydroxylation reaction, especially for the formation of suprarenal hormones.

The organism can synthesise its own enzymes and hormones, but not vitamins (or not in sufficient amounts). The supply must therefore come from the diet, either from vitamins contained in a free or bound form in foods, or else, as is the case for vitamin A and D, in the form of precursors called pro-vitamins. The latter can easily be converted into the actual vitamins by the body. At the present time about 20 vitamins are known, which have distinct chemical properties and specific effects in intermediate metabolism. They cannot replace each other in this function, although their actions may be interrelated. For example, the antioxidant properties of vitamin E (tocopherol) protect the vitamin A in the living organism and this means that an adequate supply of vitamin E in the body is necessary before vitamin A can be stored in the liver.

The importance of vitamins with their specific catalytic properties for the control and regulation of metabolism is immediately obvious when a vitamin is lacking in the diet or is present in insufficient amount. Symptoms soon arise because of specific reaction in the intermediate metabolism which follow from a greater or lesser lack of vitamins. Typical illnesses (deficiency symptoms in the sense of avitaminoses and hypovitaminoses) occur, which can be traced back to inadequate supply of the respective natural or synthetic vitamin. (Pharmacological effects result from high vitamin doses given medicinally).

To safeguard and preserve the vitamins contained in food, the food chemist and technologist needs a thorough knowledge of their physical and chemical properties – sensitivity to air (oxygen), acids, alkalies, trace metals, temperature, light, ionising radiation, solubility in water or lipids – and also of their behaviour (loss) during production, processing and preservation.

7.1 CLASSIFICATION AND NOMENCLATURE OF VITAMINS

Vitamins are usually divided into the following two groups:

fat soluble vitamins	**water soluble vitamins**
vitamin A	B vitamin group
vitamin D	ascorbic acid
vitamin E	
vitamin K	

This is an arbitrary division from the point of view of chemical composition with only solubility as a criterion. Nevertheless, this grouping is useful – for lack of a better – because the solubility in fat or water determines certain important biological and analytical properties, as for example, presence in foods, whether the organism can store it, possible methods of isolation, extraction from natural products and the principles of analytical separation and determination.

Table 13 Nomenclature of vitamins

Nomenclature from the International Union for Pure and Applied Chemistry	Common nomenclature	Old name
retinol	vitamin A	anti-infection vitamin
ergocalciferol	vitamin D_2	anti-rachitic vitamin
cholecalciferol	vitamin D_3	
tocopherols	vitamin E	anti-sterility vitamin
no official decision	vitamin K	anti-haemorrhagic vitamin coagulation vitamin prothrombin factor
thiamin	vitamin B_1 aneurin	anti-neuritic vitamin anti-beriberi vitamin
riboflavin	vitamin B_2 lactoflavin	
nicotinamide	niacin nicotinic acid nicotinic acid amide	PP vitamin PP factor
no official decision	pyridoxin	vitamin B_6
pantothenic acid	pantothenic acid	
biotin	biotin	vitamin H
no official decision	folic acid pteroylglutamic acid	
cobalamin	cobalamin vitamin B_{12}	
ascorbic acid	vitamin C	anti-scorbutic vitamin

7.2 ANALYSIS OF VITAMINS

The analysis of individual vitamins will be found in each respective section. The general problems involved however, are introduced here.

Tests for the detection and determination of vitamins are based on:

1. **chemical** methods using more or less specific reactions.
2. **physical** methods by determining constants, for instance, extinction at the absorption spectra maxima.
3. **biological** methods.

Chemical determination is frequently affected by other substances, chemically closely related, but biologically inactive. Its use is naturally limited, although the use of absorption chromatography permits the removal of many undesired materials. When **physical** methods are used to determine vitamins, purification is normally necessary as, for example when measuring light absorption.

So-called **Standard Addition** tests are frequently used in both chemical and physico-chemical determination of vitamins. Here parallel determinations are done where a known amount of the vitamin to be determined is added to the sample. The amount must be approximately equal to the expected vitamin content. The main and Standard Addition tests must, however, be carried out at the same time and under the same conditions. In this manner, any disturbances caused on the concentration and sample effect (food) which cannot be controlled, can be eliminated. The Standard Addition test also gives a control for possible losses of vitamins during analysis.

Biological determination of vitamins is still the basic method of determining the nature and quantity of a vitamin. It can be carried out as a **prophylactic** or **curative** test on animals, as in the form of the growth test, rickets test, scurvy test, or prothrombin test. Biological tests (animal experimentation) are very time consuming, have a wide margin of error (± 30%) and also use a large amount of material. **Microbiological** tests are widely used today, especially for the B vitamins. The microorganisms (bacteria, yeasts, moulds) which need a supply of vitamins for growth and will only flourish when the appropriate (test) vitamin is added to the nutrient medium. The amount of added vitamin is proportional to the increase in cell numbers.

Microbiological methods of testing are much more specific than chemico-physical methods, and require less time, material and personnel. However, biological and microbiological determination of vitamins can only be carried out in well equipped food analysis laboratories. In many cases a combination of chemical and microbiological methods must be used.

Before the chemical constitution of vitamins was known, quantitative information on the amounts of vitamins found were quoted in international units (I.U.), determined biologically (animal testing) under precisely laid down standard conditions. Now that their constitution is known and the vitamins can be prepared in a pure form, it is possible to give a figure for the actual amount of the vitamin contained in an international unit. Vitamin dosage is almost always given in a weight unit, or a particular quantity (g) of the pure material. (The World Health Organisation – WHO – provides standard preparations of particular vitamins, for instance D_3, B_{12}, and the content of these standard substances is given in USP units. USP units correspond with the International Units).

7.3 FAT SOLUBLE VITAMINS

Vitamin A

The best known natural compound with vitamin A activity is retinol (vitamin A_1 alcohol). It is an alcohol derivative from an unsaturated long chain hydrocarbon built up from isoprene.

The side chain contains four conjugated double bonds, in conjugation with the double bond in the *β*-ionone ring.

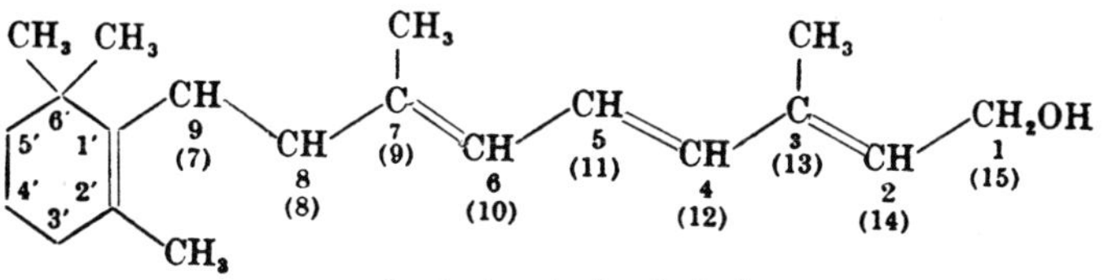

retinol, vitamin A_1 alcohol

The all-*trans* form is mainly found in nature. Synthetic products are also usually in the all-*trans* form. As they are alcohols, all forms can form esters; the commonest natural vitamin A esters are with the higher fatty acids. The esters are more stable to chemical and physical action (for instance atmospheric oxygen, heat, light catalysts) and synthetic products are prepared mainly as acetates or palmitates.

Compounds of the A_2 series have a second double bond in the ring and are found in the liver oils of fresh water fish.

If the alcoholic group of vitamin A is oxidised, vitamin A aldehyde or vitamin A acid is obtained. Both molecules exhibit limited biological vitamin A activity, since the *β*-ionone ring – essential for vitamin A activity – is still present.

Vitamin A alcohol ($C_{20}H_{30}O$), retinol, has a melting point of 62–64°C, is insoluble in water, but can be brought into colloidal solution with materials which facilitate solution (e.g. Tweens). Vitamin A alcohol dissolves easily in oils and fats, light petroleum, ether, benzene, methanol, acetone, chloroform and other lipophilic solvents.

Vitamin A has a specific absorption in the ultra-violet region at about 328 nm. The position of the maximum varies a little according to the solvent.

Vitamin A_1 acetate, $C_{22}H_{32}O_2$, m.p. 57–60°C has a solubility similar to that of vitamin A_1 alcohol. Absorption in the UV is exactly the same as for vitamin A_1 alcohol.

Vitamin A palmitate, $C_{36}H_{60}O_2$, m.p. 28–29°C, has a solubility similar to the alcohol and acetate.

Some carotenoids are **provitamins** of vitamin A. Vitamin A can be formed in the human and animal organism from the provitamins taken in the diet. The most important provitamins are α-carotene, β-carotene, γ-carotene and cryptoxanthine (see p. 123). The provitamins found in nature are all in the all-*trans* form.

When β-carotene is transformed into vitamin A in the animal or human organism, the molecule is split exactly in the centre as shown in the formula:

$$C_{40}H_{56} + 2H_2O \rightarrow 2C_{20}H_{29}OH$$

Details of the mechanism of the reaction are not known at present, but it seems certain that the transformation occurs in the mucous membrane.

It is important to know that vitamin A is present in the liver of fishes and mammals, in milk and milk products, in egg yolk and also that β-carotene occurs as a provitamin in green plants and vegetables.

Commercial production of vitamin A, in the form of vitamin A concentrates, can be carried out by molecular distillation from natural sources, especially fish liver oils. The chemical synthesis of the complicated vitamin A molecule is now carried out on a commercial scale so that there need be no future lack of vitamin A as well as with vitamin D.

During the autoxidation of edible fats, the vitamin A present is destroyed due to the susceptibility of its double bonds to O_2. Vitamin A is also completely destroyed during the hardening of fats, as in the hydrogenation of whale and fish oils. Direct sunlight should be avoided in drying plant foods, and in artificial drying oxygen should be excluded as far as possible, because loss of carotene can be considerable in the presence of O_2 and at elevated temperatures. Enzymatic fat oxidation often accompanies the destruction of vitamin A or provitamin A (carotenoids), and takes place in the drying of vegetable products if the enzymes are not denatured (inactivated) quickly enough by treatment. According to a thorough investigation by SCHEUNERT the vitamin A or carotene content of fruit and vegetables is not greatly affected by the usual domestic processing, nor by preservation in bottles or cans.

The **utilisation** of vitamin A and carotene in the body depends to a large extent on the fat content and its distribution in the diet. When it is fed with a large amount of fat the degree of absorption is higher than in the absence of fat, and the effect is even more marked for the carotenes than for vitamin A itself. Absorption from raw vegetables is often poor; in carrots, which have a very high carotene content, 98% is excreted. When lecithin and other emulsifiers are present, the body absorbs and utilises both vitamin A and provitamin A better because of the uniform distribution. Feeding liquid paraffin reduces the utilisation of carotene and vitamin A considerably, because liquid paraffin in a good solvent for carotenoids but is barely absorbed by the gut. Hence, the body is deprived of the fat soluble vitamins present in the gut. The absorption of carotene is also promoted by the presence of bile and by protein in the diet. The vitamin A absorbed is carried along the lymphatic pathways with the other fat soluble materials, and deposited preferentially in the liver.

The consumption of autoxidised edible fats impairs the vitamin A balance because it destroys the vitamin. Tocopherol (vitamin E) prevents the enzymatic and also the non-enzymatic oxidation of vitamin A in the body because it is an **antioxidant**.

Vitamin A is central to many metabolic processes, for instance it is important for certain sterol transformations. In addition the synthesis of cholesterol in the organism is arrested by the lack of vitamin A. The best known function of vitamin A is its participation in the visual processes. Symptoms of lack of vitamin A, such as inflammation of the skin and mucous membrane, retarded growth, changes in bones and nerves, indicate that the vitamin has additional effects on the metabolism. There is a close correlation between the role of vitamin A in the metabolism and its effect on growth.

The daily requirements for adults based on considerable research is given as 5000 International Units per day. When the organism is under stress as in heavy manual work, heat cold, pregnancy and illness, the requirement increases, see Table 17, p. 234.

The quantities of vitamin A and carotene are often given in International Units (I.U.). An I.U. of vitamin A corresponds to 0.3 μ.g. vitamin A alcohol and is calculated on 0.344 u.g. of a standard preparation of crystalline vitamin A acetate.

An I.U. of carotene corresponds to 0.6 μg of a standard preparation of all trans-β-carotene.

It is most important to exclude oxygen when processing foods containing vitamin A, either by addition of antioxidants or by working in an inert gas atmosphere. Because vitamin A is sensitive to acids, it is preferable to use alkaline extraction processes in treating foods before determination of vitamin A, and analyses should not be done in direct sunlight.

D Vitamins

Vitamin D_2 or ergocalciferol and vitamin D_3 or cholecalciferol, are the two most important D vitamins.

Both compounds can be produced from sterols by suitable irradiation with UV rays: vitamin D_3 is formed from **7-dehydrocholesterol**, whereas vitamin D_2 is formed from **ergosterol**. Continued irradiation leads to the formation of additional compounds, including some of unknown structure such as the so-called supra-sterol and toxisterol. The following formulae show the transformation of the provitamins D, 7-dehydrocholesterol and ergosterol, by irradiation into vitamins D_3 and D_2.

7-dehydrocholesterol → (ultraviolet light, isomerisation) → vitamin D_3

ergosterol → (ultraviolet light, isomerisation) → vitamin D_3

The D vitamins are colourless and odourless substances, very stable to oxygen in an oil solution, even at 100°C (in contrast to vitamin A). They are insensitive to alkalis, but very sensitive to light.

Vitamin D_2 is insoluble in water, slightly soluble in fats and oils, and dissolves readily in organic solvents. The compound exhibits optical activity, the value depending on the solvent used. A solution of vitamin D_2 in organic solvents has a characteristic absorption spectrum with a maximum absorption around 265 nm with extinction value 475 nm (1% solution and, 1 cm cell path thickness) measured in alcohol. Vitamin D_3 has similar solubility to vitamin D_2. The characteristic absorption spectrum has a maximum around 265 nm (solution in alcohol).

Occurrence: Vitamin D has not been found in large quantities in nature. Among foods, fish oils seem to have the largest vitamin D content. In human food the main sources are milk and milk products, eggs and ox and pig's liver. As far as is known, the actual vitamin (vitamin D_3) is only found in fish liver oils.

A number of foods contain a considerable amount of the D provitamins, especially milk and milk products, eggs and fungi. The D provitamins are deposited primarily in the skin; pig's skin is a rich source.

Provitamin D_3 (7-dehydrocholesterol) formed from cholesterol in the organism, is therefore a normal metabolic product and is laid down mainly in the skin. The human body does not have enzyme systems able to catalyse the transformation of provitamin D_3 into vitamin D_3. This transformation only occurs when the skin is exposed to sunlight or UV light. In cow's milk the vitamin D content is higher if the cows are out on pasture.

The D vitamins are considered in general to have the following **physiological** activity: they influence the calcium and phosphate absorption from the gut, the excretion of calcium and phosphate from the kidneys and the deposition of calcium and phosphate in bones.

It is difficult to determine the absolute daily requirement of vitamin D, because it is normally formed in the human body with the help of sunlight. For infants a daily requirement of 400–500 IU is suggested. A suitable amount must be fed to the body in the absence of sunlight otherwise signs of vitamin D deficiency appear, called rickets or the English disease. The first signs of rickets in the infant are changes in the normal bone growth and a delay in the ossification of the bones of the skull. The rib cage also shows the typical rachitic rosary.

Vitamin D is the only vitamin where deficiency leads to serious permanent damage, for instance, there may be deposition of calcium in the tissues, in kidneys, lungs and stomach. Hence the addition of the D vitamins to dietetic products and food must be carefully scrutinised.

An International Unit (IU) of vitamin D corresponds to the activity of 0.025 g of crystalline vitamin D_3.

Analysis: Vitamin D activity is determined by various standard tests using rats and chicks as test animals.

Recently a new test to determine vitamin D activity has been adopted in which radioactive phosphate or calcium is laid down in the bony skeleton of experimental animals. Both these methods are very exact and can also be carried out much more quickly than the classical methods.

The chemical determination of vitamin D in foods is relatively difficult, because the D vitamins are only found in very small concentrations and are present with a number of other chemical substances which react in a very similar way. There is no general recommended chemical method laid down for the determination of vitamin D. Most methods use a reaction between vitamin D and antimony trichloride in the presence of acetyl chloride, which gives a yellow colour. The reaction does not work well when larger amounts of vitamin A are present. For this reason, when determining vitamin D in preparations which contain concentrated vitamin D and A, a preliminary separation of vitamin A by thin layer chromatography is first carried out.

E Vitamins (tocopherols)

A number of substances in nature have been found to have vitamin E activity. They are named α-tocopherol, β-tocopherol, γ-tocopherol and so on. They are all derivatives of chromane and are derived from tocol:

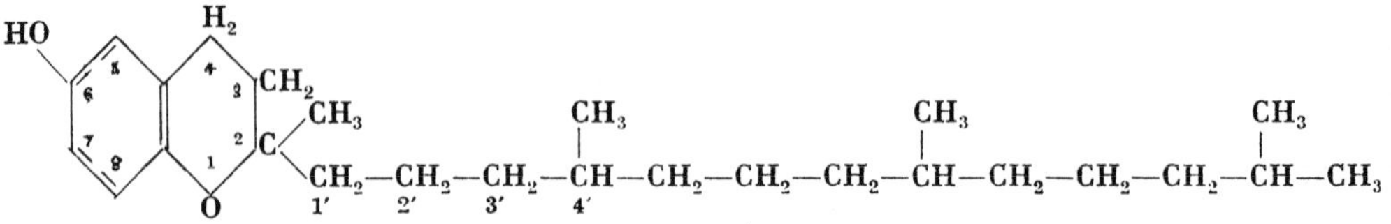

tocol
(2-methyl-2-[4′, 8′, 12′-trimethyldecyl]-6-oxychromane)

The individual tocopherols differ in their various substituents in the benzene ring: α-tocopherol = 5, 7,8-trimethyltocol, β-tocopherol = 5,8-dimethyltocol, γ-tocopherol = 7,8-dimethyltocol, δ-tocopherol = 8-methyltocol, ε-tocopherol = 5-methyltocol.

α-Tocopherol, $C_{29}H_{50}O_2$, can be distilled under vacuum without decomposition (boiling point 0.003 = 140–170°C). At room temperature it is a viscous, yellow oil. Its solubility in fat solvents and oils is relatively good; it is insoluble in water.

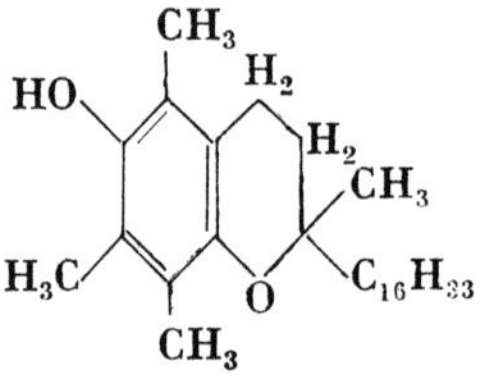

α-tocopherol

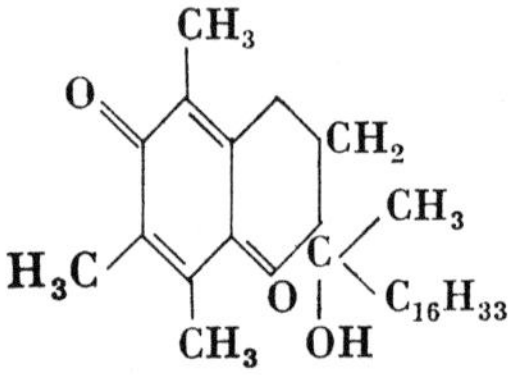

α-tocopherylquinone

α-Tocopheryl acetate, the acetic acid of α-tocopherol, is similar to α-tocopherol in solubility. Tocopherol esters are biologically more active than free tocopherols. Probably this is because the esters withstand oxidative attack better.

Occurrence: As far as is known, the tocopherols only occur in the vegetable kingdom in free form. The tocopherols taken in the form of esters, are transformed into the free form in the animal and vegetable organism. The following table gives a summary of the tocopherols found in natural products.

Table 14 Tocopherols in natural products (after H. KUBIN and H, FINK 1961)

Material investigated	Total tocopherol mg/100 g dry material	α %	β and γ %	δ %	ζ2 %
algae (*Scenedesmus obliquus*)	8.0	37	52		11
peas (yellow Viktoria)	7.1		100		
pike (muscle flesh)	0.96	74	26		
egg yolk	6.1	64	20		16
cultivated mushroom	0.91	28	30		42
Boletus edulis fungus	5.50	7	18	10	65

The presence of tocopherol in leaf vegetables, milk, butter, egg yolk and in a number of vegetable oils is important for human nutrition.

In animals, the liver, the suprarenal glands and the heart are rich in tocopherols. Significant amounts of tocopherol are also stored in body fat and according to their tocopherol content, give a greater or lesser degree of protection against autoxidative deterioration when these fats are stored.

Tocopherol is first oxidised to tocopherylquinone (formula p. 206), in the organism and this is closely connected with the antioxidative effects of vitamin E. Tocopherylquinone is however not stored in the organism; after the transformation (SIMON Metabolite) it is conjugated by glucuronic acid and excreted as a water soluble compound.

Relatively little is known at present about the biochemical activity of vitamin E. It is certain that it takes part in some of the oxidation processes in the living cell; probably there is a connection with the trace element selenium. There are some interrelationships in the body between the unsaturated fatty acids present and vitamin E which can be traced back to the antioxidative properties of α-tocopherol and the prevention of the formation of fat peroxides. From this point of view, tocopherol should be regarded as a specific physiological antioxidant.

Their natural antioxidant action makes the tocopherols important for the food industry: they prevent or delay the formation of peroxides in unsaturated fats and fatty acids during processing and storage.

An adequate supply of vitamin E is necessary for the body because of the interrelationship between vitamin E and the unsaturated glycerides or fatty acids in our diet. For each 1 g of linoleic acid in the diet, about 1 mg tocopherol is needed. It is important that there is an adequate supply of tocopherol because the consumption of autoxidised and heated fats increases the demand for tocopherol.

The Internationa Unit (IU) of vitamin E corresponds to the activity of 1 mg synthetic (racemic) α-tocopherol acetate. Daily requirement approx. 30 IU.

Analysis: Biological determination of vitamin E is by the resorption sterility test, which depends on the fact that rats with vitamin E deficiency are not able to bear their young to term, but reabsorp the foetus after a short time. The haemolysis test, which gives an accurate measure of vitamin E activity, is based on the greater tendency to haemolysis of erythrocytes in animals with vitamin E deficiency.

There are no specific reactions available for chemical analysis of the tocopherols. The total tocopherols in foods are carefully separated and freed from accompanying interfering materials. The individual tocopherols are then isolated by chromatography without difficulty. The determination of tocopherols is usually done by the methods of EMMERIE and ENGEL, where trivalent iron is reduced to divalent iron with the presence of tocopherols, and the Fe^{++} reacted with α, α'-dipyridyl to give an intensely red coloured complex whose extinction can be measured photometrically and compared with a standard calibration curve.

Fluorometric and the recent elegant polarographic methods are the commonest physical methods of determination.

K vitamins

The K vitamins are derivatives of naphthoquinone; vitamin K_1 and K_2 only differ in the isoprene side chains. The side chain of vitamin $K_{2(30)}$ is a squalene residue, that of K_1 a phytyl residue. The 30 refers to the number of C atoms in the side chain.

$$CH_2-CH=\underset{\displaystyle CH_3}{C}-(CH_2)_3-\underset{\displaystyle CH_3}{CH}-(CH_2)_3-\underset{\displaystyle CH_3}{CH}-(CH_2)_3-\underset{\displaystyle CH_3}{CH}-CH_3$$

vitamin K_1
(2-methyl-3-phytyl-1,4-naphthoquinone α-phylloquinone)

Vitamin K_1 is a yellow, viscous oil, dissolving readily in fats and oils but insoluble in water. It is sensitive to light and unstable with alkali, strong acids and oxidising agents.

It occurs in liver fat (pig, chicks) and more plentifully in green plants, such as spinach, cabbage, lucerne and so on. Bacteria also form vitamin K_2.

Vitamin K is responsible for the normal coagulation of the blood which is why it is called the coagulation or antihaemorrhagic vitamin. According to recent research it is also the hydrogen carrier in the respiratory cycle.

Biological methods of determination are usually employed (control of the prothrombin mirror, blood coagulation test). Chemical determination is unspecific because of the presence of other quinone bodies in biological material.

In recent years new materials have been isolated from mitochondria, plants and microorganisms which also (like the K vitamins) have a quinone structure. As they are very widespread, for instance in yeasts and heart muscle, they have been given the name of **ubiquinones** (or coenzyme Q). Chemically they are tetra-substituted benzoquinone derivatives with a long isoprenoid side chain. (Following the nomenclature of the K vitamins, the ubiquinones are also given an index number which indicates the number of the C atoms in the side chain).

The first ubiquinone was isolated from heart muscle. Its side chain has 50 C atoms (10 isoprene residues, and is therefore given the name ubiquinone (50) or coenzyme Q_{10}: H).

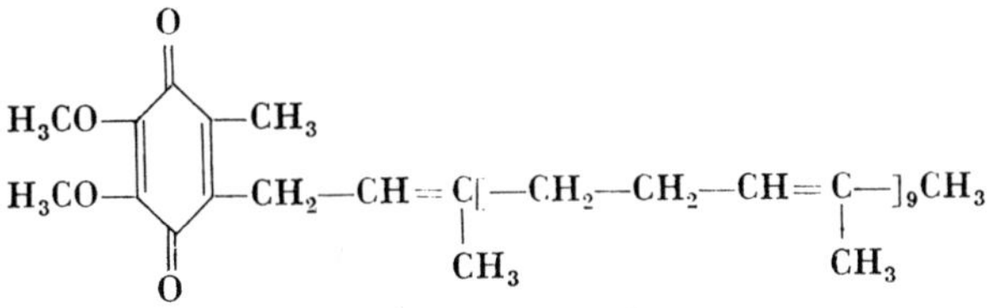

$$H_3CO,\ H_3CO,\ CH_3\text{-benzoquinone}-CH_2-CH{=}\underset{CH_3}{C}[-CH_2-CH_2-CH{=}\underset{CH_3}{C}-]_9CH_3$$

ubiquinone 50 (coenzyme Q_{10})

The ubiquinones probably function in the transport of electrons in the respiratory cycle, because of their redox systems.

It is thought that the same function is carried out by the structurally similar quinone bodies, vitamin E quinone (tocoquinone) and vitamin K_2. Therefore these ubiquinones are considered to have an action similar to vitamins.

Vitamin F

Formerly the unsaturated fatty acids linoleic, linolenic and arachidonic acids were collectively called vitamin F (see p. 81). Normally the body fat contains these fatty acids in appreciable amounts, so that they are no longer included among vitamins, but among **essential nutrients** (see p. 279). When the organism is deficient in these fatty acids, skin diseases appear and they are also involved in the synthesis of the various prostaglandins.

7.4 WATERSOLUBLE VITAMINS

The B vitamin group

Because of the historical course of research into vitamins, individual vitamins were grouped together in the B group, but this is only because they occurred together, are water soluble and have coenzyme activity. Chemically the B vitamins differ greatly, and each B vitamin has a specific physiological activity. Lack of a vitamin leads to specific avitaminoses. It is therefore useful to know them by their individual names.

Table 15 gives the vitamins with their names; the activity and function in metabolism indicate their differences and specificity.

Vitamin B_1 (thiamin, aneurin)

Thiamin is built up from a substituted pyrimidine ring and a substituted thiazole ring; both rings are coupled through a CH_2 group. Salts are formed on the amino group of the pyrimidine ring and on the N atom of the thiazole ring.

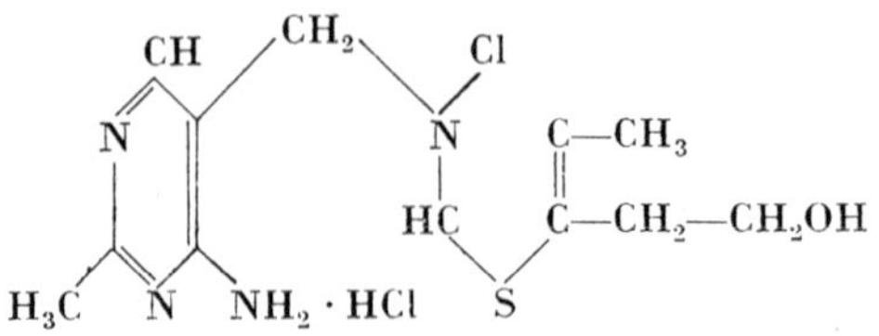

vitamin B_1 hydrochloride

Table 15 Activity and functions of the B vitamins

Vitamin	active form	metabolic function
B_1 (thiamin, aneurin)	cocarboxylase	metabolism of α-keto acids
B_2 (riboflavin, lactoflavin	isoalloxazine nucleotide	Transport of hydrogen or electrons
nicotinic acid (niacin, PP factor)	diphosphopyridine nucleotide (DPN, NAD) triphosphpyridine nucleotide (TPN, NADP)	dehydrogenation of substrates
B_6 (pyridoxin)	pyridoxal-5-phosphate	decarboxylation of amino acids, transamination, tryptophan metabolism
pantothenic acid	co-enzyme A	transacetylation, activation of fatty acids, formation of citric acid leading to endo-oxidation of nutrients
biotin	'activated CO_2'	carboxylation
folic acid (pteroylglutamic acid)	N^{10}-formyltetrahydro-folic acid	inclusion of C_1 compounds in the biosynthesis of complicated substances e.g. purines
vitamin B_{12} (cobalamin)	cobamide co-enzyme	biosynthesis of labile methyl groups, nucleoside formation

The white crystals with a yeast odour are readily soluble in water, less so in alcohol and insoluble in ether, benzene and oils and fats. The solutions decompose rapidly in neutral and alkaline pH, especially if air is present. The vitamins are very stable in an acid environment. International standard : 1 IU = 3 μg thiamin.

Thiamin is widely distributed in the animal and vegetable kingdom: rice cereals, vegetables, potatoes, yeast, meat and milk etc. It was discovered in connection with a deficiency disease, Beriberi, which occurs frequently in the Far East, but which does not depend on vitamin B_1 deficiency alone. The vitamin is found free or as a diphosphate (for instance in yeast). Only plants seem to be able to synthesise it completely, whereas in animals the body is able, to some extent, to carry out the synthesis from the two components (thiazole residue and pyrimidine ring).

This vitamin is now prepared commercially following its synthesis (by the two German chemists (ANDERSAG and WESTPHAL, 1936) and is available in any required quantity.

In the organism, thiamin pyrophosphate, is the active agent. It is formed into cells from thiamin and ATP:

$$\text{thiamin} + \text{ATP} \xrightarrow{\text{thiamin kinase}} \text{thiamin diphosphate} + \text{AMP}$$

Its physiological functions are mainly concerned with the enzymatic metabolism of carbohydrates – thiamin diphosphate is the co-enzyme of the enzyme involved in the decarboxylation of α-keto acids. In animals and plants the most important reaction is oxidative decarboxylation of pyruvic acid, which is in the main pathway of oxidative carbohydrate metabolism:

$$\text{pyruvate} \rightarrow \text{acetyl co-A} + CO_2 (+ 2H^+ + 2e^-)$$

Vitamin B_1 is an essential growth compound for certain yeasts, which are therefore used for microbiological estimation. Carboxylase is particularly important in alcoholic fermentation by yeast.

The daily requirement of vitamin B_1 depends on the composition of the diet; it is increased by carbohydrates and reduced by fats. In a balanced diet the requirement is 1-2 mg in adults, see Table 17, p. 234.

Vitamin B_1 is very stable in normal cooking and preservation (canning) of vegetables and meat, although its solubility can lead to considerable losses in certain circumstances, especially in the boiling and blanching of vegetables (p. 293). When milk is boiled, the loss depends on duration and especially on the presence of air.

Although vitamin B_1 is widely distributed in nature, it is a vitamin which is not always consumed in adequate amounts, even in those countries with an adequate diet. It is not that there is insufficient vitamin B_1 available in the food but there is a movement towards other foods (containing less thiamin). Contributing factors are the decrease in heavy manual labour which has led to a decrease in the volume of food consumed (and that automatically decreases the total supply of vitamins), poor choice of foods, increased consumption of white bread and foods made from white flours in the place of wholemeal breads, decrease in the total consumption of cereals, the replacement of beef with pork and an increase in sugar consumption. Because vitamin B_1 is not stored in the organism, these changes in diet can lead to a latent lack of vitamin B_1.

Analysis: Thiamin is determined chemically or physico-chemically, by measurement of the ultra-violet fluorescence of thiochrome (oxidation product of thiamin) or by azo dye reactions. Vitamin B_1 can also be determined microbiologically because of the activity of the vitamin as a growth material in *Phycomyces* (Phycomyces test = increase in dry weight of the fungus with a fixed amount of vitamin B_1, however the test is not specific). Tests involving animals have also been used, for instance the beri-beri pigeon test.

Vitamin B_2 (riboflavin, lactoflavin)

Vitamin B_2 belongs to the flavin group (compounds with three condensed rings, derivatives of isoalloxazine). It can be produced synthetically on a commercial scale.

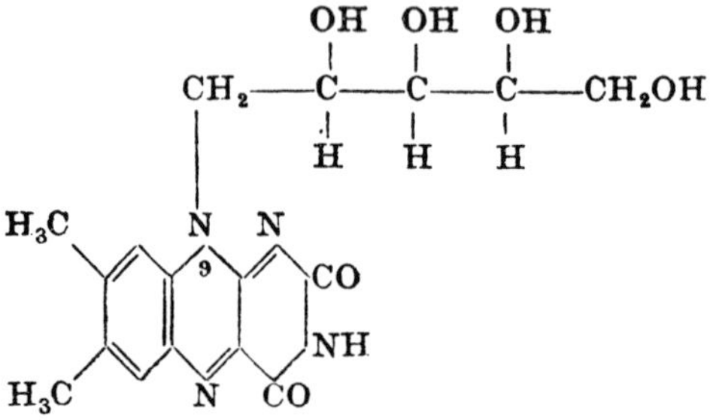

OH OH OH
CH_2—C—C—C—CH_2OH
H H H
H_3C N N
9 CO
NH
H_3C N CO

vitamin B_2 (riboflavin, lactoflavin)

OH
CH_2—O—P=O
HOCH OH
HOCH
HOCH
CH_2
H
H_3C C C N N C=O
C C C
C C N C NH
H_3C C C
H O

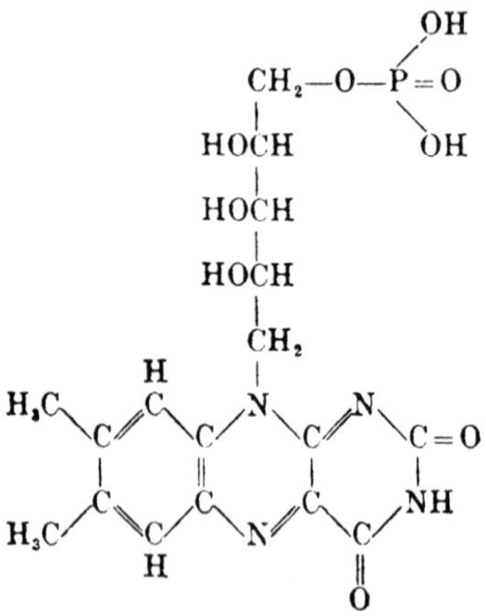

riboflavin-5′-phosphate
flavin mononucleotide (FMN)

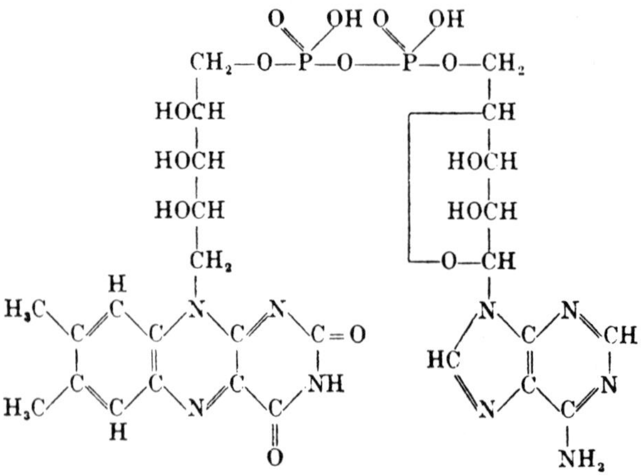

O OH O OH
CH_2—O—P—O—P—O—CH_2
HOCH CH
HOCH HOCH
HOCH HOCH
CH_2 O—CH
H
H_3C C C N N C=O N N
C C C C CH
HC
C C N C NH N
H_3C C C N C C
H O NH_2

dimethylisoalloxazine-adenine-dinucleotide
flavin-adenine-dinucleotide (FAD)

The nitrogen at the 9 position carries ribitol, a super alcohol residue corresponding to ribose (see p. 147). Vitamin B_2 is difficult to dissolve in water, giving a yellow green fluorescence. The solubility can be increased by the addition of many substances such as urea, p-aminobenzoic acid. Vitamin B_2 is amphoteric and forms salts with acids and bases, hence its improved solubility with dilute acid and bases. Riboflavin is stable to heat and fairly resistant to oxygen in the air, but it decomposes in light. In an alkaline solution, irradiation with light transforms it to lumiflavin which has a methyl group instead of a ribitol residue in the 9 position. Lumichrome is formed from riboflavin with the splitting off of the ribitol residue by light irradiation in an acid solution. Acid solutions of riboflavin are stable when light is excluded.

The vitamin B_2 content of milk and cheese, muscle, liver and kidney from warm blooded animals, of hen's eggs, of yeast and of certain vegetables is important in human nutrition. Riboflavin is found in practically all living cells, either as a phosphoric acid ester or as a dinucleotide bound to protein. Free riboflavin is thought only to be present in larger amounts in milk (yellowish colour of whey), in urine and in the retina of fishes. The active form of riboflavin in the organism is in coenzymes: riboflavin-5′-phosphate (flavin mononucleotide, FMN) and flavin-adenine-dinucleotide (FAD), see p. 214).

These enzymes are called flavin enzymes or sometimes 'yellow ferments', as the flavin residue gives the protein a yellow, or sometimes greenish to brownish colour.

The flavin enzymes act as hydrogen carriers in various redox reactions in the intermediate metabolism. Recently semi-quinoid forms have been postulated to take part in so-called single electron steps. These redox reactions occur in the riboflavin portion of the enzyme. The oxidised and reduced form of the flavin enzyme can be shown as follows:

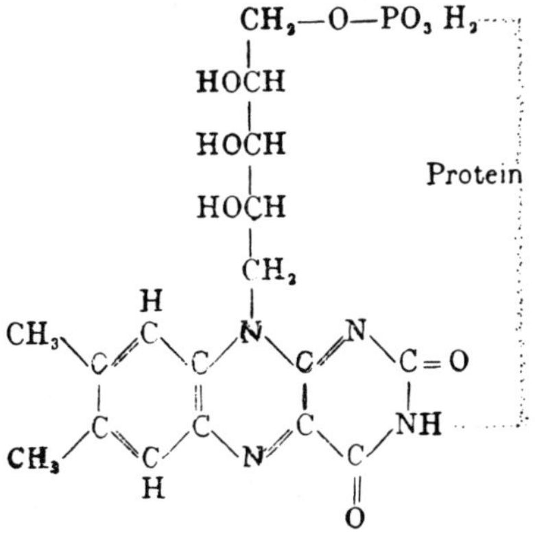

CH_2—O—PO_3H_2 ⋯ Protein
HOCH
HOCH
HOCH
CH_2
(isoalloxazine ring: CH_3, CH_3, H, H, N, N, N, C=O, NH ⋯ Protein, O)

oxidised form of a flavin enzyme

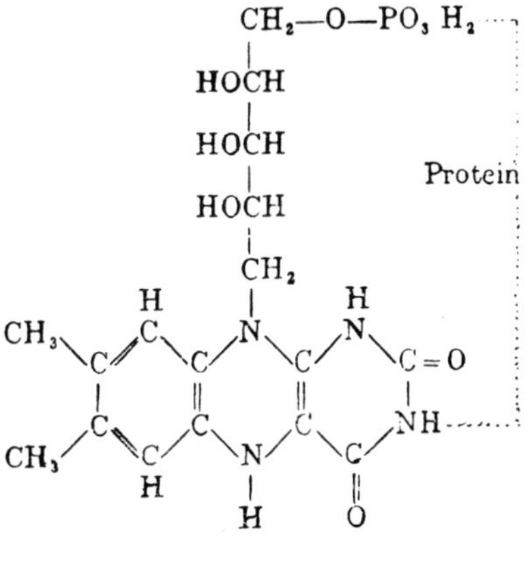

CH_2—O—PO_3H_2 ⋯ Protein
HOCH
HOCH
HOCH
CH_2
(isoalloxazine ring: CH_3, CH_3, H, H, N, N–H, N–H, C=O, NH ⋯ Protein, O)

reduced or leuco- form of a flavin enzyme

The most important physiological enzymes containing riboflavin in the coenzyme are the DPNH-cytochrome-c-reductases, TPNH-cytochrome-c-reductases, succinic acid dehydrogenase, D-amino acid oxidase, L-amino acid oxidase, xanthine oxidase and acetyl coenzyme-A-dehydrogenases. Xanthine oxidase is also used in food analysis (see SCHARDINGER reaction p. 214).

Riboflavin is clearly of basic importance for nutrition and metabolism in the respiratory cycle.

The riboflavin taken up in the diet is absorbed in the small intestine and probably phosphorylated to riboflavin-5′-phosphate in the mucosa. This compound reaches the liver via the bloodstream and is transformed there into riboflavin-adenine-dinucleotide (FAD).

Riboflavin deficiency manifests itself by a number of non-specific symptoms: general lethargy and disinclination to work, changes in the mucosa of the lips, mouth (fissures at the corners of the mouth) and tongue, dystrophy (breaking) of the fingernails and finally in changes in the skin of the whole body. Reduced resistance to infection (typhus, pneumonia) is often a sign of vitamin B_2 deficiency. Whereas severe B_2 avitaminosis does not normally occur in our life style, latent vitamin B_2 deficiency is common in central Europe. Due, not to a lack of food, but to a badly balanced diet, the body may only receive 60% of its actual riboflavin requirements. Mild deficiency in vitamin B_2 may also result from insufficient absorption of the vitamin, due to disturbances in the digestive tract and to liver disease.

The recommended daily dosage in the latest report (1979) from the Department of Health and Social Security in the United Kingdom are as follows: men, of all ages 1.6 mg; women of all ages 1.3 mg; growing boys and girls 0.4 to 1.7 mg depending on the age. During preganancey the figure should be 1.6 mg and during lactation 1.8 mg (see also Table 17, p. 234).

Vitamin B_2 is not stored in the organism, any surplus taken in food or supplements is excreted (like vitamin C).

Analysis: riboflavin can be determined microbiologically, or chemically. As long as the concentration of riboflavin is high enough a chemical method such as the so-called lumiflavin method is preferable. It depends on the fact that riboflavin is transformed into limiflavin by light. The intensity of fluorescence can then be measured in UV light. Polarographic determination is particularly suitable because of its specificity and the simplicity of the method. If the total content of vitamin B_2 in a food is to be measured the vitamin, which is present as a phosphoric acid ester, as a dinucleotide, or bound to protein must first be liberated, by using, for example an enzyme which splits phosphoric esters such as clarase; only then can the actual determination be carried out either by microbiological or chemical methods.

Niacin (Nicotinic acid, nicotinic acid amide, vitamin PP).

Niacin is the general term for nicotinic acid and nicotinic acid amide. Both substances have the same action, because they can be transformed into one another in the organism. In the human body the amino acid **tryptophan** (see p. 67) can be transformed into nicotinic acid. This amino acid is therefore called **provitamin PP**. PP factor = pellagra prevention factor.

Nicotinic acid: melting point around 234-237°C (sublimation); soluble in water and alcohol. Because it is amphoteric, the compound dissolves better in acids and alkalis than in pure water. A weak solution has an acid reaction. Acid, alkaline and neutral solutions of nicotinic acid are very stable.

Nicotinic acid amide melts at 128-131°C; it is readily soluble in water, and also in 95% alcohol and in glycerol. Nicotinic acid and nicotinic acid amine have a characteristic UV absorption spectrum with a maximum about 260 nm.

O C—OH N O C—NH₂ N O C—NH₂ ⊕N H $X^{\ominus}$

nicotinic acid niacin | nicotinic acid amide | nicotinic acid amide pyridinium salt

Niacin is widespread in food from animal and vegetable sources. It is active in the human body as two coenzymes: Nicotinamide adenine dinucleotide (NAD) and nicotinamide-adenine-dinucleotidephosphate (NADP). Both are coenzymes of **dehydrogenases** (see p. 256) and therefore transport hydrogen from substrates to hydrogen acceptors, such as the flavin enzymes (see pp. 212 and 214).

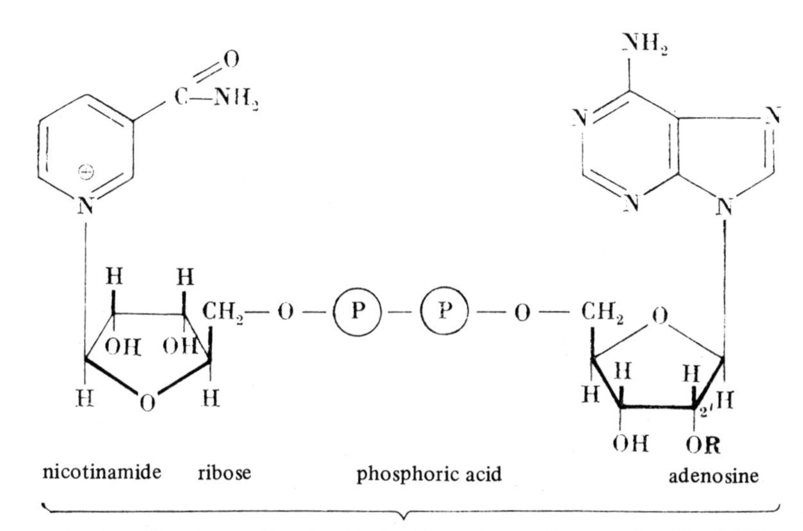

nicotinamide adenine dinucleotide (= diphospho pyridine neculeotide, DPN) NAD

R = H in NAD; R = PO_3H_2 in NADP

In the coenzymes, the pyridine ring (see formula) is linked to ribose as an N-glycoside. Such a bond is only possible with the pyridinium cation, with an H atom on the nitrogen. Nicotinic acid amide is bound to adenosine via pyrophosphoric acid, so that the dinucleotide (= NAD) has the formula.

The nomenclature 'diphospho-pyridine-nucleotide' (DPN or Codehydrase I) used formerly, is incorrect and misleading. It is not a nucleotide of diphospho-pyridine but rather a dinucleotide, whose base is a pyridine **derivative**. IUPAC has suggested the name **nicotinamide adenine dinucleotide**, abbreviation NAD. The coenzyme previously called 'triphosphopyridine nucleotide' (TPN or Codehydrase II), in which the OH group in the 2′ position of the adenosine ring is esterified with phosphoric acid, is now called **nicotinamide adenine dinucleotide phosphate** (abbreviation NADP). This book uses the new internationally agreed nomenclature.

To indicate that NAD and NADP are pyridinium salts, the abbreviations $NAD^{\oplus}$ and $NADP^{\oplus}$ should be used for the reducing forms NADH and NADPH.

The importance of nicotinic acid amide is related to the mechanism for hydrogenation of the coenzyme.

H O C—NH₂ ⊕ N rib–(P)–(P)–ad

$$NAD^{\oplus} + 2[H] \underset{-2[H]}{\overset{+2[H]}{\rightleftharpoons}} NADH + H^{\oplus}$$

This reaction with nicotinic acid amide occurs in many cases of substrate hydrogenation where protein, fat, carbohydrate, amino acids and so on are dehydrogenated. The apoenzyme protein determines the specificity.

Niacin prevents human pellagra, a complicated avitaminosis, caused by deficiency of this and other B vitamins. This illness manifests itself in the skin, the digestive and nervous systems. Nicotinic acid is frequently given therapeutically as a 'liver protection' material. Presumably the effect results from an increase in the coenzyme content (NAD and NADP) of liver cells.

A daily intake of 4.7 mg of nicotinic acid per 1000 cal. of food is thought to be adequate to prevent the appearance of deficiency. However, a minimum of 8.8 mg should be consumed. The human body can make part of the niacin required from tryptophan (see p. 215). 60 mg L-tryptophan yields on average 1 mg niacin. Therefore when the protein intake is normal, sufficient niacin is always available.

Bound nicotinic acid, is formed in cereals, particularly maize; it cannot be utilised by the body unless special extraction procedures are used. Here niacin is bound to a polypeptide and is stable to acids but labile to alkalis; the enzymes in the human digestive tract cannot split the niacin free and hence it is unavailable.

Pellagra caused by nicotinic acid deficiency, is still a serious nutritional problem in many parts of the world, especially Asia Minor, Iran, China, Egypt, Sudan, Latin America and Puerto Rico. In most areas where pellagra is endemic, the diet consists mainly of maize because animal protein and legumes are in short supply. Maize not only contains vitamin PP in the bound form, but is lacking in tryptophan (the provitamin of niacin). Only in Mexico, where the maize, before being made into 'tortillas', is treated with lime, which breaks down the bound form, is pellagra less common. Niacin deficiency is also important in connection with **amino acid imbalance**:

When the essential amino acids (see p. 229) are not present in the diet in the optimal relationship to each other, 'amino acid imbalance' can occur which is seen in retarded growth in young rats. The niacin requirement rises considerably with amino acid imbalance. Maize, with its low tryptophan and lysine content and its unusually high leucine content produces such an imbalance in the body and leads at the same time to pellagra – because of the lack of trytophan.

Niacin can be determined by animal tests or microbiologically. The two compounds, nicotinic acid and nicotinic acid amide, must however be present as such and must first be freed (by acid or alkali) from their derivatives. Chemical methods for determination of nicotinic acid and nicotinic acid amide almost all depend on KONIG's pyridine reaction with cyanogen bromide or chloride and an aromatic amine. The pyridine ring of nicotinic acid is broken to form yellow polymethine dyes.

Vitamin B_6 group (pyridoxin)

Pyridoxin is the name given in the new international nomenclature to three substances which are closely related to one another and are easily transformed. They are **pyridoxol, pyridoxal** and **pyridoxamine**. In the vegetable kingdom they are usually found together. In the animal organism for practical purposes only pyridoxal and pyridoxamine are found. Foods containing vitamin B_6 are milk, egg yolk, offal, yeast, cereals and green vegetables.

The physiologically active form is **pyridoxal-5-phosphate** as co-factor for enzymes which are able to catalyse a great variety of reactions.

The interrelationships and transformations of the compounds in the body are shown in the following scheme.

Phosphorylation of pyridoxal in the cells is carried out by pyridoxal kinase: pyridoxal + ATP → pyridoxal-5-phosphate + ADP. The coenzyme **pyridoxal phosphate** is present mainly in enzymes which effect the amino acid metabolism, for instance in amino acid decarboxylases, amino transferases and amino acid racemases. Pyridoxal-5-phosphate is also necessary for the transformation of linoleic acid into arachidonic acid. As the same coenzyme can catalyse totally different reactions one can assume that all these reactions have the same intermediate (condensation) product. At present it is considered that a Schiffs base can be involved in different pathways in the reaction and lead to different end products, with the apoenzyme controlling the final product (see p. 239).

Pyridoxin deficiency rapidly leads to acrodynia in the body due to disturbance of the amino acid metabolism. In young animals vitamin B_6 deficiency leads to severe disturbance of growth, atrophy of the reproductive glands and changes in the kidneys. The most conspicuous symptom, which led to the discovery of the vitamin, is a particular type of inflamation of the skin. Disturbance of the protein metabolism from B_6 deficiency leads to atrophy of the thymus gland, muscular dystrophy and disturbance of the ability to form antibodies.

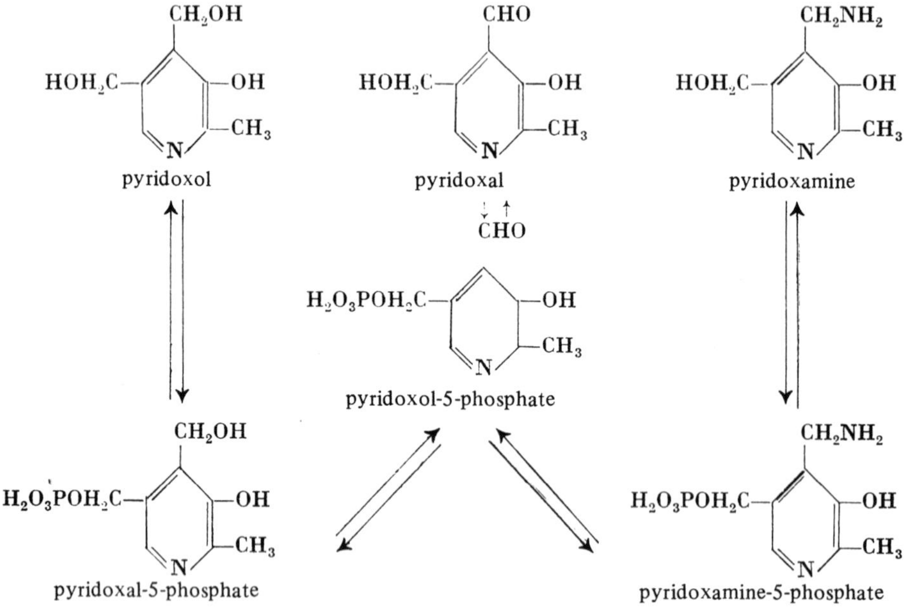

CH_2OH ; HOH_2C ; OH ; CH_3 ; N
pyridoxol

CHO ; HOH_2C ; OH ; CH_3 ; N
pyridoxal

CH_2NH_2 ; HOH_2C ; OH ; CH_3 ; N
pyridoxamine

CHO ; $H_2O_3POH_2C$; OH ; CH_3 ; N
pyridoxol-5-phosphate

CH_2OH ; $H_2O_3POH_2C$; OH ; CH_3 ; N
pyridoxal-5-phosphate

CH_2NH_2 ; $H_2O_3POH_2C$; OH ; CH_3 ; N
pyridoxamine-5-phosphate

Pyridoxin deficiency does not occur in man with a normal adequate diet. Experimental avitaminosis with definite symptoms can be induced when pyrodoxin is withdrawn from the diet.

The human requirement of pyridoxin is calculated at 2–3 mg per day with the usual diet supplying this amount. When heavy demands are made on the body, as in illness or increased rate of metabolism, the need for vitamin B_6 increases. An increase in protein consumption in particular leads to a greater need for pyridoxin.

Chemical and biological methods of determination of vitamin B_6 are relatively difficult. Microbiological determination is particularly difficult as the various organisms used do not have the same requirements for pyridoxol, pyridoxal and pyridoxamine.

Pantothenic acid

Pantothenic acid was first discovered as a growth requirement for yeasts. Only later was it found to have vitamin activity for humans. (+)-pantothenic acid is the only form possessing vitamin activity.

$$HO{-}CH_2{-}\overset{\displaystyle CH_3}{\underset{\displaystyle CH_3}{\overset{|}{\underset{|}{C}}}}{-}CH(OH){-}CO{-}NH{-}CH_2{-}CH_2{-}COOH$$

pantothenic acid

[D(+)-α, γ-dihydroxy-β, β-dimethylbutyryl-β-alanine]

The alcohol corresponding to pantothenic acid, pantothenol, also acts as a vitamin (but has no growth effect for microorganisms.)

Pantothenic acid is a viscous oil dissolving readily in water, acetone and absolute alcohol. It crystallises as the calcium salt, which is useful in making vitamin preparations. Pantothenic acid, in neutral solution is stable to light and atmospheric oxygen, but is very sensitive to heat; it is hydrolysed into β-alanine and the lactone of pantoic acid in an acid or alkaline solution or by heat.

The active form of pantothenic acid in the organism is as coenzyme A:

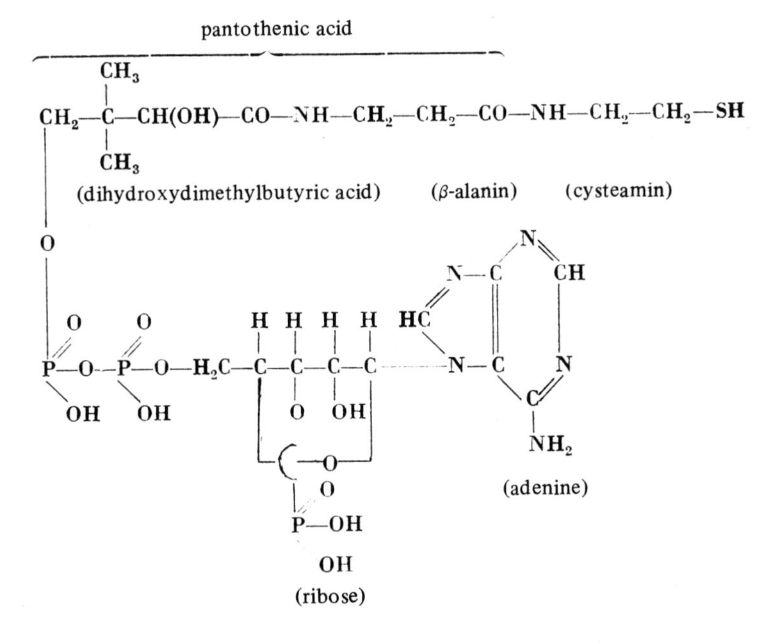

Coenzyme A

Coenzyme A holds a key position in the breakdown of carbohydrates, amino acids and neutral fats and in the synthesis of neutral fats, phosphatides, sterols (e.g. cholesterol) and in the acylation processes of metabolism; see ligases p. 251. Reactive acetic acid, so-called 'activated acetic acid' is an acetylated Coenzyme A.

Due to the importance of Coenzyme A to life, it is found in small amounts in all natural products, especially yeasts, green plants, fruits, cereals, animal organs (liver, kidneys, muscle), egg yolk and milk. The diet therefore supplies ample amounts and there is no problem with pantothenic acid deficiency under the conditions we live. For this reason no figures for the requirements of pantothenic acid are normally quoted.

The determination of pantothenic acid is carried out microbiologically or with animal experimentation.

Folic acid group

Compounds with folic acid activity derive from folic or pteroyl glutamic acid

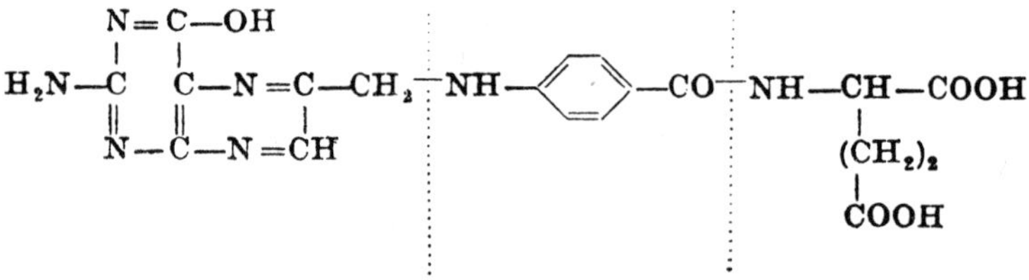

```
        N=C—OH
        |  |
 H2N—C   C—N=C—CH2 ⋮ NH—⟨benzene⟩—CO ⋮ NH—CH—COOH
     ||  ||    |                            |
     N—C—N=CH                             (CH2)2
                                            |
                                           COOH

    pteridine        p-aminobenzoic acid       glutamic acid
                         folic acid
```

Folic acid is made up from the heterocyclic ring system of 2-amino-4-hydroxy-6-methylpteridine, p-aminobenzoic acid and glutamic acid.

Compounds of the folic acid group are found in almost all living animal and vegetable tissues, but only in very small concentrations. A relatively high folic acid content (50–100 μg/100 g) is present in dark green leaf vegetables, in fresh green vegetables, in the liver and kidneys of animals used as food; small amounts are found in beef, cereals, roots, vegetables, tomatoes, cheese and milk.

Folic acid is only slightly soluble in cold water and alcohol, but more in hot water. Folic acid dissolves readily in acids, because it forms salts with the NH_2 group. It is stable to heat in neutral alkaline and acid environment. Folic acid is destroyed by light.

After absorption in the body folic acid is transformed into the so-called 'activated formic acid' and 'activated formaldehyde' via a series of intermediate steps. Both these forms of the coenzyme of folic acid take part in a series of enzymatic reactions, in which methyl groups, hydroxymethyl groups or formyl groups are transported, (see transferases p. 257). This occurs in protein metabolism, in the biosynthesis of purines and nucleic acids and for the formation of the porphyrin ring and blood pigments.

Folic acid deficiency produces changes in the normal blood picture and the development of a special type of anaemia.

In a normal diet there is no folic acid deficiency, but it is possible for a folic acid deficiency to develop on an unsuitable diet, (for instance, when the wrong foods are chosen or when the food is wrongly prepared). In intestinal tract diseases or after an operation on the stomach (gastrectomy) it sometimes happens that folic acid is not released from food and insufficient is absorbed.

Analytical determination of folic acid can be carried out satisfactorily by microbiological methods. Because of the minute amounts involved, chemical methods are not used.

Biotin

Biotin has long been known as a growth substance for plants. It was isolated from liver extract and named vitamin H as one of the yeast growth factors in egg yolk.

Chemically, biotin is a cyclic urea derivative containing a thiophan ring.

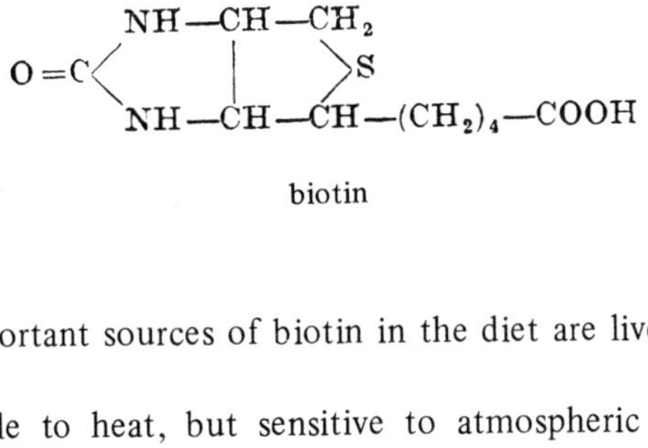

```
        NH—CH—CH2
O=C<       |    >S
        NH—CH—CH—(CH2)4—COOH
```

biotin

The most important sources of biotin in the diet are liver, milk, meat and vegetables.

Biotin is stable to heat, but sensitive to atmospheric oxygen and light including UV.

Biotin is concerned in a number of important biochemical processes, in which carbon dioxide is built into or split off from various molecules. The CO_2 which is bound to biotin inside the enzyme is called 'activated CO_2' (activated carbonic acid).

The best known of these reactions is the carboxylisation of pyruvate to oxalacetic acid (WOOD-WERKMAN reaction). Biotin is also involved in the condensation (carboxylation) of acetyl Coenzyme A with CO_2, with formation of malonyl Coenzyme A; this reaction is the key reaction for the biochemical synthesis of long chain fatty acids from smaller fragments.

Because biotin is of vital importance to many biochemical reactions for as in protein metabolism, fatty acid metabolism and purine metabolism, its deficiency exhibits itself in a severe disturbance to metabolism.

The biotin in the diet is supplemented by the biotin synthesised by the intestinal flora, but this biosynthesis depends on the composition of the food ingested.

If much raw egg white is included in the diet, a component of egg protein, the so-called avidin, binds to biotin and other biotin type proteins in a complex, which cannot be broken down by enzymes in the stomach and intestinal tract. Hence the biotin is unavailable and the biosynthesis of fatty acids is inhibited by avidin.

The biotin requirement of man is estimated at 150–300 μg. This dosage is generally met by synthesis from the intestinal bacteria, so that the supply of biotin is largely independent of the diet. However, when the intestinal flora is disturbed, or in other diseases of the intestinal tract, the organism needs biotin in the diet.

The detection and estimation of biotin is carried out microbiologically or by animal experimentation.

The vitamin B_{12} group (cobalamin)

The most important compound in this group is vitamin B_{12} or cyanocobalamin. This complicated molecule, has a structure similar to a porphyrin ring, the basic material of red blood pigment, but contains trivalent cobalt as its covalently bound central atom bound in a complex.

Man's best source of vitamin B_{12} is offal, such as liver and kidneys. In industrial production, use is made of the fact that numerous microorganisms form vitamin B_{12}. The vitamin is either the main or by-product of many fermentation processes. Use is made of both crude and purified forms in animal feeds and in pharmaceutical and medical preparations.

Vitamin B_{12} taken orally in the diet, combines in the human intestinal tract with a substance secreted by the mucous membrane of the stomach, the so-called 'intrinsic factor'. The formation of this complex is vitally important for the absorption of vitamin B_{12}. Without it man cannot absorb vitamin B_{12} in adequate amounts.

Vitamin B_{12} is involved in many biochemical processes in higher organisms. Its participation in the biosynthesis of labile methyl groups is important, as these are involved in the biosynthesis of important substances. These include the purine and pyrimidine rings, nucleic acids, methionine, choline and creatine. There is some indication that vitamin B_{12} is also involved in many other important metabolic processes, such as carbohydrate metabolism and in amino acid and protein metabolism.

Vitamin B_{12} deficiency seldom occurs in man on a normal diet. Only strict vegetarians may exhibit clinical signs of the onset of vitamin B_{12} deficiency. When vitamin B_{12} is not absorbed from the intestinal tract because of lack of intrinsic factor due to gastrectomy, chronic anaemia may develop, an illness greatly feared and not understood before the discovery of vitamin B_{12}.

The daily requirement of vitamin B_{12} is estimated to be 5 μg per day for a healthy man.

Analysis: Combined biological and chemical methods are usually employed to determine vitamin B_{12}. In many cases chemical determination is only successful after suitable enrichment and the removal of interfering substances. In the modern chemical determinations, radioactive labelled vitamin B_{12} is used as an internal standard and the total loss of uptake of vitamin B_{12} from the diet or the vegetable and animal material is calculated from the loss of activity. Microbiological methods are generally so sensitive that the analysis can be carried out without the need for special enrichment.

Vitamin C L(+)-ascorbic acid

Rich sources of ascorbic acid are rose hips, sea buckthorn (*Hippophae Rhamnoides*), lemons and blackcurrants although it is also found in animals. Potato is the main source of vitamin C in our diet (6–35 gm/100 g).

Superficially L-ascorbic acid, a lactone of a hexo acid, shows a great similarity to the hexoses.

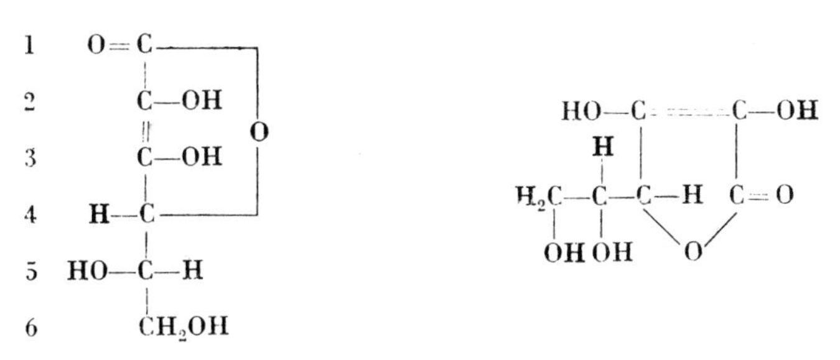

L(+)-ascorbic acid
m.p. 192°C, specific rotation $[\alpha]_D^{20} = +23°$ (in water)

The presence of hexoses and vitamin C in all green plant cells indicates that there is a link between them. Experimentally it has been shown that glucose is the mother substance for the biogenesis of vitamin C, with 2-ketogulonic acid (= α-ketogulonic acid) as an intermediate stage. There is no enediol grouping in α-ketogulonic acid in contrast to vitamin C. In aqueous solution there is no equilibrium between the keto-, alcohol- and the enediol form; the lactone ring of ascorbic acid is not opened, so that the carboxyl group can be esterified. Mineral acids catalyses the transformation of 2-ketogulonic acid into ascorbic acid by enolisation.

The commercial synthesis of vitamin C starts from D-glucose via D-sorbitol → L-sorbose → 2-ketogulonic acid directly to ascorbic acid. Hydrogenation of D-glucose yields the first step D-sorbitol. This alcohol, on biochemical dehydrogenation with *Acetobacter* yields L-sorbose. *Acetobacter* oxidises only those specific OH groups adjacent to a second configuratively similar group. This occurs at the C atoms 4 and 5 of sorbitol and the specificity is so marked that it has been used to establish the configuration in sugar chemistry.

The oxidation to 2-keto-L-gulonic acid is achieved either directly with oxygen over platinum or via a diacetone derivative with potassium permanganate. Acid hydrolysis splits off the acetone residue. Enolisation followed by lactonisation yields vitamin C.

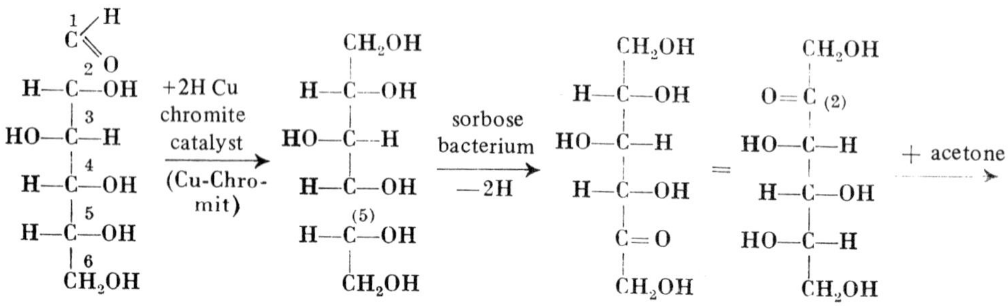

$^{1}C(H)=O$ / $H-{}^{2}C-OH$ / $HO-{}^{3}C-H$ / $H-{}^{4}C-OH$ / $H-{}^{5}C-OH$ / ${}^{6}CH_2OH$

D-glucose

+2H Cu chromite catalyst (Cu-Chromit) →

CH_2OH / $H-C-OH$ / $HO-C-H$ / $H-C-OH$ / $H-C^{(5)}-OH$ / CH_2OH

D-sorbitol

sorbose bacterium, −2H →

CH_2OH / $H-C-OH$ / $HO-C-H$ / $H-C-OH$ / $C=O$ / CH_2OH = CH_2OH / $O=C$ (2) / $HO-C-H$ / $H-C-OH$ / $HO-C-H$ / CH_2OH

L-sorbose

+ acetone →

CH_2OH / $(CH_3)_2C<(O-C, O-C-H)$ / $H-C-O$ / CH / H_2C-O $>C(CH_3)_2$, ring O

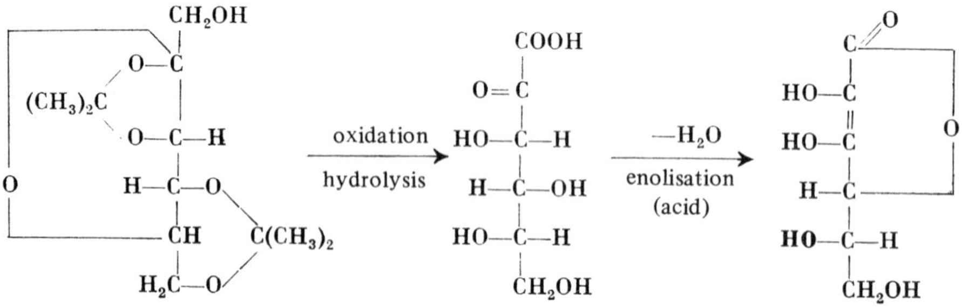

diacetone-L-sorbose

oxidation, hydrolysis →

$COOH$ / $O=C$ / $HO-C-H$ / $H-C-OH$ / $HO-C-H$ / CH_2OH

2-keto-L-gulonic acid

$-H_2O$, enolisation (acid) →

$C=O$ / $HO-C$ / ‖ / $HO-C$ / $H-C$ (ring O) / $HO-C-H$ / CH_2OH

L-ascorbic acid

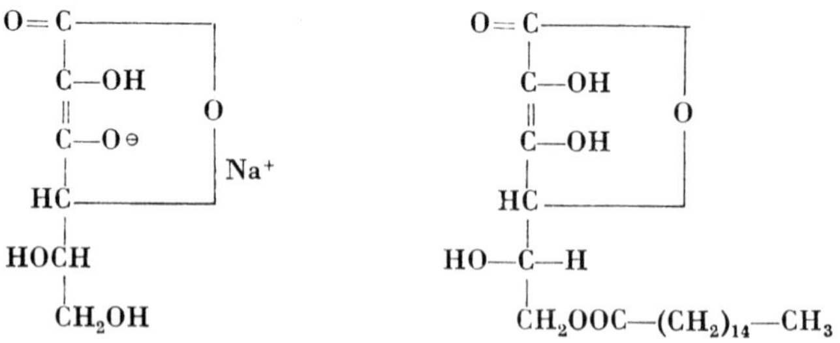

$O=C$ / $C-OH$ / ‖ / $C-O^{\ominus}$ Na^{+} / HC (ring O) / $HOCH$ / CH_2OH

sodium ascorbate

$O=C$ / $C-OH$ / ‖ / $C-OH$ / HC (ring O) / $HO-C-H$ / $CH_2OOC-(CH_2)_{14}-CH_3$

ascorbyl palmitate

The alcoholic OH groups of ascorbic acid can be esterified with fatty acids; **ascorbyl palmitate** (see formula) is fat soluble and can be used in fats as an antioxidant and synergist and has applications in the food industry.

The physiological **activity** of ascorbic acid in the human body is concerned primarily with the prevention of scurvy, defence against infection of all kinds and mobilisation of the body's defence mechanisms. It also acts as a redox system in the transfer of hydrogen in cell metabolism and its function in the biosynthesis of corticoids has recently been confirmed. The daily requirement of vitamin C (50-75 mg) for the prevention of scurvy and all pre-scorbutic symptoms (spring lassitude) can be supplied from commercial sources (synthetic production). 1 IU = 0.05 mg crystallised vitamin C, (see also table 17, p. 234).

The anti-scorbutic effect of vitamin C is mainly the result of its configuration. Of the four optically active ascorbic acids – vitamin C has two asymmetrical C atoms (at C atoms 4 and 5) – the biologically active form is dextro-rotatory L-ascorbic acid, true vitamin C.

D-isoascorbic acid, one of the four stereoisomeric forms of ascorbic acid, is also important as a water soluble antioxidant.

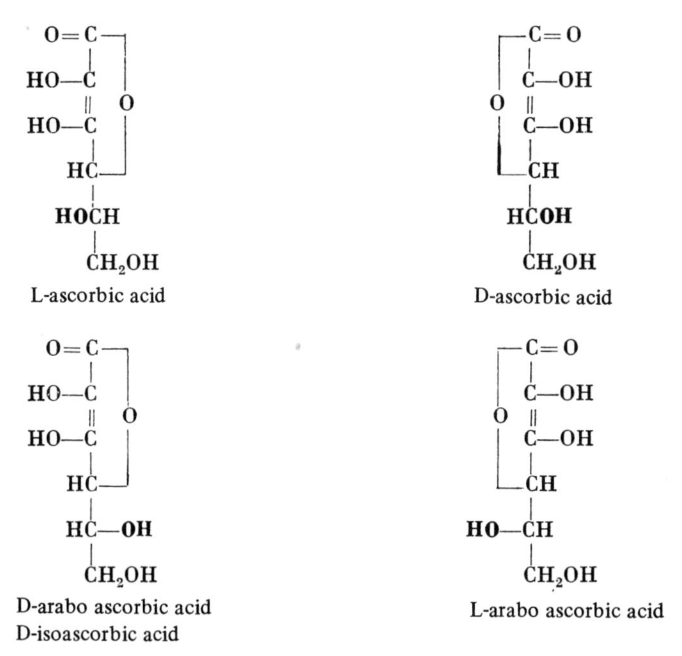

stereoisomeric forms of ascorbic acid

Vitamin C forms colourless crystals, which are stable when protected from light. They dissolve readily in water with a marked acid reaction. Ascorbic acid is a relatively strong acid and this acid property is due to the presence of the acid enol hydroxyl on the C_3, whose acidity is increased by the neighbouring lactone carbonyl to an acidity greater than usual for an enol (see formula of Na ascorbate). A 2% ascorbic acid solution in water has a pH of 2,8. The degree of dissociation of ascorbic acid depending on pH can be seen from Figure 12.

The properties of a solution of ascorbic acid in water (e.g. sensitivity to oxygen, tendency to autoxidation, and enzymatic oxidation) depend on the pH of the solution and therefore on the types of ions present.

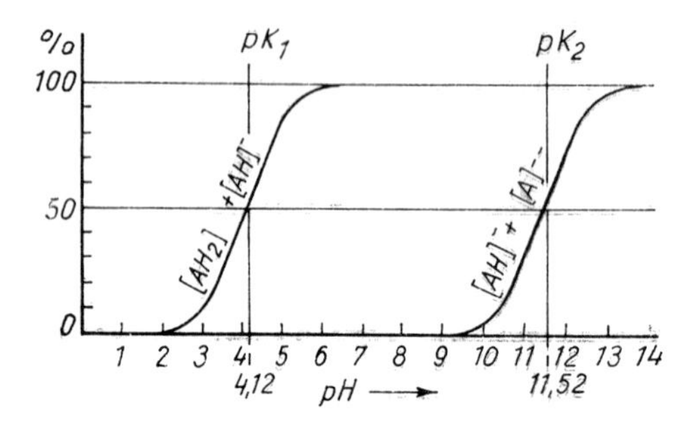

Fig. 12. Dissociation of ascorbic acid in relation to pH (expressed as %). Dissociation constants in water; $pK_1 = 4.1$ $pK_2 = 11.5$.

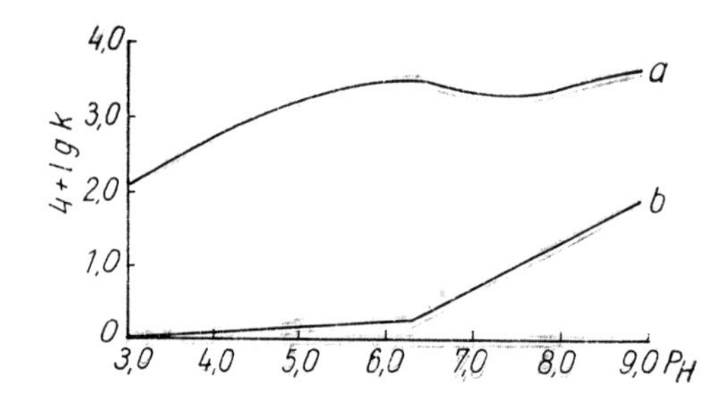

Fig. 13. Changes in speed of reaction of oxidation of L-ascorbic acid in relation to hydrogen ion concentration in presence of (a) and in absence of (b) copper ions (according to SCHULTE & SCHILLINGER).

In a strongly acid solution, vitamin C is very stable when oxygen is excluded. Even in presence of O_2 (air) pure solutions of vitamin C are surprisingly stable in an acid environment and not very sensitive to oxidation; on the other hand a neutral or alkaline environment or the slightest traces of heavy metals (in particular Cu and Fe) greatly speed up decomposition by oxygen. Even in concentrations down to 10^{-9}, Cu^{2+} ions are effective. The oxidation of ascorbic acid depends on the conversion of $Cu^{2+} \rightarrow Cu^{+}$ and reoxidation of the autoxidisable

Cu^+ by oxygen. In general the speed of autoxidation of ascorbic acid and the oxidation catalysed by Cu increases with increasing dissociation (see Fig. 13).

In food, in addition to autoxidation and oxidation by heavy metal catalysts, there are also the **enzymatic** processes of oxidation of ascorbic acid. Enzymatic oxidation can account for loss of vitamin C during the harvesting, transportation and storage of vegetable foods containing vitamin C (spinach, peas, beans etc.) as well as in household and industrial processing (see Tables, p. 231).

There are a series of **oxidases** oxidising ascorbic acid which contain as their coenzyme, either copper (ascorbic acid oxidase, phenolase) or iron (peroxidase, cytochromoxidase). Only ascorbic acid oxidase is able aerobically to oxidise vitamin C directly, the other enzymes (e.g. phenolases, peroxidases) attack ascorbic acid indirectly. With the polyphenols, for instance, such as catechin, quercetin, chlorogenic acid, they act as agents or electron transporters in the reactions between the oxidase enzyme and vitamin C.

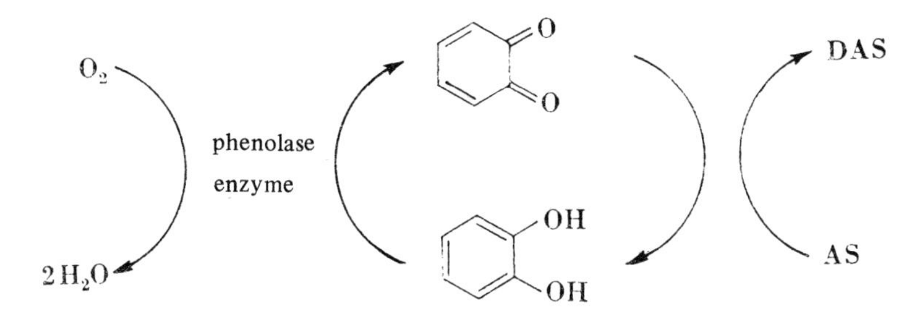

Chain of reaction: O_2-phenolase-polyphenol-ascorbic acid

The cyclic process continues with the intact enzyme systems in vegetables as long as there is any unchanged ascorbic acid present. Because the animal organism has not polyphenol oxidase, this reaction chain cannot occur. In a similar way an enzymatic system which destroys ascorbic acid is also possible with peroxidase.

The most marked characteristics of vitamin C are its **ready oxidation** and its strong power of reduction. Many methods of determination of vitamin C are based on these characteristics. (TILLMAN's reagent, iodine solution etc.) These properties are related to the enediol grouping in the ascorbic acid molecule. The reductones (see p. 138) and reductinic acid are other enediol compounds of interest to the food chemist as possessing strong reducing properties. During oxidation the vitamin with the loss of 2H atoms is first transformed into the oxidised form, dehydroascorbic acid (DAS) which still possesses anti-scorbutic activity. This in turn, due to the instability of the lactone ring in neutral or alkaline solution, readily oxidises further to inactive products. The stages are via 2,3-diketogulonic acid (and other enediol forms) which after decarboxylation, is finally oxidised to oxalic acid and L-threonic acid. In acid solution, dehydroascorbic acid is destroyed. Ascorbic acid on boiling forms reductone-like bodies, which behave similarly to ascorbic acid itself.

Oxidation to dehydroascorbic acid is reversible with the addition of hydrogen in vivo and in vitro. This is used analytically in food chemistry to determine **total vitamin C** (ascorbic acid + dehydroascorbic acid). Because hydrogen is released and taken up readily, ascorbic acid and dehydroascorbic acid (both forms are present in tissue) can form a redox system in the organism (hydrogen transfer), which is of biological importance in metabolism.

It is particularly important to protect vitamin C from oxidation during the storage and processing of foods, because of its nutritive importance. Oxygen can be excluded by storage under vacuum or by working in an inert gas atmosphere.

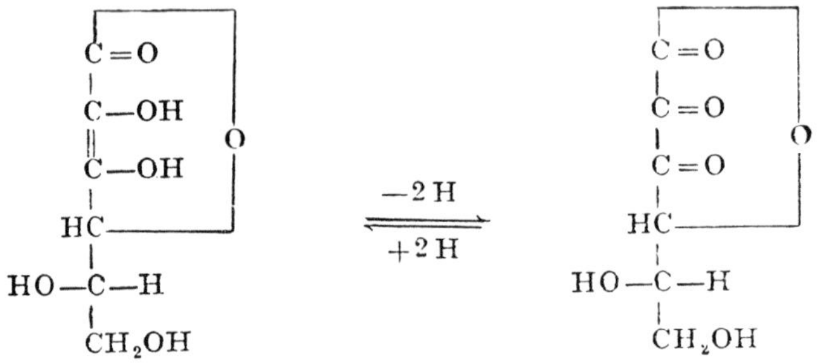

```
      ┌────────────┐                         ┌────────────┐
      C=O          │                         C=O          │
      |            │                         |            │
      C—OH         │                         C=O          │
      ||           O                         |            O
      C—OH         │        —2 H             C=O          │
      |            │      ═════════          |            │
     HC────────────┘        +2 H            HC────────────┘
      |                                      |
  HO—C—H                                 HO—C—H
      |                                      |
      CH2OH                                  CH2OH

  ascorbic acid                        dehydroascorbic acid
```

Vegetable foods containing vitamin C should be transported at low temperatures, for instance by using crushed ice during the time between harvesting and processing (spinach, beans, peas). Ascorbic acid destruction in foods, catalysed by heavy metals, may be prevented by substances naturally present in food:

amino acids:	glycine (glycocoll), creatine, creatinine, asparginic acid, glutamic acid, cysteine
proteins	
hydroxy acids:	citric acid, tartaric acid
dicarboxylic acids:	oxalic acid
mono- and polysaccharides pectins	
flavonoids, tannins etc:	quercitol, anthocyanins, catechins

It must not be overlooked that protein, sugar and pectin solutions tend to help stabilisation in a purely physical fashion by reducing the rate of oxygen diffusion as well as by the formation of hydrogen bridges between the vitamin C molecule and the sterically favoured OH-, > NH-, > CO and COOH groups of the protein, sugar or pectin molecules.

Enzymatic destruction of vitamin C in foods is best prevented by inactivation of the oxidation enzymes with heat as in blanching, see p. 227 and 293.

Analysis: The strong reducing property of vitamin C is used in almost all methods for determination. Vitamin C can be determined by using an oxidising agent, such as iodine solution, methylene blue solution, or 2,6-dichlorophenol-indophenol solution (TILLMAN's reagent). The latter is particularly important as it can be made specific for the determination of ascorbic acid. 2,6-dichloro-phenolindophenol is a dye, red in acid solution, blue in neutral or alkaline solution and colourless in its reduced form.

1

Cl, Cl: O=C₆H₂Cl₂=N—C₆H₄—OH (Na) $\underset{-2H}{\overset{+2H}{\rightleftharpoons}}$ HO—C₆H₂Cl₂—NH—C₆H₄—OH

2,6-dichlorophenol-indophenol
(TILLMAN's reagent)

leuco form
(colourless)

Because it is sensitive to oxygen, vitamin C must be extracted from natural material in an acid environment, using metaphosphoric acid or oxalic acid to stabilise the solution. In certain cases substances which interfere or reduce tanning agents, colouring matter, reductones, glutathione etc.) should first be removed, preferably by paper chromatography. Polarography can also be used for the determination of vitamin C.

Enzyme chemistry provides a specific method, based on ascorbic acid oxidase from pumpkin or cucumber. The most specific method for determination, although it is the most time consuming, is biological testing with animals, using the prophylactic and curative scurvy test.

7.5 OTHER ACTIVE DIETARY COMPOUNDS

There are other organic active agents in addition to vitamins, which the body receives with its food and which either cannot be synthesised, or else not synthesised in sufficient amount by the body. These materials differ from vitamins either because the quantity needed daily is greater than for vitamins, or else because they are only essential under certain circumstances.

It has been mentioned earlier that essential items of the diet needed in larger quantities than vitamins include **essential amino acids** (see p. 44 and **essential fatty acids** (previously called 'vitamin F', see p. 81).

There are also substances which the organism normally synthesises from food in the diet, but which improve the body functions if administered under certain conditions. Here it is assumed that the body is no longer able to synthesise these materials in sufficient quantity to maintain its normal functions. Animal feeding tests have shown that the following can be included in this group:

lipoic or **thioctanic acid** (once included in the vitamin B group); **orotic acid** (uracil-4-carboxylic acid); **myo-inositol** (previously called meso-inositol); **uni-quinone** or **Coenzyme Q** (a quinone similar to vitamin **K**, see p. 208); an organic **selenium compound**; so-called **glucose tolerance factor**, a compound which contains chromium bound in a complex; **bifidus factor**; **rutin** or **vitamin P**; **choline**.

This short summary indicates the connection between these materials and vitamins and enzymes on the one hand and trace elements on the other. Myo-inositol, rutin (vitamin P) and choline are of particular interest in food chemistry.

Myo-inositol

Inositol is chemically a hexahydroxycyclohexane. Of the 9 possible isomeric forms only the optically inactive myo-inositol (previously called meso-inositol) is active biologically.

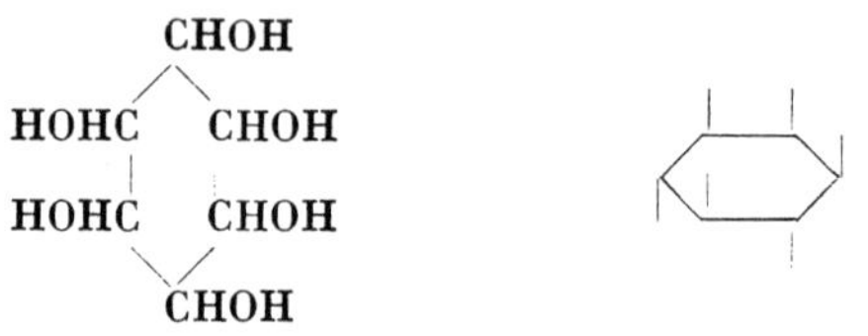

```
        CHOH
       /    \
HOHC        CHOH
  |
HOHC        CHOH
       \    /
        CHOH
```

myo-inositol
(the vertical lines indicate the position of the OH groups)

Inositol in the bound form is widely distributed in the vegetable kingdom. In the animal organism and in plants it is present in inositol phosphatides as a diphosphoric acid or monophosphoric acid ester.

The most important compound nutritionally is the hexaphosphoric acid ester of inositol, phytic acid, often simply called **phytin**. About 80% of the total phosphoric acid content of cereals, for instance in wheat or rye grains, consist of phytin. Cereal grains contain 0.2–0.27% phosphorous as phytin. Phytin inhibits the absorption of calcium which must be considered in a nutritional evaluation; the utilisation of the phytin phosphorus as a source of phosphate is particularly important (see p. 119). Phytin as such cannot be absorbed by the human gut. It is also not known to what extent it is enzymatically split in the gut by phytases, a group of low specificity mono-esterases. Monophosphoric acid ester is absorbed more rapidly than free inositol.

At present it is not yet known whether man needs to obtain part of his supply of inositol from outside sources (diet), or whether the amount synthesised in the body is sufficient.

'Vitamin P'

These compounds, formerly called vitamin P (permeability factors consist of a series of bioflavonoids, which are frequently found as glycosides, for instance in

Table 16 Vitamin content of some foods. (The values refer to the edible portion of 1000 g of raw food bought)

	Retinol µg	Carotene µg	Vitamin D µg	Thiamin mg	Riboflavin mg	Nicotinic acid mg	Vitamin C mg	Vitamin E mg	Vitamin B_6 mg	Vitamin B_{12} µg	Folic Acid Free µg	Folic Acid Total µg	Pantothenic acid mg	Biotin µg
Meat (beef)	Tr	Tr	Tr	0.70	2.4	52	0	1.5	3.2	20	40	100	7.0	Tr
(pork)	Tr	Tr	Tr	8.9	2.5	62	0	0	4.5	30	Tr	50	11	30
(lamb)	Tr	Tr	Tr	1.4	2.8	60	0	1.0	2.5	20	Tr	50	7.0	20
(veal)	Tr	Tr	Tr	1.0	2.5	70	0	–	3.0	10	Tr	50	6.0	Tr
(chicken)	Tr	Tr	Tr	1.0	1.6	78	0	1.0	4.2	Tr	100	120	12	20
Liver (veal)	146,000 (83,000–317,000)	1,000	2.5	21	31	124	180	2.4	5.4	1,000	1,900	2.400	84	390
(ox)	165,000 (116,000–240,000)	15,400	11.3	23	31	134	230	4.2	8.3	1,100	2,200	3,300	81	330
(pig)	92,000 (56,000–142,000)	0	11.3	3.1	30	148	130	1.7	6.8	250	590	1,100	65	270
Hen's eggs	1,400	Tr	17.5	0.9	4.7	0.7	0	16	1.1	17–29	250–390	250–390	18	250
Cow's milk	260–350	130–220	0.13–0.30	0.4	1.9	0.8	15	0.7–1.0	0.4	3	40	50	3.5	20
Butter, salted	7,500 (5,200–9,700)	4,700 (3,500–6,500)	7.6 (6.3–10)	Tr	Tr	Tr	Tr	20	Tr	Tr	Tr	Tr	Tr	Tr
Cheese, Cheddar type	3,100	2,050	2.61	0.4 (0.2–0.8)	5.0 (3.0–8.0)	1.0 (0.1–2.0)	0	8	0.8 (0.5–1.4)	15	–	200 (100–400)	3.0 (1.0–7.0)	17 (4–23)
Vegetable oils	0	Tr	0	Tr	Tr	Tr	0	10–800	Tr	0	Tr	Tr	Tr	Tr
Lard	Tr	0	Tr	Tr	Tr	Tr	0	Tr	Tr	Tr	Tr	Tr	Tr	Tr
Wheat flour 72% extraction rate	0	0	0	3.1	0.3	20	0	Tr	1.5	0	140	310	3.0	10
Potatoes	0	Tr	0	1.1	0.4	12	80–200	1.0	2.5	0	100	140	3.0	1.0
Peas	0	3,000 (2,500–4,000)	0	3.2	1.5	25	250 (150–350)	Tr	1.6	0	–	–	7.5	5.0
Apples	0	300	0	0.4	0.2	1.0	20–300	2.0	0.3	0	20	50	1.0	3.0
Granulated sugar	0	0	0	0	0	0	0	0	0	0	0	0	0	0

The figures given are taken from tables prepared in the 4th Edition of the Composition of Foods, McCance and Widdowson, by A. A. Paul and D. A. T. Southgate, H.M.S.O., London 1978.

The upper line gives the amount of the vitamin normally present and the lower line, variations which may be encountered.

citrus fruits. They include the turinosides, **rutin** (aglycone, quercetin), **hesperidin** (aglycone hesperetin), **naringin** (aglycone naringenin)

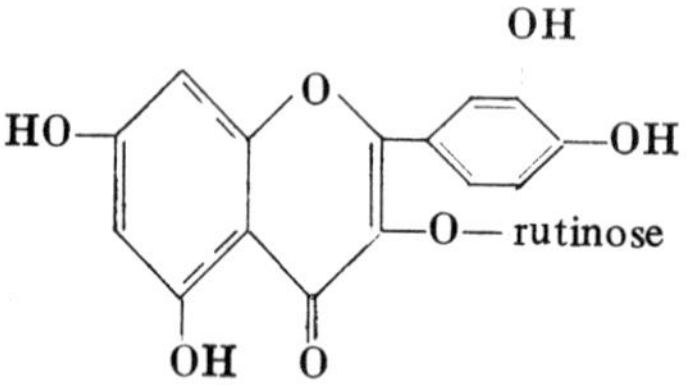

rutin (flavonol glycoside)

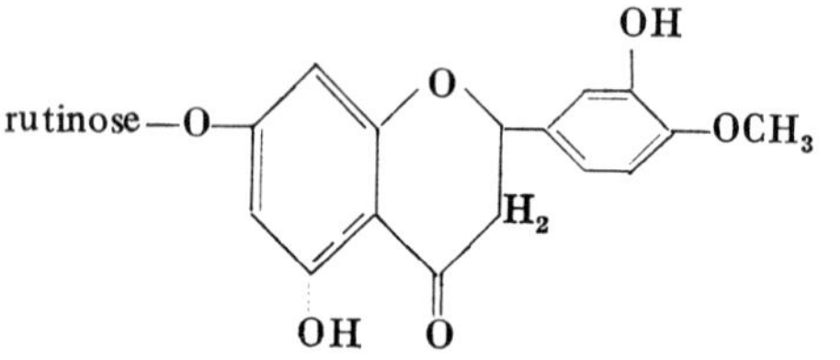

hesperidin (flavanone glycoside)

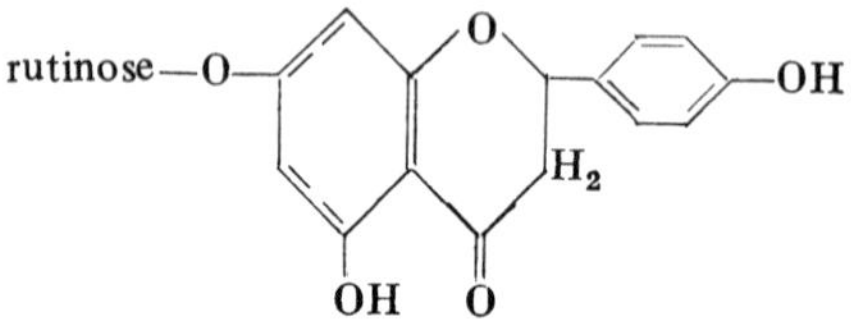

naringin (flavanone glycoside)

rutinose = disaccharide from glucose and rhamnose

7.6 VITAMIN CONTENT OF VARIOUS FOODS

The polyphenolic flavonides (previously called vitamin P) are however, neither vitamins nor essential nutrients, as their deficiency does not necessarily cause nutritional diseases. They have a specific pharmacological action and medical opinion ascribes to them a beneficial effect on the reduction of permeability of the blood vessels – especially when given in conjunction with increased vitamin C. They have no known task in the normal functions of the body.

Choline

Chemically, choline is trimethyl ethanolamine (see p. 119). Animals can form choline by methylation from ethanolamine, the methionine (activated methionine) as the principal methyl donator. Ethanolamine (colamine) itself is formed by decarboxylation of the amino acid serine and is used by the body in practice only for the formation of choline or as a basis for phosphatide synthesis, for instance cephalins.

Choline has a number of metabolic tasks. The three most important are: providing the basic building blocks for the synthesis of lecithin, providing the methyl groups for tansmethylation and formation of acetyl choline. Before lecithin can be synthesised biologically in the organism, there must be sufficient choline available. Most of the choline absorbed in food or formed by transmethylation in metabolism is normally used for the formation of lecithin.

Choline deficiency symptoms in animals are retarded growth and in man, fatty degeneration of the liver, which frequently develops into cirrhosis of the liver. Fatty degeneration of the liver can be prevented by administering 'lipotropic' substances of which choline is the most active. Its lipotropic action depends on the fact that it is a building block for lecithin synthesis. The fatty acids formed constantly in the liver by biosynthesis, are normally passed into the blood in the form of lecithin (that is, phosphatides). When choline is deficient the transport of the fatty acids from the liver is disturbed. The fatty acids are then deposited in the liver in the form of triglycerides resulting in fatty degeneration of the liver.

7.7 VITAMIN REQUIREMENTS

Compared with the quantities of calorific nutrients and structural materials needed by man, the amounts of vitamins are very small. The desirable daily requirement which from experience is thought to be necessary for complete health and full capacity for work, has been evaluated by various committees in the form of recommendations given in Table 17.

Vitamin requirements in fact cannot be a fixed figure, because they will vary according to external influences. In general the requirement rises with the demands made on the body, for instance with heavy manual work, heat, cold, during illness, during pregnancy and during lactation. In the special case of thiamin (vitamin B_1) the requirement depends on the supply of carbohydrates (see cocarboxylase and carbohydrate metabolism, p. 209); vitamin B_6 requirements are similarly dependent on the amount of protein consumed, see p. 217.

The vitamins available to us in foods are listed in Table 16. However, the amount of vitamin remaining in the food after preparation is the decisive figure rather than the 'native' content, or better still, the amount which is effectively absorbed.

7.8 INFLUENCE OF STORAGE AND PREPARATION ON THE VITAMIN CONTENT OF FOOD

Serious vitamin loss may occur during preparation and preservation of foods, if suitable methods are not used. This loss occurs because of the sensitivity of many vitamins to light, for instance vitamin A, B_2, B_6, E and K. Other vitamins are heat labile, such as vitamin B_1, folic acid and antothenic acid. In the presence of oxygen, especially if accompanied by oxidative enzymes and traces of heavy metals, vitamins liable to oxidation such as vitamin A, D, E, and C, suffer heavy losses or are completely destroyed. As is shown in Table 18, some vitamins may be destroyed by irradiation.

A decrease in vitamin content dependent on temperature can be seen immediately after harvesting, during **transport** and **storage**, unless special protection systems (stabilisers) are present in the cell. Citrus fruits are an example where the destruction of vitamin C is inhibited. Spinach, green beans and asparagus, stored at room temperature, lose 50% of their vitamin C content after

Table 17 Recommended Daily Amounts of Vitamins for Population Groups in the United Kingdom

		Thiamin	Riboflavin	Nicotinic acid	Total Folate	Ascorbic acid	Retinol equivalents	Vitamin D
		mg	mg	mg equivalents	μg	mg	μg	μg
Boys under	1	0.3	0.4	5	50	20	450	7.5
	1	0.5	0.6	7	100	20	300	10
	5– 6	0.7	0.9	10	200	20	300	Depends on exposure to sunlight
	9–11	0.9	1.2	14	200	25	575	Depends on exposure to sunlight
	15–17	1.2	1.7	19	300	30	750	Depends on exposure to sunlight
Girls under	1	0.3	0.4	5	50	20	450	7.5
	1	0.4	0.6	7	100	20	300	10
	5– 6	0.7	0.9	10	200	20	300	Depends on exposure to sunlight
	9–11	0.8	1.2	14	300	25	575	Depends on exposure to sunlight
	15–17	0.9	1.7	19	300	30	750	Depends on exposure to sunlight
Men	18–34	1.2	1.6	18	300	30	750	Depends on exposure to sunlight
	35–64	1.1	1.6	18	300	30	750	Depends on exposure to sunlight
	65–74	1.0	1.6	18	300	30	750	Depends on exposure to sunlight
	75+	0.9	1.6	18	300	30	750	Depends on exposure to sunlight
Women	18–54	0.9	1.3	15	300	30	750	Depends on exposure to sunlight
	55–74	0.8	1.3	15	300	30	750	Depends on exposure to sunlight
	75+	0.7	1.3	15	300	30	750	Depends on exposure to sunlight
Pregnancy		1.0	1.6	18	500	60	750	10
Lactation		1.1	1.8	21	400	60	1200	10

These figures have been taken from the Department of Health and Social Security. Report on Health and Social Subject 15, Recommended Daily Amounts of Food Energy and Nutrients for Groups of People in the United Kingdom, Report by the Committee on Medical Aspects of Food Policy, H.M.S.O. 1979.

24 hours; the loss is less during cold storage. Young leaf vegetables are particularly sensitive; in spring the loss of vitamin C in spinach is three times as high as in autumn spinach. These losses demand fast transportation and careful storage before processing, for instance storage under cold conditions or possibly even in a controlled gas atmosphere.

Table 18 Destructability of vitamins (according to K. LANG)

Vitamin	thermo labile	oxidisable	light sensitive	sensitive to irradiation
A		+	+	+
D		+		
E		+		
K			+	+
B_1 (thiamin)	+			+
B_2 (riboflavin)			+	+
B_6 (pyridoxin)			+	
pantothenic acid	+			
pteroylglutamic acid (folic acid)	+			
C		+		+

Even in vegetables normally stored, such as potatoes or root vegetables, the loss of sensitive vitamins is considerable. By spring, potatoes only contain half or even less of the vitamin C present when they were harvested. Better preservation of vitamins, for example vitamins B and C, can be ensured when maize and dried beans are cold stored.

Vitamins are subject to many influences because they are compounds which react readily and are present in small amounts in a complex mixture of substances in food. Individual findings about vitamin loss should not be taken as a general rule. The figures given in table 19 only act as a guide. However, they show clearly the possible vitamin loss which can occur in **food preparation in the home**.

Table 19 Average vitamin loss during household preparation of foods (according to K. LANG; deviations may be considerable in individual cases)

thiamin	30%
riboflavin	15%
niacin	20%
ascorbic acid	35%
pantothenic acid	35%
folic acid	40–50%

In general, slow heating, long cooking, exposure to oxygen (air) and slow cooling have a bad effect on vitamin preservation. Cooking in a pressure cooker under increased pressure is more advantageous than ordinary cooking. The metal of which the cooking pot is made must also be considered; heavy metas (Cu and brass vessels) quickly destroy all vitamin C in foods. When cooking or blanching water is thrown away there is a serious loss of water soluble vitamins, especially vitamin C, B_1, B_2 and pantothenic acid. Modern food industry tries to apply this knowledge by careful processing, in the preparation and preservation of vegetables and fruits.

Loss of vitamins after the preparation of food should also be considered. This occurs when food is kept warm for a long time. It is better to keep cooked food in a cool place to preserve the vitamins. Further loss of vitamins during storage can also occur with frozen and with sterilised tinned or bottled foods. The losses depend both on how they are treated in the first place, for instance inadequate inactivation of the oxidation enzymes during blanching, and also on the storage temperatures.

CHAPTER 8

Enzymes

The metabolism of a living organism consists of a complicated series of interconnected chemical reactions. Almost without exception, biochemical synthesis and decomposition are catalysed and directed by enzymes – formerly called ferments in the German literature. In the English literature the expression 'fermentation' is only used for microbiological fermentation. Materials on which enzymes act are called **substrates**. From many points of view, enzymes are of central importance to the food chemist and technologist.

Native or **endogenous enzymes** are present in the food itself. Our food derives mainly from living animal and vegetable organisms and many reactions caused by enzymes continue in it – both the natural enzymatic reactions in life and those due to the onset of the postmortem state. Some of these reactions improve the quality and are therefore desirable, for example the ripening of fruit or the late maturing or conditioning of meat. However, the majority of the processes catalysed by enzymes in food reduce its value by nutrient decomposition or changes in aroma, taste, structure and colour. Finally these enzymatic processes lead to decay and decomposition of the food. Man therefore tries to delay or prevent the activity of these native enzymes by suitable preservation methods, such as cooling, freezing, drying, sterilisation or pasteurisation.

Many processes of food preparation and manufacture use **added enzymes**. Such enzymatic procedures have been used empirically for a long time, in the brewing of beer, wine, vinegar, in the preparation of fermented vegetables, cabbage, cucumber and so on, in milk products, yoghurt, cheese, and in baking processes using yeast and sour dough. In these cases the added enzymes are derived from microorganisms. Nowadays such processes, suitably controlled, are widely used in industrial practice. Hence commercially produced enzyme preparations are becoming increasingly important in the food industry, for instance in baking, for fruit juice preparation, and in confectionery, see p. 266.

Microbial spoilage of food is due to foreign or exogenous enzymes. Microbial enzymes catalyse decomposition reactions, often resulting in obnoxious, unpalatable and even poisonous products which can reduce or completely destroy the edibility of the food. For centuries, man has used methods such as drying, smoking, pasteurising and sterilising, souring, sugaring and chemical preservation to prevent the growth of microorganisms and therefore the spoilage of foods by microbial enzymes.

Enzymes are being used to an increasing extent in **food analysis** (p. 269). They can be used as indicators, in that their presence or absence in foods permits conclusions about the normal constitution of the food or about attack by microorganisms. In addition, enzymes are subject to chemical or physical influences (e.g. O_2, CO_2, preservatives, pest control agents, temperature, humidity) during the production, processing and storage of foods; they therefore mirror the state and quality of many foods during technical processing – sterilisation, pasteurisation or blanching. On the other hand, enzymes are used for the qualitative and quantitative determination of food components, for instance of sugar, alcohol, glycerol, lactic acid, sorbitol and citric acid.

8.1 CHEMICAL AND CATALYTIC NATURE OF ENZYMES

Enzymes can be defined as proteins which act as catalysts. Their chemical and physical properties are similar to proteins, in that they are denatured physically by heat and chemically by various reagents. Their molecular weight can vary widely between 10^3 and 10^6. Enzymatic activity is not distributed over the whole of the molecule, but localised in particular, sharply delimited areas, at the so-called 'active centres'. Their activity is expressed in enzyme units. An enzyme unit is the amount of enzyme which acts on one micromole of the substrate per minute under optimum conditions. The so-called 'specific activity' is used to define the purity of an enzyme. It is expressed as the number of enzyme units per milligram protein.

The following classification is based on structure:

1. Enzymes with a **native protein**. Amylases, pepsin, trypsin and urease belong to this group.
2. Enzymes with a **protein** and **cofactor**. This group includes enzymes with a prosthetic (active) group which usually has a low molecular weight. When the prosthetic group can be readily separated off, it is called a **coenzyme**. The enzyme complex, comprising protein + coenzyme as a functional unit is called a holoenzyme.

These enzymes are subdivided according to the nature of their prosthetic groups as follows:

(a) Enzymes whose prosthetic groups contain metals, for instance Cu, Fe, Mg. Peroxidase, catalase, cytochrome oxidase, phenol oxidase, ascorbic acid oxidase and phosphates are of importance in food chemistry.

(b) Enzymes whose coenzymes are organic compounds, usually without metals. Frequently they are vitamin derivatives for instance NAD, NADP, FAD.

These coenzymes have a chemical role in the catalytic process, as they act as donors and acceptors for electrons of certain molecules. They are not themselves changed during the reaction and must therefore return to their original state by a further enzymatic reaction.

Table 20 lists the known coenzymes, the groups transferred by them and their relationship to the vitamins.

Table 20 Known coenzymes

Coenzyme	usual abbreviation	Group transferred	Vitamin
1. Coenzymes of the transferases			
S-adenoxylmethionine*		$-CH_3$	
pyridoxal-5-phosphate	PAL	$-CH_2OH$ $-NH_2$	pyridoxin (adermine)
biotin		CO_2	biotin
tetrahydrofolic acid	CoF	HCOO– (formyl residue)	folic acid
coenzyme A*	CoA	acetyl (CH_3CO-)	pantothenic acid
thiamin pyrophosphate	TPP	aldehyde or keto group	thiamin (aneurin)
cytide diphosphate	CDD	phosphoryl choline etc.	
adenosine triphosphate	ATP	H_3PO_4 and AMP^{2+}	
phosphoadenylic acid sulphate	PAPS	H_2SO_4	
uridine diphosphate†	UDP	sugar	
2. Coenzymes of oxidoreductases			
nicorinamide-adenine-di-nucleotide	NAD (DPN. CoI)	H_2	niacinamide
nicotinamide-adenine-dinucleotide phosphate	NADP (TPN, CoII)	H_2	niacinamide
nicotinamide-mono-nucleotide	NMN	H_2	niacinamide
flavine-mononucleotide	FMN	H_2	riboflavin
flavine-adenine-dinucleotide	FAD	H_2	riboflavin
haems of cytochrome		electrons	
3. Coenzymes of isomerases			
uridine diphosphate	UDP	sugar isomerisation	pyridoxin
pyridoxal phosphate	PAL	racemisation of the α-amino acids	pyridoxin
nicotinamide-adenine-	NAD	sugar isomerisation	nicotinamide
B_{12} coenzyme		carboxyl displacement	cobalamin

*Adenosylmethionine = 'active methyl' in metabolism (see pp. 43, 119, 239).
formyl-tetrahydrofolic acid = 'active C_1 residue' in metabolism.
acetyl CoA = 'activated acetic acid' = active C_2 residue in metabolism.
glucose-uridine diphosphate = 'activated glucose' in metabolism (for instance for the synthesis of glucosides, disaccharides, etc.) see p. 157.
†AMP = adenosine monophosphate; adenosine diphosphate.

Table 20 Known coenzymes–*continued*

Coenzyme	usual abbreviation	Group transferred	Vitamin
4. Coenzymes of pyases			
pyridoxal phaosphate		decarboxylation	pyridoxin
thiamin pyrophosphate		decarboxylation	thiamin
guanyl triphosphate	GTP	decarboxylation	
adenosine triphosphate	ATP	decarboxylation	

8.2 ENZYMES AS CATALYSTS

Most chemical reactions are reversible and tend to reach equilibrium, so that the original materials are still present in measurable quantities, in addition to the products. An example is the formation of an ester from an acid and alcohol.

$$\text{acid} + \text{alcohol} \rightleftharpoons \text{ester} + H_2O$$

This dynamic equilibrium state is expressed by the proportionality constant K of the Law of Mass Action. In simplified form the reaction is as follows:

$$A + B \underset{k_{-1}}{\overset{k_{+1}}{\rightleftharpoons}} C + D \qquad K = \frac{k_{+1}}{k_{-1}} = \frac{[C] \cdot [D]}{[A] \cdot [B]},$$

where k_{+1} and k_{-1} are rates of the reversible reaction forwards and backwards. Attainment of equilibrium depends on the distribution of energy between the components, for instance the reaction acid + alcohol $\rightleftharpoons$ ester + water can take a very long time.

The time taken to reach equilibrium can be reduced by adding catalysts, for instance to provide the proton H^+ or by adding an esterase (carboxyl ester hydrolase). The state of equilibrium, that is the magnitude of the constant K, will not be changed by this addition. The magnitude depends only on the concentration of the components [A], [B], [C] and [D] with their respective energies. If the equilibrium lies well towards the formation of C and D, then K will be large. If the reverse reaction is too small to be measured, that is if the limit of change of [A] or [B] $\rightarrow 0$ then K tends towards infinity.

If a large amount of water is added to the system

$$\text{acid} + \text{alcohol} \rightleftharpoons \text{ester} + \text{water}$$

the energy relationship is disturbed and the equilibrium will lie on the side of acid and alcohol. If a great excess of alcohol is present a large amount of ester will be formed provided that depletion of the acid does not occur.

The equilibrium constant K is linked with a quality ΔG°, which represents the 'change of free energy of the system'. However the quantity ΔG°, meaning the 'standard free energy change' is used as in the relationship

(2) $$\Delta G^\circ = -RT \log_e K$$

or $$\Delta G^\circ = 2.3 \; RT \log_{10} K$$

where R represents the gas constant (1.98 cal/mol/degree), T is the absolute temperature and ΔG° represents the maximum useful work of a reaction under standard conditions, that is with a concentration of reactants of the order of 1 mol/litre.

If the reaction is not carried out under **standard conditions** as is normally the case, then the following formula applies:

(3) $$\Delta G = \Delta G^\circ = 2.3. \; RT \log_{10} \frac{[C] \cdot [D]}{[A] \cdot [B]}$$

the magnitude of ΔG is defined as the difference between the sum of the free energies of the reactants and the sum of the free energies of the reaction products

$$\Delta G = \Sigma G_{Pr} - \Sigma G_{Re}$$

According to the law of Thermodynamics the following applies to reversible isothermic processes

$$\Delta G = \Delta H - T\Delta S$$

where ΔH is the change in enthalpy of the reaction and ΔS is the entropy or measure of change in the state of organisation during the reaction.

For equilibrium conditions $\Delta G = 0$ as can easily be seen by inserting the equations (1) and (2) into equation (3). An exothermic reaction can only give energy as long as it is not in a state of equilibrium and it will only continue as long as energy can be set free. This indicates $G < 0$ and the energy given off is given the negative sign for conventional purposes. In endothermic processes, energy must be supplied continuously since $G > 0$ and the energy is given the positive sign.

Although there is a larger or smaller gain in energy, most exothermic reactions do not occur spontaneously. H_2 and O_2, for instance, can remain in a mixture in the absence of catalysts, without a reaction occurring. In the same way, the equilibrium for most organic compounds in the presence of oxygen remains far to the CO_2 and H_2O side. The partners in the reaction must first be brought to an agitated or activated state by the application of energy (e.g. heat), so that the reaction may take place. Under normal conditions, such mixtures of materials are in a so-called metastable condition. If all exothermic reactions like those mentioned above occurred spontaneously all materials would soon find themselves in a state of equilibrium. There would be no energy differentials and life would no longer be possible.

The energy at the beginning of an exothermic reaction is named **activation energy** (E_a). If the rate constants k_1 and k_2 are known for reaction at temperatures T_1 and T_2, the activation energy required can be calculated according to the ARRHENIUS equation:

$$E_a = 2.303\,R\frac{T_1\,T_2}{T_2\,T_1}\log\frac{k_2}{k_1} \qquad (4)$$

where R is the general gas constant (1.98 cal/mol/degree).

If a reaction is to proceed from state A to state C (see Fig. 14), the activated state B must first be reached, in order that the reaction may proceed to C. If a

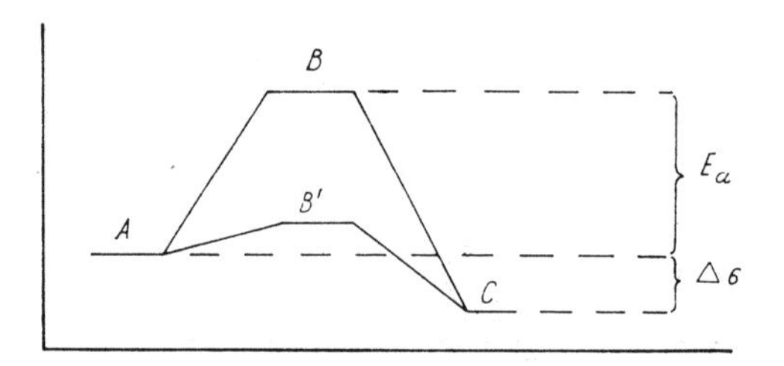

Fig. 14. Reduction of the activation energy by a catalyst (enzyme)

catalyst or an enzyme is added a state B′ need only be reached for the reaction to continue. This means a reduction in the amount of energy needed to activate the reaction, and a new pathway involving a different reaction mechanism is opened up to reach the same end product.

Many chemical reactions, for instance ionic reactions, need almost no activation energy. They are therefore spontaneous and very fast. This is also true for the reverse reaction, that is when both k_1 and k_2 tend towards ∞; $\frac{k_1}{k_2}$ move towards one, $\log\frac{k_1}{k_2}$ tends towards zero, so that the activation energy E_a is, for practical purposes zero.

The phosphorylation of glucose to glucose-6-phosphate is an example of a reaction, in which the coenzyme (ATP) is involved. This is important in food chemistry as the determination of glucose is based on it. It also demonstrates how nature makes it possible for endothermic reactions to take place. The reaction (esterification)

$$\text{D-glucose} + H_3PO_4 \rightleftarrows H_2O + \text{glucose-6-phosphate} \quad (\Delta G^\circ = +3\text{kcal/mol})$$

is **endothermic**. The equilibrium is almost completely on the left side. In order that the reaction may occur in spite of this, it is coupled to a strong **exothermic** reaction, the cleavage of adenosine triphosphate (ATP, see p. 62) to adenosine diphosphate (ADP) and H_3PO_4

$$\text{ATP} \longrightarrow \text{ADP} + H_3PO_4 \quad (\Delta G^\circ = -7\text{kcal/mol})$$

The phosphosphoric acid bond in ATP is called an 'energy rich bond' (often indicated by the symbol ~). Energy rich bonds are those where more than 5kcal/mol is set free when they are split. Groups linked by such bonds are highly reactive (activated) and react readily with other molecules in exothermic reactions.

If the phosphorylation of glucose is carried out with the enzyme hexokinase with the energy rich ATP as its co-substrate, then

$$\text{ATP} + \text{D-glucose} \xrightarrow[\longleftarrow]{\text{hexokinase}} \text{ADP} + \text{D-glucose-6-phosphate} \quad (\Delta G^\circ = -4\text{kcal/mol})$$

occurs in an exothermic reaction. The adenosine diphosphate (ADP) produced by this reaction is converted back to ATP by a further coupled exothermic reaction, through the enzymatic oxidation of 3-phosphoglyceraldehyde to 3-phosphoglyceric acid (Oxidative phosphorylation). The dehydrogenation or oxidative phosphorylation of 3-phosphoglyceraldehyde provides the necessary energy:

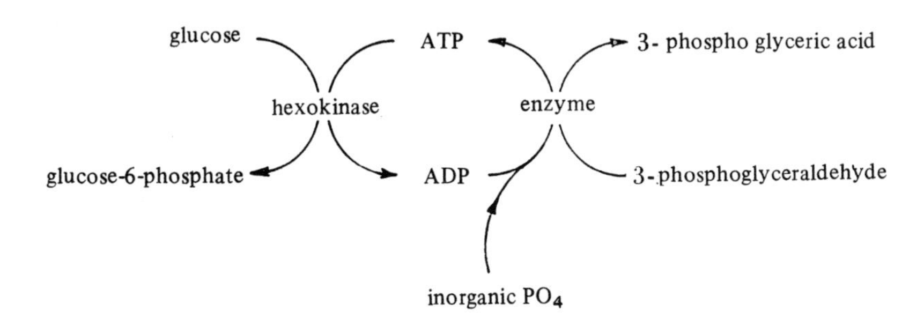

This example shows that nature uses 'energy rich bonds' as a means of storing the energy which later can be freed to be used again for a great variety of endothermic reactions (biosyntheses).

In the same way, the number of moles of ATP formed or consumed in such processes can be used directly as a measure for ΔG.

There are a number of other compounds, as well as ATP (the most frequent in nature), which contain energy rich bonds. These include acetyl-coenzyme A

(CoA, see p. 220), activated carbonic acid (p. 221), phosphoeneol pyruvic acid, creatine phosphate, adenosyl methionine, uridine diphosphate (see p. 159), see also p. 239 footnote*. The figures given in the literatnre for the free energy of these bonds are **standard values**, as is obvious from the definition of ΔG. The amount of energy which actually is available to the cell (ΔG) is considerably influenced by the relative concentration of the two partners in the reaction (Equation 3).

8.3 ENZYME SPECIFICITY

The action of enzymes is usually specific and a distinction is made between those which are **substrate specific** and those which are **specific in their action.** The phenomenon of specificity can be explained in the following way; as a first step in the reaction the substrate in solution with the enzyme forms a complex. In doing this the substrate attaches itself to particular groupings on the surface of the enzyme molecule. Enzyme specificity can be visualised as a lock and key; the substrate is the key and must fit exactly onto the active centre on the enzyme which is the lock, to enable the reaction to start.

The importance of the **apoenzyme** for the reaction specificity and for the substrate specificity is now known. Formerly the protein portion of the enzyme was thought only to act as a non-specific colloidal carrier of the effective 'active group'. Frequently, one finds the same coenzyme occurring in different enzymes and catalysing different reactions. The apoenzyme determines which substrate is chosen and converted; the coenzyme determines which of the possible reactions will take place on the substrate – oxidation, dehydrogenation, decarboxylation and so on.

The property of an enzyme which permits it to catalyse only one of several thermodynamically possible reactions on a substrate is called its **activity** or **reaction specificity**. For example, an α-amino acid requires a specific enzyme for each of the most diverse reactions: (a) oxidative deamination with formation of an α-oxo acid, (b) reactions with another oxo acid with exchange of the C=O– and C–NH_2 groups (transamination), (c) decarboxylation with formation of a primary amine and CO_2:

$$\underset{\alpha\text{-amino acid}}{R{-}CH(NH_2){-}COOH} \xrightarrow[-2H;\ +H_2O]{\text{(a) oxidoreductase}} R{-}C({=}O){-}COOH + NH_3$$

$$R{-}CH(NH_2){-}COOH \xrightarrow[+R'{-}C({=}O){-}COOH]{\text{(b) aminotransferase}} R{-}C({=}O){-}COOH + R'{-}CH(NH_2){-}COOH$$

$$R{-}CH(NH_2){-}COOH \xrightarrow[-CO_2]{\text{(c) carboxylase}} R{-}CH_2{-}NH_2$$

Enzymes can be divided into three groups, based on their **substrate specificity.**

1. Enzymes with only **slight specificity.** In these it is the type of bond between the two components which is decisive. For instance many carboxylesterhydrolases split ester bonds independent of the nature of the acid and the alcohol.
2. Enzymes with **group specificity** are specific to the type of bond and structure of one of the components. Examples are, glycosidehydrolases with their specificity for the glycosidic link (α or β) and for the sugar residue. (The type of aglycone has no determining effect on this enzymatic reaction).
3. Enzymes with **absolute specificity** only attach themselves to a particular substrate. For example, in optical antipodes only the D or the L form is active and in compounds with –C=C bonds only the cis or the trans-configuration is active. (See lipoxygenases p. 255). Urease for instance only splits urea.

Isoenzymes catalyse the same reaction, but vary in their activity because their protein structures differ, for instance by having different molecular weights, see p. 55. They can be separated electrophoretically.

8.4 ENZYME KINETICS

Factors which influence the reaction rate

The rate of an enzymatic reaction, that is the **enzyme activity** is influenced by various factors: **Substrate concentration – concentration – temperature – pH – ionic strength – redox potential – activators – inhibitors.**

Effect of substrate concentration. If more substrate is added continuously to a given (constant) amount of enzyme, the rate of the reaction increases until a limiting rate for the reaction is reached (Fig. 15).

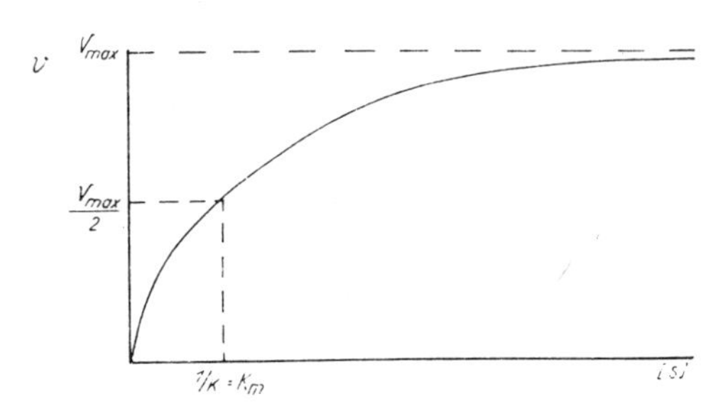

Fig. 15. Dependence of rate (v) of the enzymatic reaction on the concentration of the substrate (s). Determination of $1/K = K_m$.

MICHAELIS and MENTEN proposed a kinetic theory of enzyme activity which has proved very fruitful in enzymatic chemistry:

Enzyme and substrate form an ‘activated’ intermediate compound in solution, the ‘enzyme substrate complex’. The actual enzymatic reaction takes place within this complex, the reaction products from the reaction then split off and the freed enzyme can react again.

(5) $$\text{enzyme (E)} + \text{substrate (S)} \underset{k_{-1}}{\overset{k_{+1}}{\rightleftharpoons}} \text{enzyme-substrate-complex (ES)} \underset{k_{-2}}{\overset{k_{+2}}{\rightleftharpoons}} \text{end products (P)} + \text{enzyme (E)}$$

k_{+1}, k_{+2}, and k_{-1}, k_{-2} are the rate constants for the reversible reaction. The law of mass action applies as usual to the concentration of the partners in the reaction

(6) $$\text{Equilibrium constant } K = \frac{k_{+1}}{k_{-1}} = \frac{[ES]}{[E] \cdot [S]}$$

further, when [Eg] is the total enzyme concentration then:

(7) $$[E] = [Eg] - [ES]$$

from equations (6) and (7) it follows that

(8) $$K = \frac{[ES]}{([Eg] - [ES]) \cdot [S]} \quad \text{or, transformed, } [ES] = \frac{[Eg]\ [S]}{1/K + [S]}$$

According to the MICHAELIS theory, the actual enzymatic reaction takes place within this enzyme substrate complex and the reaction products (P) are split off.

(9) $$ES \underset{k_{-2}}{\overset{k_{+2}}{\rightleftharpoons}} E + P$$

The freed enzyme returns to a state of equilibrium (5). If it is assumed that the rate of formation of the complex is distinctly greater than the rate of transformation – which in most enzymatic processes is clearly the case –, then this slower monomolecular step (9) is the rate determining stage. The concentration of the complex [ES], (8) which follows from the law of mass action, is maintained continuously during the reaction, and the rate of reaction v (increase in reaction products formed P per unit of time), is proportional to ES

(10) $$v = \frac{\Delta[P]}{\Delta t} = k_{+2}\,[ES] = k_{+2}\,[Eg] \cdot \frac{[S]}{1/K + [S]}$$

The manner in which the rate of the reaction v substrate concentration, depends on the concentration of the substrate S, when the **enzyme concentration** is constant is sown in Fig. 15. At a reaction rate (v_{max}) the total enzyme is present as an enzyme substrate complex: [ES] = [Eg]. Then

(11) $$v_{max} = k_{+2}\,[Eg]$$

Once this rate of reaction v_{max} has been achieved, the size of K (association constant of the enzyme substrate complex) or 1/K = K (**Michaelis constant,** dissociation constant of the enzyme substrate complex) can easily be determined. From equations (10) and (11) is derived

$$\frac{v}{v_{max}} = \frac{[S]}{1/K + [S]} = \frac{[S]}{K_m + [S]} \tag{12}$$

When only enough substrate is added to reach half the maximum reaction rate $\frac{v}{v_{max}} = \frac{1}{2}$, then from equation (12) becomes $\frac{1}{K} = K_m = [S]$, that is **the substrate concentration which exactly reaches half the maximum reaction rate is equal to the Michaelis constant of the enzyme substrate complex.**

In order to determine K it is possible to proceed as follows by determining the speed of reaction of the enzyme with a series of increasing substrate concentrations and then determining K graphically by means of a curve analogous to the curve in Fig. 15. In this case the **total concentration of substrate** is usually substituted for the **concentration of free substrate** [S], because the amount of substrate bonded to the enzyme is usually so low that it can be ignored. The LINEWEAVER-BURK procedure for determining K_m is very elegant.

The **Michaelis constant** has a **characteristic magnitude for the enzyme-substrate pair**. If it is high, a large amount of substrate must be added in order to reach half saturation of the enzyme: the enzyme has only slight affinity for the substrate. If two enzymes which act on the same substrate are compared, then the one with the smaller Michaelis constant will transform the substrate more rapidly. The Michaelis constant of most enzymes is between 10^{-2} and 10^{-5} mol/litre. The determination of the Michaelis constant is used in food chemistry for the characterisation of certain enzymes and substrates.

Effect of enzyme concentration. If the **enzyme concentration** is varied, while the **substrate concentration** is kept the same, with other conditions being constant, then it can be shown experimentally as in Figure 16, that **the reaction rate v is proportional to the enzyme concentration [E]**. This only applies on condition, that the enzyme concentration remains small in comparison to the substrate concentration.

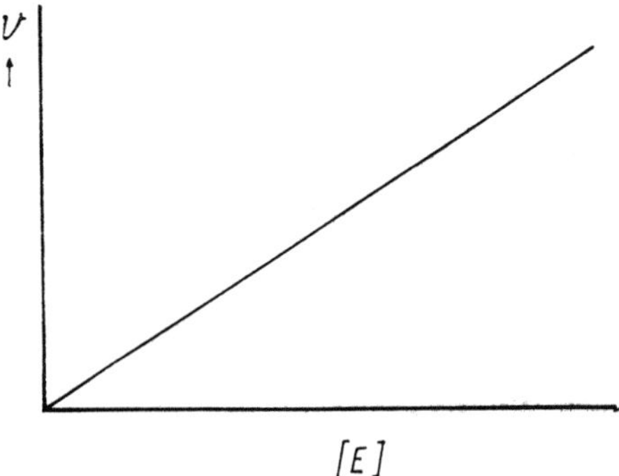

Fig. 16. Dependence of the reaction rate (v) on the enzyme concentration [E].

This relationship only applies to the **initial velocity** of the reactions, because the speed of an enzymatic reaction usually changes during the reaction (it usually decreases). The reasons for this may be, for instance, impoverishment of the substrate, loss of activity of the enzyme during the reaction, inhibition of the enzyme by the reaction products and by allosteric feedback inhibition (control).

Effect on temperature. The action of enzymes is, in general, dependent on temperature. Each enzyme has an **optimum temperature**, for activity which usually lies between 30 and 50°C, as shown in Fig. 17. Very few enzymes, except those from thermophilic bacteria, have an optimum activity at above 50°C.

Above 60°C enzymes are more or less rapidly inactivated (denaturation of the enzyme protein). Therefore undesired enzymes in food can be rendered harmless by heat. **Heat inactivation** of enzymes is used in many types of food processing: blanching of vegetables, high frequency heating, pasteurisation, sterilisation, drying by heating.

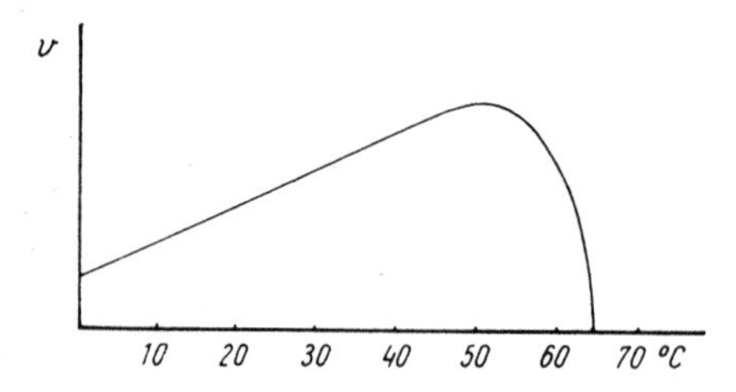

Fig. 17. Dependence of enzyme activity on temperature.

Inactivation of enzymes by heat depends to a large extent on the water content and pH of the food and on the presence of accompanying materials, (salts, sugar, acids and so forth). An enzyme may be more stable within the cell, that is in its natural environment, than when it has been purified. The food technologist must take into account that the degree and rate of inactivation is largely determined by the **humidity** or moisture content of the environment. Some esterases in carefully dried flour (0% H_2O) are stable up to 105°C, whereas these enzymes are inactivated at 50°C when the humidity of the flour is 25%. Research has shown that water absorption and swelling of food (substrates) has a decisive influence on the thermal resistance and activity of the enzyme. Effective inactivation of enzymes in foods and microorganisms as well as the destruction of microbes is best carried out in damp heat.

After the enzymes in food have been inactivated by heat, reactivation may occur, as sometimes happens with the phosphatase in milk and with peroxidase, in tinned peas and frozen vegetables. Although no enzymatic activity can be detected immediately after heat treatment, it can frequently be shown to occur later on. It is considered that the tertiary structure (see p. 50) of the enzyme protein, which was destroyed by the heat treatment, reforms itself to some extent with simultaneous regeneration of the enzymatic activity.

Below 50°C the Van't HOFF relationship for enzymatic reactions applies, according to which, when the temperature rises by 10°C the rate of reaction is increased to double or treble ($Q_{10}°C = 2$–3). With falling temperature the rate of the enzymatic reaction decreases in the reverse way. The effect of temperature on an equilibrium is quantitatively expressed by the Van't HOFF relationship

$$\frac{d \ln K}{dT} = \frac{\Delta H°}{RT^2}$$

This is useful in food preservation (p. 291); in cold and frozen storage the lowering of the temperature reduces the activity both of the native enzymes contained in foods and in microorganisms. At low temperatures, even down to −30°C, the activity of enzymes is greatly reduced, but it is often not completely stopped. The formation of ice in the cell leads to a withdrawal of the free water, which enzymes need for activity, similar to that in dried foods, (see drying). However, even at −20°C, only 70–90% of the water is frozen in most foods. Enzyme activity, even if it continues at a very low rate, sometimes leads to undesirable deterioration of the foods when they are stored for a long time at normal frozen storage temperatures (about −20°C). This may include discoloration, the formation of hay-like, liver-oily smell and taste because of an increase in the products of metabolism, as observed in cauliflower, peas, beans, peaches, mirabelle plums. Many varieties of strawberry develop a 'metallic' or freezer flavour. Other forms of deterioration include hydrolytic and oxidative decomposition of fat, bleaching or browning of green vegetables due to the destruction of chlorophyll (phaeophytin formation) and changes in the carotene as well as enzymatic oxidation of vitamin C and other substances. For this reason, the enzymes in vegetables are inactivated where appropriate before freezing, by heating as quickly as possible for a short but adequate time **(blanching)**, (p. 293).

In foods which are normally eaten raw, for instance certain types of fruit (apricots, strawberries, peaches, tomatoes), the blanching process cannot be used to inactivate the enzymes. One therefore tries to exclude the oxygen in the air – which is essential for the activity of oxidation enzymes (oxidases) – by freezing and storing some foods under vacuum or in an inert gas atmosphere such as nitrogen or CO_2. The oxygen in the cell interstices can also be driven out by adding sugar solutions or solid sugar which dissolves to some extent before freezing. The enzymes, because they are proteins, are affected osmotically and their activity is inhibited by the loss of water (dehydration effect of sugar on the enzyme protein), for instance in apricots, peaches, strawberries and cherries.

Effect of pH. Enzymes are ampholytes, like all proteins. The pH value of the solution therefore has a large influence on enzyme activity. The optimum pH of an enzyme is usually near its isoelectric point, adjusted to the conditions of its isoelectric point, adjusted to the conditions of its natural environment. Pepsin, for instance, has an optimum between pH 1.5 and 3, trypsin between pH 7 and 9. The pH optima of plant enzymes mostly lie between pH 4 and 6.5, those of the tissue enzymes of higher animals between pH 6.5 and 8.

Specific enzyme activators. Enzymes which belong to the metalloprotein group need the presence of certain metal ions. These are called activators because they can easily be removed from association with the enzyme but if they are added back the enzyme is reactivated. Take amylase, for instance, needs one cobalt ion per enzyme molecule for its enzyme activity to develop and arginase needs cobalt, manganese or iron ions. Others may require magnesium, zinc, copper or calcium ions. The effectiveness of these ions indicates the importance of trace elements in human nutrition, in animal nutrition, and in the application of fertilizers for plants, because without them, certain enzymatic reactions could not take place.

Many enzymes are only formed by the living cell as inactive precursors or so-called proenzymes. This process can be considered as a protective mechanism of the cell against autolysis. These proenzymes are only activated under certain conditions by specific activators, some of which are themselves enzymes. Pepsinogen is activated in the stomach by H^+ ions or by pepsin which is already present, and only then develops its proteolytic activity. The proenzyme trypsinogen produced by the pancreas is transformed in the gut into active trypsin by enterokinase (enteropeptidase) or by trypsin itself.

Enzyme inhibitors. There are three types of inhibition in enzyme chemistry, **competitive, non-competitive** and **allosteric**.

Competitive inhibitors are substances able to combine reversibly with an enzyme and so exclude the actual substrate. Because they are similar in structure to the substrate, they can attach themselves to the enzyme molecule, but are not themselves transformed. This means that the action of the enzyme on the actual substrate is inhibited to a greater or lesser degree according to the concentration of the inhibitor present, the key fits the lock, but can no longer be turned. For example, succinic acid dehydrogenase, which is closely connected with the respiration cycle, is competitively inhibited by substances structurally similar to succinic acid, such as malonic acid, oxalacetic acid, oxalic acid and so on.

Non-competitive inhibitors combine irreversibly with the enzyme and the enzyme activity decreases with increasing concentration of the inhibitor. This inactivation analogous to the poisoning of a catalyst cannot be reversed by increasing the concentration of the substrate. This means that enzymes with heavy metals as their active groups, can be activated by adding heavy metal complex formers such as citrates. In the food industry trypsin inhibitors in the seeds of Leguminosae must be treated before use. This inhibitor protein, which inactivates the digestive enzyme trypsin, is rendered harmless by heat, by sprouting of the seeds or by using microorganisms (mould cultures). Trypsin inhibitors have also been found in egg white and milk.

In **allosteric** inhibition the conformation of the enzyme protein is changed by reversible bonding of an inhibitor which sterically prevents bonding to the substrate or makes it more difficult.

8.5 CLASSIFICATION AND NOMENCLATURE OF ENZYMES

The nomenclature of the more than 700 enzymes so far known and described has been carried out from many different points of view; according to origin, substrate or reaction etc. The **International Union of Biochemistry** proposes in their Recommendations (1964) on the Nomenclature and Classification of Enzymes, that, for the systematic designation of an enzyme, the **total reaction** catalysed by it should serve as a basis for the nomenclature. The ending -ase may only be used for individual enzymes and not for enzyme systems. The nomenclature of the substrates follow the rules laid down by IUPAC. Well-known trivial names such as trypsin and so on, are retained.

At the same time the system evolved, makes it possible to classify each enzyme by four numbers. The first number allocates the enzyme to one of the **six main groups** into which all enzymes can be divided:

1. oxidoreductase
2. transferase
3. hydrolases
4. lyases
5. isomerases
6. ligases

Lyases are decomposing enzymes, which split off groups with formation of double bonds, or take up groups at double bonds; ligases are enzymes, which catalyse the combining of two molecules with simultaneous splitting of energy rich phosphate bonds (coupled reaction, synthesis, see p. 264).

Each main group is divided into several sub-groups. This subdivision depends on the chemical nature of the bonds, or on the grouping of the substrates which are attacked by the enzyme. This is indicated by the second figure.

Examples:

Hydrolases which split groups of esters are given the number 3.1.

Oxidoreductases, which attack CH(OH) groups 1.1. These sub-groups are further divided into sub-sub-groups, indicated by the third number. This division is also carried out on a chemical basis: for instance; oxidoreductases, which attack carbinol groups, with NAD or NADP as hydrogen acceptor: 1. 1. 1. with cytochrome as acceptor 1. 1. 2. with oxygen as acceptor 1. 1. 3.

Esterases which split carboxyl esters: 3.1.1. Esterases which split thiol esters: 3.1.2. and so on.

In the sub-subgroups all known enzymes are numbered in order. Example: Acetyl-CoA-hydrolase (a thiolesterhydrolase) 3.1.2.1.

The word '**desmolase**', which formerly described all those enzymes which are non-hydrolytic, is now no longer used. The enzymes which decompose are classified as **lyases**. If the equilibrium is completely on the synthesis side, the enzyme can also be called a 'synthase'. On the other hand, '**synthetases**' are enzymes which catalyse the bonding of two molecules with the help of ATP or similar nucleoside triphosphates, that is, by the splitting of an energy rich phosphate bond. The systematic name for this main group is **ligase**, in order to avoid confusion with the synthases mentioned above.

8.6 INDIVIDUAL ENZYMES

As this is a book for food chemists and food technologists, only those enzymes which are important for them will be treated here.

Oxidoreductases

A large number of enzymes catalyse both oxidation and reduction processes. The main event in redox reactions is always a transfer of electrons. A molecule which can be oxidised, an **electron donor**, giving up one or more electrons to an **electron acceptor**, which is thereby reduced (an oxidation is always coupled with a reduction). The tendency of a substance to take up or give up electrons is called its **redox potential**, measured in mV (millivolts). According to the reducing or oxidising effect of the substance relative to the normal hydrogen electrode, the redox potential will be **positive** or **negative**.

When two redox systems with different redox potentials are present together, the reaction which is thermodynamically possible is frequently hindered, that is, the rate of the reaction is very slow or practically non-existent. Ascorbic acid, for instance, can be kept in **pure water** for a long time in spite of the presence of O_2.

The reaction can be speeded up if a **reversible redox system** is introduced whose redox potential lies between the two systems and whose components (oxidised and reduced forms) react more readily with both the other systems. In the example quoted above, the oxidation of ascorbic acid is considerably increased by adding traces of certain heavy metal ions (cu^{2+} or Fe^{3+}) or ascorbic acid oxidase (see p. 254).

In this sense oxidoreductases can be defined as **redox catalysts** which link two redox systems (a, B) causing as it were, equal reactivity, so that the substrates are reduced by the electron or hydrogen donator and at the same time oxidised by the electron or hydrogen acceptor (see p. 253).

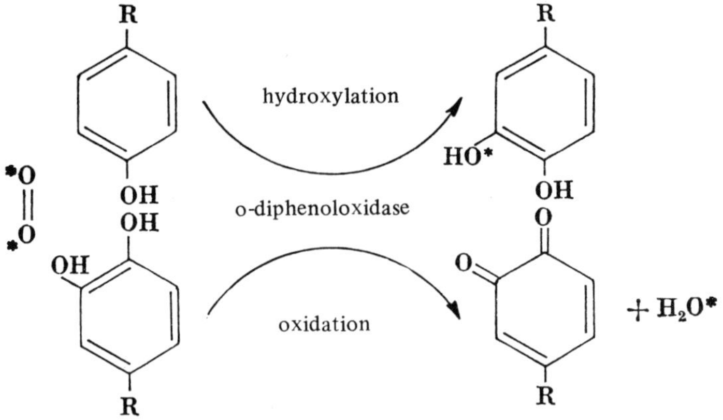

In nature, many coenzymes and other compounds act as electron (or hydrogen) acceptors, for instance NAD = (cozymase, coensyme I, DPN) NADP (c enzyme II, phosphocozymase, TPN), cytochromes, quinones, disulphides, oxygen, H_2O_2. A number of oxidoreductases show little or no specificity as far as the hydrogen acceptor is concerned. In these cases synthetic acceptors can be used, which simplify the proof and study of enzymatic reactions in vitro by producing characteristic colour changes, as for instance methylene blue, resazurin ('reductase test' in milk), Janus green, cresyl blue, pyocyanine, m-dinitrobenzene, tetrazol compounds.

Phenoloxidases. Many oxidases are important in food chemistry: the **diphenol oxireductases** (1.10.3) are widespread in plants, fruits and fungi, and affect a large number of compounds. Chemically they are copper proteins. They can be classified, according to their substrates, as **o-diphenol: oxygen oxidoreductases** (polyphenolases) and as **p-diphenol: oxygen oxidoreductases (laccases)**. In both reaction processes, oxygen from the air acts as hydrogen acceptor with formation of water. The reaction chain is: $O_2 \rightarrow$ enzyme $\rightarrow$ substrate.

o-Diphenol: oxygen oxidoreductases (pyrocatechin oxidase, polyphenolase) only oxidise o-diphenols. At the same time a hydroxylation of phenols can occur, (mixed function enzyme system).

Catechins, **anthocyanidins**, **chlorogenic** and **caffeic acids**, **flavonols** and so on can be used and are natural substrates important in food chemistry and technology. The o-quinones, which are formed during oxidation, are very unstable and polymerise in a reaction which is not yet understood, to further brown and black pigments. Examples are the **browning** which occurs on freshly cut surfaces of apples, potatoes, celery, asparagus and in freshly pressed fruit juices, such as apple juice and grape juice etc. The coloured products of the reaction are grouped together under the heading **phlobaphenes**. The formation of melanin (cephalopods, polyps) is similar: 3,4-dihydroxyphenylalanine (dopa) is formed primarily from tyrosine by hydroxylation and is then oxidised to the o-quinone which polymerises further to dark coloured pigments. These enzymes and reactions also play a role in the fermentation of tobacco, coffee, tea and cocoa and result in 'enzymatic browning'.

For **enzymatic browning**, the interaction of three components is necessary: oxygen – intact enzyme – substrate. If one of these factors is missing it does not occur. To prevent enzymatic browning infood processing, it is necessary to exclude oxygen by vacuum or inert gas, or to inactivate the enzyme by heat or chemically by sulphite (dried fruit, wine). The enzyme may also be checked by reducing the pH (putting peeled potatoes in a solution of vinegar and water, or fruits in 0.5% citric acid solution), so that the enzymes no longer work at their optimum pH and the speed of the reaction will be reduced (see p. 249).

The role of ascorbic acid in this system is of interest to the food chemist and technologist. The quinones formed enzymatically from polyphenols are coupled with ascorbic acid in a further oxidation to **dehydroascorbic acid** in a second purely chemical oxidation. The quinones are reduced to phenols and at the same time again serve as substrates for the phenoloxidases. When oxygen is present (as thermal H acceptor), ascorbic acid can be continuously oxidised in this system. On the other hand, if oxygen is excluded or there is a surplus of ascorbic acid, the formation of quinones and hence the resulting **enzymatic browning reaction** can be prevented.

Ascorbic acid oxidase. The enzyme **ascorbic acid oxidase** (L-ascorbate: oxygen oxidoreductase) is very similar to the diphenoloxidases, and it oxidises ascorbic acid to the dehydro form. Like the diphenoloxidases, this enzyme is a copper protein. The oxidation of ascorbic acid is usually irreversible, because the dehydro form is very unstable and is easily oxidised further to other products. **Browning reactions** in the juices of citrus fruits, especially in lemon and grapefruit juice concentrates, can be attributed to the fact that the dehydroascorbic acid is transformed to the open chain 2,3-diketo-L-gulonic acid when the lactone ring is broken. With the splitting off of CO_2 and water, this forms various decomposition products, which readily polymerise to brown resins

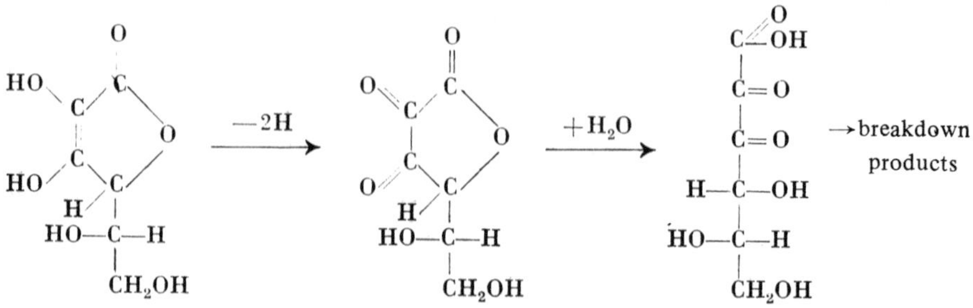

L-ascorbic acid $\xrightarrow{-2H}$ dehydroascorbic acid $\xrightarrow{+H_2O}$ 2,3-diketo-L-gulonic acid → breakdown products

Aldehydodehydrase. Another oxidoreductase, which is important in food analysis, is xanthine: oxygen oxidoreductase, also called **xanthine** or **aldehydoreductase**, or **Schardinger's enzyme**. It occurs mainly in liver, spleen and milk and catalyses the reaction xanthine + O_2 + H_2O → uric acid + H_2O_2. Purine derivatives are usually oxidised. Chemically, the enzyme is a flavoprotein containing molybdenum. The proof for its presence in milk is by adding formaldehyde as a substrate with methylene blue as H acceptor (Janus green, tetrazolium compounds, resazurin, neutral red etc. can also be used.) The colour is lost during the reaction (**Schardinger's reaction**). For this reason this test is suitable for use with raw or boiled milk.

Lipoxygenases (Lipoxidases). **Lipoxygenases** (linoleate: oxygen oxidoreductases) are enzymes which are found in many plants, especially in the soya bean. They catalyse the direct addition of oxygen to the cis, cis-1,4-pentadiene system $-CH=CH-CH_2-CH=CH-$, that is to isolated double bonds of polyunsaturated fatty acids such as linoleic, linolenic or arachidonic acid forming fat hydroperoxides. Carotenoids are bleached in a coupled reaction. Oleic acid is not affected. These oxidation processes play a part in the development of rancidity in fats, in the decomposition of carotene, in changes in flavour in frozen peas and so on. Enzymatic oxidation is a radical chain reaction and proceeds in a similar manner to fat oxidation.

Glucose oxidase. **Glucose oxidase** (β-D-glucose: oxygen oxidoreductase) is a flavoprotein catalysing the reaction

$$\beta\text{-D-glucose} + O_2 \xrightleftharpoons{\text{enzyme}} \text{D-glucono-}\delta\text{-lactone} + H_2O_2$$

The enzyme is used by itself in food technology or combined with catalase, which decomposes the H_2O_2 which is formed:

Foods which contain both sugar and protein have a tendency for undesirable browning (**Maillard reaction**) when they are processed, dried or stored. In foods which are rich in protein (e.g. meat, eggs) but contain little glucose, this can be prevented by removing the glucose by introducing glucose oxidase. (The gluconic acid-δ-lactone which is formed does **not** react in the Maillard reaction).

The glucose oxidase + catalase system can also be used to reduce the oxygen content of the headspace in bottled products like beer (see p. 256).

The oxidoreductases which use H_2O_2 as acceptor include catalase (hydrogen peroxide: hydrogen peroxide oxidoreductase) and peroxidase (donor: hydrogen peroxide oxidoreductase).

Peroxidase. **Peroxidase** a haem protein, is widely distributed in animal and vegetable matter, for instance in fruits, microorganisms and milk. Peroxidase sets oxygen free from H_2O_2 and from certain other peroxides, and transfers it directly onto organic substrates (AH_2) such as mono- and diphenols, aromatic amines etc.

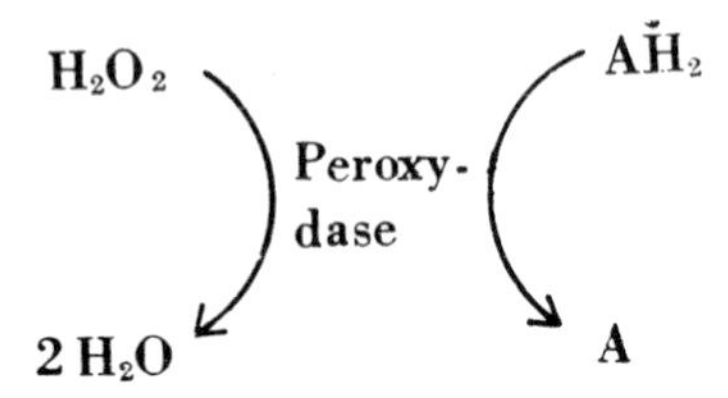

The oxidised compounds, for instance quinones (A), lead to other reactions yielding polymerisation products, which discolour and change the flavour of foods. It is therefore necessary to inactivate the peroxidase by heat treatment in foods before storage or processing, and special care must be taken to prevent a possible regeneration of this enzyme (see p. 297). Peroxidase is particularly **resistant to heat**, compared with other enzymes. This property is used in the **peroxidase test** for adequate enzyme inactivation in blanching vegetables before freezing or drying.

In food processing involving heat (blanching, sterilisation, drying), the elimination of peroxidase by adequate heating (negative peroxidase test) also guarantees the inactivation of other undesirable native enzymes (polyphenol oxidases, catalases, lipoxygenase, ascorbic acid oxidase, lipases) which could reduce the palatability of the food by colour changes and alteration in aroma and taste on further processing and storage. Proof of the presence of the enzyme can be easily demonstrated by the peroxide test in the presence of H_2O_2.

Catalase. Catalase is found almost exclusively in living cells. The reaction it catalyses,

$$2\,H_2O_2 \rightarrow 2\,H_2O + O_2$$

can take place extraordinarily rapidly. One molecule of the enzyme splits 4.2 10^4 molecules H_2O_2/second at 0°C. Catalase, like peroxidase, is a haem protein. It is distinguished from peroxidase by its greater substrate specificity. Hydrogen donators, other than H_2O_2 are only attacked very slowly.

Catalase preparations from animal organs, bacteria and fungi (usually in combination with glucose oxidase) are now used in the food industry to remove O_2 and H_2O_2 compounds (pp. 267 and 268).

Dehydrogenases. The dehydrogenases belong to the group of oxidoreductases and include NAD (nicotinamide adenine dinucleotide) which was originally known as coenzyme I and later as DPN (diphosphopyridine nucleotide). NADP (nicotinamide adenine dinuclotide phosphate) originally known as coenzyme II and later as TPN (triphosphopyridine nucleotide) also belong here.

Dehydrogenases of importance in food technology include the following:
(a) **Alcohol dehydrogenases** (alcohol: NAD oxidoreductase, ADH) which catalyse the reaction

$$\text{alcohol} + \text{NAD} \overset{\text{ADH}}{\rightleftharpoons} \text{aldehyde (or ketone)} + \text{reduced NAD}$$

Primary and secondary alcohols or semi-acetals can act as substrates. In this reaction, NAD, as hydrogen acceptor, is reduced to NADH. This enzyme is widely used to determine blood alcohol (ADH method), to determine the alcohol content of spirits, fruit juices and so on. The acetaldehyde, which is formed during the dehydrogenation of ethanol, is reacted with semicarbazide and the stoichiometric reaction is displaced to the right.

$$C_2H_5OH + \text{NAD} \rightarrow CH_3CHO + \text{reduced NAD}$$

The reduced NAD which is formed can easily be estimated spectrophotometrically at 340 nm as a direct measure for the amount of alcohol converted, (see the WARBURG optical test, p. 269).

Glucose-6-phosphate dehydrogenase (D-glucose-6-phosphate: NADP-oxidoreductase, G-6-P-OH), catalyses the following reaction:

$$\text{glucose-6-phosphate} + \text{NADP} \overset{\text{G-6-P-OH}}{\rightleftharpoons} \text{D-glucono-}\delta\text{-lacton-6-phosphate}$$

$$+ \text{reduced NADP}$$

This reaction, which proceeds stoichometrically, is used analytically for quantitative determination of glucose. Glucose is first phosphorylated with hexokinase (ATP: D-hexose-6-phosphate-dehydrogenase) and ATP to glucose-6-phosphate. This is then dehydrogenated using glucose-6-phosphate dehydrogenase and NADP as coenzyme. The resultant reduced NADP (formerly TPNH) is again spectrophotometrically determined in an optical test and is a direct measure of the glucose.

Transferases

Those transferases which catalyse the transfer of a residue R from one molecule to another according to the general formula

$$\text{AR} + \text{B} \overset{\text{enzyme}}{\rightleftharpoons} \text{BR} + \text{A}$$

are of fundamental importance to food chemistry and technology, in fermentation, fusel oil formation, development of aroma, ripening of fruit and in many other biological systems. The following Table gives a summary. Certain transferases have recently been introduced with great success in analytical food chemistry. Examples are the phosphotransferases for determining glucose, fructose and glycerol using the enzymes hexokinase and glycerokinase respectively.

Table 21. Summary of important transferase systems

	Enzyme	Example	Reaction catalysed
2.1	which transfer C_1 units methylhydroxy-methylformyl, carboxyl-	serine hydroxymethyl transferase	L-serine + tetrahydrofolate = glycine + 5.10-methylene tetrahydrofolate
2.3	which transfer acetyl residues	choline acetyl transferase	acetyl-CoA + choline = CoA + 0-acetyl choline
		acyl-CoA: acetyl-CoA-acyl transferase	acyl-CoA + acetyl-CoA = + 3-oxoacyl-CoA + CoA
2.4	which transfer glucosyl	UDP-glucose: D-fructose-2-glucosyl transferase	UDP-glucose + D-fructose = UDP + sucrose
2.6	which transfer nitrogen (NH_2-) groups	L-alanine: 2-oxoacidamino-transferase	L-alanine + 2-oxoacid = pyruvic acid + L-amino acid
2.7	which transfer phosphoric acid groups	hexokinase	ATP + D-hexose = ADP + D-hexose-6-phosphate
			a) ATP + D-glucose = ADP + D-glucose-6-phosphate b) ATP + D-fructose = ADP + D-fructose-6-phosphate
		adenosinekinase	ATP + adenosine = ADP + AMP (adenosine mono-phosphate)
		pyruvatekinase	ATP + pyruvic acid = ADP + phosphoenolpyruvic acid.
		Glycerokinase	ATP + glycerol = ADP + glycerol-1-phosphate

The **phosphotransferases** (2.7), usually still classified as kinases, are important for metabolism in the building and decomposition of vital compounds. They control the formation of the energy-rich phosphate compounds of many coenzymes, sugar phosphates and so on.

The **acyl transferases** (2.3) take part, for instance, in the citric acid cycle and in the synthesis and decomposition of fatty acids (see p. 93) and the formation of acetylcholine.

Aminotransferases (2.6 transaminases) are important in all the biological processes in the cell and in bacteria. The play a role in cheese ripening and in fusel oil formation by yeasts during alcoholic fermentation (see p. 257).

Phosphorylases belong to the glucosyl transferases. In the phosphorylation of glycogen, for instance, they transfer glucose residues to phosphoric acid with the formation of glucose-1-phosphate.

Hydrolases

Hydrolases are important in food chemistry and technology. They catalyse the hydrolytic cleavage of certain molecular bonds. They can be divided according to the types of bond they attack.

3.1 ester bonds
3.2 glycosidic bonds
3.3 ether bonds
3.4 peptide bonds
3.5 non-peptide C–N bonds
3.6 acid anhydride bonds
3.7 C–C bonds
3.8 halide bonds
3.9 P–N bonds

Esterhydrolases

The carboxylesterases are the most important in this group. They include the lipases, pectinesterases, chlorophyllases and tannases. **Chlorophyllase**, which splits the alcohol phytol from chlorophyll, is involved in the yellow colour of fruit ripening. **Tannases** are enzymes which split tannins, for example gallic acid esters.

Phosphoric acid ester hydrolases are also important for food chemistry, p. 260.

Lipases. **Lipases** from the pancreas, liver, gut and other animal organs hydrolyse glycerol esters (fats). Lipases are found in quantity in all plant cells, especially in seeds (*Ricinus*) or in the flesh of oil bearing plants (olive). In the small intestine (pancreatic lipases) they control the digestion of fats. Ricinus lipase, which was formerly used in the industrial production of fatty acids and soaps, is now no longer of importance. Specific or microbial lipases cause fats or fat preparations to become sour. The ageing of flour can also be traced back to the effect of lipases. They are frequently still active at very low temperatures down to –30°C, for instance, during frozen storage of fat tissue (animal) or when frozen vegetables are not adequately blanched.

Pectin esterases (pectase). Pectin esterases convert pectins into pectic acid with the demthylation of methyl esters to give the COOH groups. Pectin esterases are used in the production of low esterified pectins (compare gelling p. 189) and in clearing fruit juices.

Phosphoric acid ester hydrolases (phosphatases) are found in all animal tissues; in milk, eggs, sour milk, cheese, in yeasts and bacteria as well as in many plants (potatoes, grain). They are involved in the synthesis and breakdown of phosphoric acid esters. They can be classified as **monoester hydrolases** (3.1.3),

diester hydrolases (3.1.4) and **triphosphoric acid ester hydrolases** (3.1.5). The optimum pH for phosphoric acid hydrolase activity varies widely according to the origin of the enzyme: they are therefore frequently called **acid, neutral** or **basic 'phosphatases'**. Phosphoric acid monoester hydrolases of yeast and of barley are involved in alcoholic fermentation. The test for the adequate heating of milk, based on the milk's natural phosphoric acid ester hydrolases is used analytically. These enzymes are normally inactivated by **pasteurisation**. However, in certain circumstances, **regeneration** may occur, as with peroxidase. When flour ages, phosphoric acid ester hydrolases may be involved.

A phosphatase, which splits phytin, is called **phytase**. It hydrolyses the inositol phosphoric acid ester bonds. Phytase is found in cereals; rice and wheat bran are rich in it, as are green and cured malt.

Those enzymes which hydrolyse phosphoric acid anhydride compounds into pyrophosphate, trimetaphosphate and polymetaphosphate, for example, were formerly also called 'phosphatases' but do not in fact, belong to this group of phosphoric acid esterases. They belong with the enzymes which attack acid anhydride bonds and are distributed in animal and plant tissue as well as in fungi and yeasts. In tests for condensed phosphates it is necessary to look out for such enzymes, because these phosphates are easily hydrolysed in the presence of the enzymes and so are withdrawn from the test.

Phospholipases. Enzymes which split phosphatides are collectively known as phospholipases. Phospholipase A and B are esterhydrolases; they split off one fatty acid residual from the phosphatide to form a lysophosphatide. Phospholipases C and D are phosphatases; they break the ester bond between phosphoric acid and the glycerol portion (C) and between phosphoric acid and the amino alcohol type (D). The following formula for lecithin shows the position of attack of the various phospholipases:

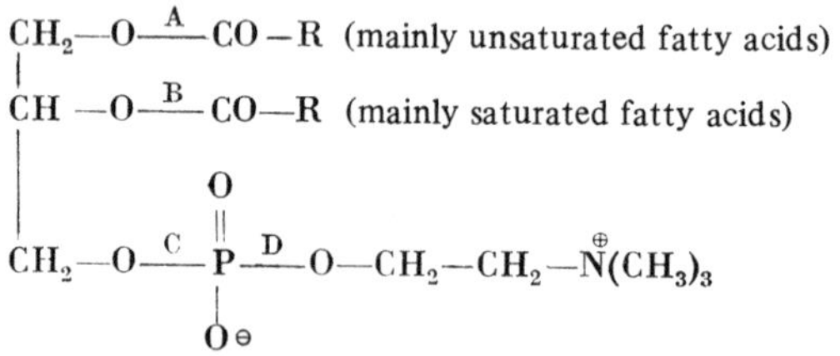

$$\begin{array}{l}
CH_2-O\overset{A}{\text{———}}CO-R \quad \text{(mainly unsaturated fatty acids)} \\
| \\
CH\ -O\overset{B}{\text{———}}CO-R \quad \text{(mainly saturated fatty acids)} \\
| \qquad\qquad\quad O \\
| \qquad\qquad\quad \| \\
CH_2-O\overset{C}{\text{———}}P\overset{D}{\text{———}}O-CH_2-CH_2-\overset{\oplus}{N}(CH_3)_3 \\
\qquad\qquad\qquad | \\
\qquad\qquad\qquad O^{\ominus}
\end{array}$$

Phospholipases of higher plants and moulds are important in food chemistry. Phospholipases can cause decomposition in egg pasta.

Glycosidohydrolases (Glucosidases).

They are usually only group specific. Only **one** of the components of the glycosidic bond determines the attack of the enzyme.

Amylases. α-1,4-glucan-4-glucohydrolase **(endo-** or **α-amulase)** splits α-1,4-glucosidic bonds **inside** the starch molecule, into glycogen and other glucans containing three or more α-1,4-glucoside bonds in series. α-1,4-glucan maltohydrase **(exo- or β-amylase)** splits off β-maltose from α-1,4-glucans at the non-reducing end. α-1,4- and α-1,6-glucan glucohydrolase only split off complete glucose units from the non-reducing end of poly- and oligosaccharides **(gluco-amylase, amyloglucosidase** or γ-amylase fungal and/or bacterial exoamylase).

The enzyme preparations formerly called **diastases** are mostly a mixture of α- and β-amylase. Both are almost always found together in the animal and vegetable kingdom, where starch and glycogen are being built up or decomposed. In the animal kingdom, and in microorganisms, α-amylase is the more abundant (for instance in the saliva, pancreas and intestines).

In **food technology**, the amylases of potatoes and sprouting seeds, especially barley, are important. The amylases of malt (green and kiln-dried) cause saccharification of the mash containing starch, or of the starch itself during the production of alcoholic drinks (beer, spirits). They are also involved in the production of ethyl alcohol, lactic acid, butyric acid, malt extract and dextrose.

Amylases from microorganisms (*Aspergillus oryzae* etc.) are used for the saccharification of rice starch in the manufacture of Japanese rice wine. In some European distilleries, amylases from moulds are also used for the saccharification of the mash. Amylases, in the form of baking powders, are used in the baking industry for dough raising. They are also used in the production of sugars and dietetic foods. Amylases, with other enzymes, help to decompose carbohydrates in the complicated process of tobacco fermentation.

Cellulases. The enzyme **β-1,4-glucan-4-glucanohydrolase** (cellulase) can break down the polysaccharide cellulose to the disaccharide cellobiose. For instance, when vegetables are stored, changes in the structure are caused by the decomposition of cellulose. Vertebrates, including man cannot break down cellulose in the digestive process but the digestion of cellulose in the rumen of ruminants is effected by the cellulases of the microorganisms present there. Cellulase preparations can be used therapeutically for disorders of the intestinal tract.

Pectinases. **Poly-α-1,4-galacturonide-glycanohydrolase** (polygalacturonase). The macromolecular pectins are split by special glycosidohydrolases (formerly called carbohydrolases), polygalacturonases, into smaller units or sometimes even into their basic building blocks (galacturonic acid). This causes reduction in viscosity and generally also in gelling properties, which are dependent on the size of the molecule. Enzymatic decomposition of pectin occurs when vegetables and fruit become soft, in the fermentation of tobacco and coffee and in the steeping of flax and hemp. The polygalacturonases of bacteria also help to decompose the pectine consumed in food in the large intestine.

Polygalacturonases are produced on a commercial scale from cultures of strains of Aspergillus and Penicillium on a nutrient base containing pectin and are sold under the trade names of Filtragol, Pectinex, Pectinol, Vinibon etc. They are used to clarify the pressed juice in the fruit juice industry and also to achieve partial decomposition of molecules in the manufacture of pectin (p. 188).

Glycoside hydrolases which split oligo- and simple saccharides. In addition to the glycosidohydrolases which attack polysaccharides, there are other enzymes which only attack oligosaccharides or simple glycosides. The **α-D-glucoside glucohydrolase maltase** (α-glucosidase) from yeasts, moulds, bacteria or malt wort, splits α-D-glucosidic bonds as in maltose, sucrose and melezitose.

β-D-glucoside glucohydrolase (β-glucosidase) attacks β-D-glucosidic bonds. A β-glucosidase is also present in emulsin isolated from bitter almonds by LIEBIG and WOHLER. β-glucosidases are involved in the formation of aromatic materials such as vanillin and mandelic acid nitrile (in bitter almonds) from the native β-glucosides in the fruit. Bitter materials are also present in spirits distilled from the Gentianaceae and are formed from glycosides by β-glucosidases. See the section on glycosides p. 157.

β-fructofuranosidase (β-D-fructofuranoside fructohydrolase) acts on bonds in which fructose in the furanoside form) is present in the β-glycosidic bonding, for instance in raw sugar. This enzyme (also called **saccharose, invertase**) is found in yeast and is used industrially in yeast preparations and in the manufacture of artifical honey, where sucrose is split enzymatically into invert sugar (a mixture of glucose and fructose). Invertase is used in the confectionery industry. It prevents crystallisation and so keeps the matrix of marzipan and fondants soft.

In the manufacture of mustard condiments, a **β-thioglucosidase** (thioglucoside-glucohydrolase, myrosinase) splits the mustard oil from the mustard glycoside during the grinding and mashing of the mustard seeds; see p. 160.

Peptide hydrolases (proteases, peptidases)
They catalyse the hydrolytic splitting on the peptide bonds (–CO–NH–) in proteins and peptides.

They are divided into exopeptidases and endopeptidases according to their function.

Exopeptidases only split off the amino acid at the end of the peptide chain while the **carboxy peptidases** only attack from the carboxyl end. The **amino peptidases** split the amino acid with a free NH_2 group at the end of the chain.

Endopepsidases only split at certain points in the middle of the peptide chain, that is they split proteins and higher polypeptides into smaller fragments. They are also called **proteinases**. The well know digestive enzymes **pepsin, rennin** (now called **chymosin**), **rennet, trypsin, chymotrypsin, enterokinase** and **cathepsin C** belong to this group, as well as the vegetable pepsidases **papain, bromelain** and **ficin**. The endopeptidases are usually only group specific. The structure of only one of the two amino acids which are linked by a peptide bond is in many cases decisive for the attack. **Pepsin** tends to attack peptide bonds adjacent to aromatic or dicarboxy-L-amino acids, and **trypsin** those in which the carboxyl group of L-lysine or L-arginine is involved.

Pepsin is the enzyme in the stomach which splits protein (proteolytic), and which decomposes all genuine proteins. It is present in the mucous membrane cells of the stomach as pepsinogen, and is only converted into active pepsin by the acid gastric juices (H^+ ions), see p. 249.

Cathepsins (organ proteases) of the mucous membrane of the stomach and the stomach secretions has an optimum pH of 4–5. They are also involved in the autolytic decomposition of the tissues (ripening and putrefaction of meat), because they are tissue proteases.

Trypsin and **chymotrypsin** from the pancreatic juices are the main enzymes which split protein in the small intestine (pH 7).

Rennin, present in calves' stomach, is included in the proteolytic enzymes, although its proteolytic action is less important than its coagulation action. It induces coagulation in the form of a calcium salt of casein, the typical protein of milk, with a characteristic serine phosphoric acid content. Rennin enzyme (rennet) from calves' stomach is important in the manufacture of cheese.

Papain is the most important protease in lower and higher plants. It is present in the sap and fruits of various plants, for instance in *Carica papaya*, as well as bacteria and yeasts. It splits proteins and peptides at the isoelectric point. A number of trade products from *Carica papaya* are used as so-called **'tenderisers'** to make old or insufficiently hung meat tender.

Bromelain is made from the juice of pineapples and **ficin** from the juice of figs.

The enzymes which split peptides are of interest to the food chemist. Papainase attacks proteins during malting and mashing. Important characteristics of beer, such as foaming, foam retention and body, depend to a large extent on proteolytic decomposition products.

When dough is prepared, the peptide hydrolases affect the constitution of the protein gluten in the flour. A gluten which is intrinsically soft, as are most European flours, may be broken down to such an extent that the swelling power of the flour is reduced and the dough no longer has any strength. A hard gluten, like that usually found in American and Canadian flours, actually needs to be partly broken down by peptide hydrolases. The strength of the dough depends on the degree of decomposition of the gluten.

The hanging (maturing) of meat also depends on the action of peptide hydrolases, which partially decompose the hard muscle proteins which had swollen during rigor mortis due to lactic acid. This makes the meat fit to eat (p. 266).

Peptide hydrolase activity is also very desirable in the fermentation of tobacco.

Peptide hydrolases in yeasts are also important for the synthesis of protein, when yeast cells are growing and dividing. They are the most effective enzymes for this synthesis and manufacture of compressed yeast and for biological protein synthesis.

Lyases

The **carboxylases** (4.1.1) are the most important lyases. These enzymes mostly have high specificity and catalyse the decarboxylation of carboxylic and amino-carboxylic acids. They are found widely distributed in animal tissue and micro-organisms. The highly specific carboxylases, those obtained from microorganisms especially, are used for the enzymatic determination of a number of amino acids.

2-keto-acid carboxylase (pyruvate decarboxylase) catalyses an important reaction in the decomposition of carbohydrates (e.g. in alcoholic fermentation). In this case thiamine pyrophosphate (vitamin B_1 pyrophosphate cocarboxylase) and Mg^{++} ions acts as co-factors:

$$CH_3-\underset{\underset{O}{\|}}{C}-COOH \xrightarrow{\text{carboxylase}} CH_3CHO + CO_2$$

Aldolase is another lyase involved in alcoholic fermentation. It splits fructofuranose-1,6-diphosphate anaerobically into glycerolaldehyde phosphoric acid (phosphoglycerol aldehyde) and dihydroxyacetone phosphate (phosphodihydroxyacetone).

Isomerases

Isomerases are enzymes, which catalyse **intramolecular** rearrangements such as recemisation and epimerisation, cis-trans-rearrangements, intramolecular redox reactions, intramolecular transfer reactions and intramolecular ring cleavage.

One of the most important isomerases in the decomposition of sugars is **D-glycerol-aldehyde-3-phosphate-ketoisomerase** (phosphotriose-isomerase, triosephosphateisomerase).

This enzyme has a particularly high exchange number ($>10^5$ mol/min). This is important because in the equilibrium glycerolaldehyde phosphate/dihydroxy acetone phosphate, glycerol aldehyde phosphate is only present at 4%. If the glycerol aldehyde phosphate is used up in a consecutive reaction, more can be made available quickly with this enzyme.

Racemases are often found in bacteria. They catalyse the racemisation of natural L-amino acids. Their presence has also been demonstrated in varieties of Lactobacillus. They are responsible for the presence of D-amino acids (D-alanine, D-glutamic acid) in ripening cheese. Some of the racemases contain pyridoxal phosphate as coenzyme (see p. 238).

Ligases

This group of enzymes, which play an important role in all endothermic metabolic processes in the vegetable and animal kingdom, catalyse the linkage of two molecules A and B, and this process is coupled with the splitting of a pyrophosphate bond in ATP (adenosine triphosphate) and so on. ATP is by far the commonest supplier of energy for enzymes of this group. Depending on the enzyme, ATP is split into adenosine monophosphate (AMP) + pyrophosphate or into adenosine diphosphate (ADP) and orthophosphate; see p. 62. The reaction follows the following pattern:

$$\text{A} + \text{B} + \text{ATP} \rightarrow \text{A–B} + \text{ADP} + \text{orthophosphate or}$$

$$\text{A} + \text{B} + \text{ATP} \rightarrow \text{A–B} + \text{AMP} + \text{pyrophosphate}$$

Acyl-coA synthetases, peptide synthetases, amide synthetases and others belong to this group.

8.7 ENZYMES IN FOOD TECHNOLOGY

Various enzymatic processes in food technology, for example the manufacture of milk products, have already been mentioned when the individual enzymes were discussed. This section is concerned particularly with the use of **enzyme preparations** in the commercial processes used in food technology – a field which is becoming increasingly important in the food industry because enzymes have the following advantages:

As proteins and therefore physiologically harmless, they are active at low temperatures and at almost neutral pH conditions. They act specifically and quickly and at relatively low concentrations and are easy to inactivate.

The enzymes used in the **food industry** are obtained from vegetable sources (especially seeds), from animal tissues (especially stomach mucosa and pancreas), but increasingly from specially cultivated microorganisms (fungus, bacteria, moulds, yeasts). The separation and subsequent isolation of enzymes is usually carried out from aqueous extracts in which there are generally a number of enzymes present. The individual enzymes are separated from these extracts although many of the extracts may be used as such. After purification procedures (by centrifugation or filtration) and careful concentration under vacuum the enzyme preparation can be obtained in a solid form by precipitation and vacuum drying. These dry products are already on the market, often ready standardised (by dilution with salt, sugar, flour etc.) according to effectiveness and activity. They keep better than the solutions of enzyme preparations which are also obtainable.

Among the **oxidoreductases**, only catalase and glucose oxidase are used industrially. There are now several **hydrolase** preparations in use – esterhydrolases (lipases, pectases), glycoside hydrolases (amylases, pectinase, maltase, saccharase, lactase, naringinase) and peptide hydrolases (rennin, pancreatin, papain, bromelain, ficin).

Enzyme preparations from plants. **Malt amylases** are the vegetable enzyme-preparations most commonly used for industrial purposes. Generally the malt extract is enriched with enzymes and also contains other glycoside hydrolases as well as phosphatases and proteases. Malt amylases are used in the baking industry, as a raising agent to supplement the flour's own enzymes; they are also used in brewing of beer and fermentation for spirits. Malt syrup and dry products containing dextrin (baby foods) are prepared by malt amylases also.

Vegetable **protease preparations** (peptide hydrolases) include **papain** from the latex of the unripe papaya, **bromelain** from the pressed out juice of pine-apples, and **ficin** from the juice of figs. Vegetable peptide hydrolases are used in the USA in brewing, to make the beer resistant to cold (**chill proofing**), so that it does not become cloudy due to protein-tannin complexes. They are used for clarification and maturing of malt drinks. Peptide hydrolase preparations are also used as **tenderisers** for meats, but the same ones are not used for malt as for meat.

The **pectin esterase preparations**, obtained from lucerne, tomatoes or citrus fruits, have industrial importance and can be used in wine and fruit juice manufacture.

Lipase preparations from the castor bean were formerly important for splitting fats. Now they are used to a small extent as aroma improvers (due to fatty acid formation) when milk low in lipases is used for the manufacture of certain types of cheese. Lipases are also used in enrobing procedures for confectionery.

A new field for the possible industrial use of vegetable enzyme preparations is as **aroma enhancers** or in the restoration of aroma in processed food. Aroma materials in fruit and vegetables are set free by the enzymes from the so-called aroma precursors. During heat treatment (sterilisation) these enzymes are destroyed, and some of the volatile aroma materials are also lost, while other components of the remaining aroma materials are changed. The precursors are preserved to a large extent so that when enzymes are added (from fresh vegetable material) the fresh aroma components are present again in the food.

Enzyme Preparations used in the Food Industry

Enzyme	Mode of action	Use
lipoxygenase (lipoxydase)	catalyses the oxidation of polyunsaturated fatty acids, carotene can be oxidised (bleached) by peroxide formation in a secondary reaction	bleaching of bread flour (wheat)

glucose oxidase (in a coupled reaction with catalase)	oxidises specifically β-D-glucose in the presence of oxygen to gluconic acid	removal of glucose, e.g. stabilising dried egg; removal of oxygen in small packets, tins and bottles to stabilise aroma.
catalase	decomposes hydrogen peroxide into water and oxygen	in preserving food, especially in the hydrogen peroxide sterilisation of milk.
lipases	hydrolyse fats	improving aroma in cheese and chocolate manufacture; improvement of quality of dried egg powder
pectase and pectinase	hydrolyse pectins: splitting off of methyl groups and decomposition of polygalacturonic acid	preparation of wine and fruit juices; removal of gelatinous skin of coffee beans
amylases in general	hydrolyse starch to dextrin and sugars in varying degrees	brewing; baking (as improvers); distilling, sugar goods, fruit juice and soft drinks industry as aids to clarification and extraction. Preparation of dietetic foods.
amyloglucosidase	hydrolyses starch to glucose	starch (mainly cornflour) industry
cellulase	hydrolyses cellulose	brewing; fruit juices and drink industry as aids to clarification and extraction. Softening and improving flavour of vegetables.
hemicellulase	decomposes various natural gum materials	prevention of gelatinisation of coffee concentrates.
maltase, lactase, invertase	hydrolyse diglucosides to the corresponding hexoses	confectionery industry, prevention of crystallisation of syrup, ice cream, artificial honey.
protease (peptide hydrolases)	hydrolyse proteins	baking (altering baking properties of wheat gluten), milk industry (e.g. prevention of oxidation flavour), cheese making (with rennet). Preparation of flavourings, special diets, cold stabilisation of beer, softening of meat (tenderizers).

Enzyme preparations from animal tissues. Enzymes such as pepsin and chymosin (rennin) are obtained on an industrial scale from animal glands and other tissues. The pig's pancreatic gland is used as a source of trypsin and chymotrypsin. The enzyme complex '**pancreatin**' is used for the production of gelatine, protein hydrolysates, peptones and also for pharmaceutical products. **Chymosin** (rennin) the chief enzyme in rennet. is extracted from the abomasum of suckling calves and is particularly important in food technology because of its specific use in cheesemaking.

Catalase preparations can be obtained from liver or erythrocytes. Used with glucoseoxidase, they are useful in removing O_2 and H_2O_2. (See p. 255 and below).

Enzyme preparations from microorganisms. Microorganisms, especially moulds, bacteria and yeasts, provide the best source of enzymes for industrial production. They are cultivated on a commercial scale under carefully controlled conditions, to obtain the highest enzyme yields possible.

Amylase preparations can be obtained from **fungal cultures** (surface and submerged cultures), especially from strains of *Aspergillus oryzae* and *Aspergillus niger.* These are used in baking, brewing, in the starch hydrolysis process for syrup and dextrin, in the confectionery industry, e.g. for chocolate syrup, liqueurs and sweets.

Protease preparations are also obtained from these fungi and are used in baking for 'chill proofing agents' in beer and fruit juices (see p. 267), and also in digestive enzyme preparations.

In the manufacture of fruit juices and wine, the so-called **'filtration enzymes'** from strains of *Aspergillus, Penicillium and Rhizopus* are widely used. These preparations contain mainly enzymes which break down pectin (pectinases and pectases) and their use leads to a higher yield and clarification of the juice. Pectolytic enzyme preparations are used in making fruit juice concentrates and fruit jellies. Recently these enzyme preparations have also been recommended for the removal of the coating on the coffee bean after fermentation.

Hemicelluloses from fungi prevent gelatinisation in coffee concentrates by acting on the gum-like polysaccharides.

Enriched glucose oxidase preparations can also be obtained from cultures of *Aspergillus niger* and *Penicillium notatum.* These enzyme preparations can be used either to remove glucose, or to bind oxygen in packaged foods. The H_2O_2 formed is removed by using catalase with the glucose oxidase (see pp. 255 and 256).

Glucose oxidation makes it possible to prevent discolouration due to the MAILLARD reaction in the production of dried egg. Removal of oxygen is already carried out in the USA in manufacture of fruit juice drinks, beer, mayonnaise and in some dry packs where oxidation reactions if not prevented would lower the nutritional value.

Naringinase, a **glycoside hydrolase**, obtained from *Aspergillus niger* can be used to remove the bitter flavour in grapefruit juice due to the naringin 7-rutinosideoglycoside of naringenin (see p. 232). Glycosidases from fungal enzymes can split superfluous anthrocyanins in fruit wines and jellies.

Amylase and **protease preparations** can also be obtained from **bacterial cultures**, usually from *Bacillus subtilis.*

Certain saccharase and lactase preparations are prepared from **yeasts** specially for use in food technology. **Saccharase** (invertase) is used in making invert sugar and artificial honey. The confectionery industry uses saccharase preparations to keep praliné fillings semi-soft (invert sugar is more soluble than sucrose). **Lactase** can prevent the crystallisation out of the milk sugar (sand sugar) in condensed milk and ice cream powder. This is achieved by the splitting of the lactose into glucose and galactose.

8.8 ENZYMES IN FOOD ANALYSIS

Enzymes are used increasingly in food analysis, because the procedures are highly specific, easy and quick to carry out.

In modern enzymatic analysis – and also in food technology – light absorption measurement is important because it is simple to carry out in cuvettes. WARBURG had already used such 'optical tests' in biochemistry. These procedures are based on the action of specific dehydrogenases, such as glucose-6-phsophate dehydrogenase, lactic dehydrogenase, alcohol dehydrogenase (ADH) among others. The specific light absorption of NADH (DPNH) at 340 nm is measured and any change is stochiometric.

In general enzymatic analyses can be divided into two groups:

Determination of enzyme activity. Testing for characteristic enzymes has long been carried out in food analysis. The **concentration** of an enzyme is determined from the speed of reaction of the enzyme under study with a surplus of suitable substrate. Enzymes serve as an indication of the state of a food; its previous treatment (sterilisation, pasteurisation, blanching), its degree of freshness, whether it has been affected by microorganisms (spoilage). In addition, enzyme systems, like other components of food, are affected by factors in the environment when the food was produced. These factors include inhibitors, preservatives, antibiotics, plant protection agents, pesticides, time of year and place of origin.

Proof for the presence of the minutest concentrations of specific enzyme inhibitors in foods can be obtained by using certain enzymes in an indicator reaction. Effector analyses are procedures in which the enzymatic reactions are influenced by inhibitors or activators (effectors). The degree of inhibition or activation permits conclusions to be drawn about the concentration of these materials using a calibration curve. Effector analysis is useful for proving the presence of certain materials used for protection against pests and as preservatives. Organic phosphoric acid esters, which have an insecticidal action (e.g. Systox or Parathion) and certain transformation products of cholinesterase specially prepared for analysis have a very strong inhibiting action. DDT can be determined quantitatively with the inhibition of carboanhydrase. The determination of fluoride in water, milk and so on, is based on the sensitive inhibition of esterases by this halogen ion. The following is a list of some enzyme activity determinations in common use.

Determination of enzyme activity

enzyme	foodstuff
diphenol oxidases	cereals, flour, milk, vegetables
xanthine-oxygen-oxidoreductase	milk
lipoxygenases (lipoxydases)	soya beans, flour
peroxidases	cereals, flour, milk, vegetables
catalase	milk and milk products
lipases	milk, milk products, flour cereals
phosphoric acid esterases	milk and milk products
amylases	honey, flour, malt, milk, bread, starch
urease	soya flour and soya products
creatinase	meat extracts, soups

Enzymatic determination of individual compounds. The following table gives those compounds which are used in food analysis together with the substrates in so far as these have been worked out. An excess of specific enzymes – or in the case of optical methods also the corresponding transfer substances (coenzyme, ATP, NAD, NADP) – must always be present.

Table 22 Some compounds which can be determined enzymatically

Group	Individual representative
alcohols	thanol, glycerol
aldehydes	acetaldehyde, glycol, aldehyde
acids or their salts	acetate, lactate, formate, malate succinate, citrate, isocitrate, pyruvate
monosaccharides and similar compounds	glucose, fructose; galactose, pentoses sorbitol, inositol
di- and oligosaccharides	sucrose, lactose, raffinose, maltose
polysaccharides	starch, cellulose, hemicelluloses
L-amino acids	glutamic acid, asparaginic acid
lipids	cholesterol

Using the simple **light absorption method** (optical test), which has already been mentioned, compounds such as ethyl alcohol, acetyl alcohol, lactate, malate, glutamate and so on, can be determined in an elegant manner in fruit juices, cider and wine, frequently without any preliminary preparation of the sample.

Part 3

Food and Man

CHAPTER 9

Nutrient Requirements

Modern studies in nutritional physiology have made it clear that the human diet may be made up from a great variety of foods, but the **nutrients** (see p. 279) contained in this food must be present in the **right** (useable) form and in calorifically adequate amounts. Today about 50 substances are known of which man needs a regular supply. These include essential amino acids, essential fatty acids, minerals, vitamins and water. Apart from mother's milk there is no food specially designed by nature as food for man, so that it is not surprising that no single food on its own satisfies all his material needs. Modern nutrition demands that the diet be varied enough for the individual foods to complement each other in a harmonious combination of nutrients which corresponds to the body's needs.

Once it is known which materials or groups of materials need to be present in the diet, the amount of each nutrient, man needs must be considered. The general answer might be given as follows: man must consume an amount of each nutrient which will keep his body in a state of normal wellbeing.

Nutrient requirement cannot be a fixed value. It depends on age, body weight, sex and type of work. Environmental conditions, for instance, the various climatic zones, and the composition of the diet also affect nutrient requirements. The USA Food and Nutrition Board and the Department of Health and Social Security in the United Kingdom, have carried out far reaching researching into food and nutrient requirements in recent decades. The main points for an optimal diet have been summarised and are now recommended in many countries as a norm.

9.1 ENERGY REQUIREMENT OF MAN AND THE ENERGY VALUE OF FOOD

During the nineteenth century, M. RUBNER demonstrated that the main nutrients, carbohydrates, fats and proteins could replace each other in the body depending on their physiological heat of combustion, (law of isodynamics). However proteins cannot be replaced by fats and carbohydrates in growth and maintenance metabolism as these compounds do not contain nitrogen.

Table 23 Recommended Daily Amounts of Food Energy for Population Groups in the United Kingdom. (Figures abstracted from the DHSS Report 1979).

Age Years	Occupational category	Energy MJ	Energy Kcal	Protein g
BOYS				
1		5.0	1200	30
2		5.75	1400	35
3-4		6.5	1560	39
12-14		11.0	2640	66
GIRLS				
1		4.5	1100	27
2		5.5	1300	32
3-4		6.25	1500	37
12-14		9.0	2150	53
MEN				
18-34	Sedentary	10.5	2510	63
	Moderately active	12.0	2900	72
	Very active	14.0	3350	84
35-64	Sedentary	10.0	2400	60
	Moderately active	11.5	2750	69
	Very active	14.0	3350	84
65-74	Assuming a sedentary life	10.0	2400	60
75+		9.0	2150	54
WOMEN				
18-54	Most occupations	9.0	2150	54
	Very active	10.5	2500	62
55-74	Assuming a sedentary life	8.0	1900	47
75+		7.0	1680	42
Pregnancy		10.0	2400	60
Lactation		11.5	2750	69

Notes:

(a) Since the recomendations are average amounts, the figures for each age range represent the amounts recommended at the middle of the range. Within each age range, younger children will need less and older children more, than the amount recommended.

(b) Megajoules (10^6 joules). Calculated from the relation 1 kilocalorie = 4.184 kilojoules.

(c) Recommended amounts of protein have been calculated as 10% of the recommendations for energy.

Therefore if one wishes to evaluate food according to its importance as a fuel (and not as building material), one gives its energy value in joules or calories. The following values, in round figures, have been calculated for the physiological energy value of the main nutrients

carbohydrates	3.75 kcal/g	16 kJ/g
proteins	4.0 kcal/g	17 kJ/g
fats	9.0 kcal/g	37 kJ/g

Using this relationship between nutrient and energy content, the necessary energy in kJ of kcal for each type of physical activity can be expressed. The total requirement for an adult lies around 2750 kcal or 11.5 MJ with moderate physical activity. The requirements of the body for the so-called main nutrients (protein, fat and carbohydrate) rise with increased physical activity.

The example of energy transformation (consumption of energy) of 2750 kcal or 11.5 MJ includes both the so-called basal metabolism and the activity metabolism. By basal metabolism is meant the energy used when lying completely still, 12 hours after last taking food. The basal metabolism varies according to sex, height, weight and age. The average is around 1500 kcal per day. Total metabolism is the sum of the basal and activity metabolism.

9.2 PROTEIN REQUIREMENT

As has already been stated (see p. 229), protein has the highest nutrient value for man if it contains the **essential** amino acids which are needed in the right amounts and in the right relationship. In general this is more often the case in animal than in vegetable protein. This is why animal protein has a higher biological value (see p. 53). Therefore animal protein should be present in adequate amounts in the diet.

The **daily protein requirement** for adults lies around 1 g/kg of body weight, that is, about 70 g per day. One third to one half of the protein consumed daily should be of animal origin. It is desirable that those who do heavy physical work should have a higher protein intake. Adequate animal protein is also important for those who do mainly mental work. Protein requirements are increased in the sick, in the convalescent, in active sportsmen, in young people and nursing mothers.

Fish, meat, milk, cheese and eggs are sources of animal protein. The soya bean is an especially valuable source of vegetable protein.

The concept 'minimum physiological protein' or minimum protein balance means the minimum amount of protein food with which the body reaches a nitrogen equilibrium between intake of nitrogen in foods containing protein and elimination of nitrogen in decomposition products (as urea in urine). It is well known from experience with mass feeding in times of crisis and from animal experiments that, even with otherwise adequate energy intake, a protein intake at or near the minimum balance cannot in the long run be reconciled with full performance, efficiency and health and resistance to infection. Even relatively light work loads or banal infections can lead to bodily breakdown in these circumstances. Protein deficiency or protein malnutrition can be observed in some parts of the world, for instance in India, South America and in tropical Africa. The syndrome which arises there through protein deficiency is called kwashiorkor. It exhibits itself in skin damage, swelling of limbs due to collection of fluid and frequently in serious liver diseases such as fatty degeneration of the liver and cirrhosis (shrinking of the liver). Mortality with this protein deficiency disease is very high. However, it is remarkable that the damage caused by kwashiorkor can be put right in a few weeks with a correct protein diet, for instance by providing milk powder.

In contrast to the fatty degeneration caused by too high an energy intake, a protein intake above normal level has no adverse effects on health and does not affect the performance of man either psychologically or physically.

9.3 IMPORTANCE OF FATS

Fat serves various purposes in the diet. It is the starting material for biosynthesis (see activated acetic acid p. 127), it provides a high value of energy, carries essential fatty acids and fat soluble vitamins. A diet lacking in fats is in practice always lacking in fat soluble vitamins, which is one reason why an adequate supply of fats is necessary in the diet. There are other reasons why fat is important and has no substitute: it has a high satiety value and has a greater energy content per unit of weight in comparison with the two other main groups of nutrients, the carbohydrates and proteins (see p. 280).

For mental work, a daily intake of fats of about 75 g, that is 1 g fat/kg is desirable. This amount includes both the fat used on bread and in cooking, the so-called 'visible' fat, and also the 'hidden' fat contained in many foods, for instance in meat, milk, cheese and so on. Intake of fats can be up to twice as high for heavy physical work. Long and thorough research by scientists all over the world has found that there is no difference in the nutritional value of the common edible fats, such as butter, margarine, olive oil, soya oil, ground nut oil etc. There is also no reason to assume, according to comprehensive nutritional research so far carried out, that animal fats or vegetable fats with similar vitamin and essential fatty acid content are superior or inferior to each other as dietary fats. The opinion, which has often been expressed in recent years, that animal fats play a part in arteriosclerosis has not been completely substantiated even after comprehensive scientific research in normal human nutrition. As to the desirable fat intake, present opinion is that in a well balanced diet, not more than about 25%–30% of the total energy requirement should be taken as fat, corresponding to about 65–70 g fat per day.

9.4 IMPORTANCE OF CARBOHYDRATES

The carbohydrates in our diet cover most of our energy requirements. Starch is the most important carbohydrate, sucrose taking second place. We consume other carbohydrates in fruit, for instance fruit sugar, fructose, and grape sugar, dextrose. Lactose (milk sugar) is the infant's only carbohydrate but adults do not utilise large amounts of it, although it is important for absorption of calcium.

In the intermediate metabolism, the body can easily form carbohydrates from fat and protein (the maximum amount of protein intake which can be converted into carbohydrate is 58%). At least 10% of the energy needs of the body should be covered by carbohydrates for physiological reasons and this is always achieved in a normal diet. Normal liver function requires an adequate supply of glycogen and therefore an adequate intake of carbohydrates. The disadvantages of such a diet are mainly the low satiety value or satisfaction of such food (important for the manual worker) and undesirable fermentation in the overloaded gut. The vitamin B balance is affected when there is excess carbohydrate. Too great an intake of sugar (and with the glycogen stores full) is in part converted to fat and deposited as body fat. This cannot be mobilised as easily as glycogen for physical work. As an average figure, an intake of about 400 g of useable carbohydrate per day is assumed which gives about 1600 kcal.

Those carbohydrates which are indigestible or difficult to digest (such as cellulose, hemicelluloses, pectins) mostly swell and pass through the gut. They fill the gut and promote peristalsis (gut movement) and defaecation. They are grouped together under the name **ballast materials**. If the diet is too poor in ballast over a longer period, the gut becomes lazy and constipation can occur. The question of intake of dietary fibre and its effects are related to these types of carbohydrates.

CHAPTER 10

Nutrient Content of Foods

Man takes his food from the animal and vegetable kingdom, apart from water, air and salt. Natural products suitable for food and nutrition are made up in a variety of ways of individual nutrients, and are therefore mixtures. Pure nutrients, such as fat, sugar, starch, casein, gelatine and albumin are only consumed separately in very small amounts. Qualitative proof and quantitative determination of the individual nutrients, protein, fat, carbohydrate, minerals, vitamins and water, can be carried out by chemical or physical analysis.

An exact knowledge of the nutrient content of animal and vegetable foods is necessary if a suitable choice of foods is to be made. Proposals for the intake of a nutritionally desirable combination of foods in the various diets and tastes for a healthy and a sick population for all lands and climates must be based on such knowledge.

The basic difference in the natural products of animal and vegetable origin is that the latter almost always contain cellulose and have a preponderance of carbohydrate, whereas animal foods contain much less carbohydrate (glycogen in meat, lactose in milk) and they act as the main suppliers of protein and fat. The water content is also important for calculating nutrient value. In meat and eggs it is 74-78%, in milk about 87%, in fresh fruit and vegetables it rises to 90% or higher and in grains and legume seeds it is only about 12-13%.

Details of the nutrient content in various vegetable and animal foods are given in the following table, where the figures are derived from the tables published in 'The Composition of Foods', A. A. Paul and D. A. T. Southgate, 4th Edition of McCance and Widdowson, published HMSO, 1978. The figures can only be taken as approximate because of the great variations caused by different varieties, by soil, climate and fertilizers, time of harvesting and so on. The table shows that animal materials (meat, eggs, milk, cheese) supply most of the proteins and fats, and that among vegetable foods, only the seeds of legumes and nuts come close to them. Flour from grain is a concentrated food because of its low water content and it contains protein and fat as well as large amounts of carbohydrate. Others are jam, honey and syrups. Fruit and vegetables only have a low energy nutrient content. However, they contribute vitamins and minerals and the aroma, taste and fibrous materials which promote digestion and so have an important function in the diet.

It is not sufficient to know just the chemical composition in order to calculate the nutritive value; even if it is obvious that a food with a low nutrient content has a low energy value, the opposite does not follow. Not every mixture is rich in nutrients and will, in fact, have a high energy value. The latter depends far more on the extent of utilisation by the body. It is not only a matter of which materials are supplied to the body, but rather of how much of them can be absorbed. The components of food have to undergo far reaching changes to reach the organs by means of the blood stream. They have to be rendered liquid and be of use nutritionally if in so far as they can be digested and absorbed.

Table 24 Nutrient and vitamin content of some foods per 100 g

Food	protein N × 6.25	fat	energy content		calcium	iron	thiamin	riboflavin	ascorbic acid	retinol
	g	g	kcal	kJ	mg	mg	mg	mg	mg	μg
pork (dressed carcase raw)	13.6	31.5	338	1397	8	0.9	0.58	0.19	0	(4)
beef (dressed carcase raw)	15.8	24.3	283	1168	7	1.9	0.05	0.20	0	(4)
veal (fillet raw)	21.1	2.7	109	459	8	1.2	0.10	0.25	0	Tr
liver (ox raw)	21.1	7.8	163	683	6	7.0	0.23	3.1	23	16.500
bacon (streaky raw)	14.6	39.5	414	1710	8	1.0	0.37	0.15	0	Tr
chicken (raw, meat only)	20.5	4.3	121	508	10	0.7	0.10	0.16	0	Tr
herring (raw)	16.8	18.5	234	970	33	0.8	(0.13)	(0.40)	Tr	45
cod (raw fillets)	17.8	0.7	76	322	16	0.3	0.08	0.07	Tr	Tr
egg (hen's)	12.3	10.9	147	612	52	2.0	0.09	0.47	0	140
milk (cow's fresh whole)	3.3	3.8	65	272	120	0.05	0.04	0.19	1.5	35
milk (fresh skimmed)	3.4	0.1	33	142	130	0.05	0.04	0.20	1.6	Tr
butter (salted)	0.4	82.0	740	3041	15	0.16	Tr	Tr	Tr	750
cheese (cheddar type)	26.0	33.5	406	1682	800	0.40	0.04	0.50	0	310
cottage cheese	13.6	4.0	96	402	60	0.10	0.07	0.30	0	32
yogurt (natural low fat)	5.0	1.0	52	216	180	0.09	0.05	0.26	0.4	8
margarine (vitamin enriched)	0.1	81.0	730	3000	4	0.3	Tr	Tr	0	900
oatmeal (raw)	12.4	8.7	401	1698	55	4.1	0.50	0.10	0	0
whole meal bread	8.8	2.7	216	918	23	2.5	0.26	0.06	0	0
white bread	7.8	1.7	233	991	100	1.7	0.18	0.03	0	0
potatoes (peeled raw)	2.1	0.1	87	372	8	0.5	0.11	0.04	12	0
cauliflower (raw)	1.9	Tr	13	56	21	0.5	0.10	0.10	60	0
beans (runner raw)	2.3	0.2	26	114	27	0.8	0.05	0.10	20	0
lettuce	1.0	0.4	12	51	23	0.9	0.07	0.08	15	0
peas (frozen raw)	5.7	0.4	53	227	33	1.5	0.32	0.10	17	0

spinach (boiled)	5.1	0.5	30	128	600	4.0	0.07	0.15	25	0
tomatoes (raw)	0.9	Tr	14	60	13	0.4	0.06	0.04	20	0
apples (eating)	0.3	Tr	46	196	4	0.3	0.04	0.02	3	0
walnuts (ripe)	10.6	51.5	525	2166	61	2.4	0.30	0.13	Tr	0
oranges (raw)	0.8	Tr	35	150	41	0.3	0.10	0.03	50	0

Figures taken from McCance and Widdowson's 'The Composition of Foods' by A. A. Paul and D. A. T. Southgate, 4th Edition, Her Majesty's Stationery Office, London, 1978.
Figures in brackets are estimates or values taken from related foods.
tr = trace

CHAPTER 11

Food Digestion

Nutrients have to be separated out from food and made water soluble before they can be absorped. This occurs by mechanical and chemical means in the digestive system. Both processes are important and occur together, because the chemical decomposition, for instance of the protein or starch macromolecule, is only possible when the food particles come into contact with the secreting and absorbing mucous membranes of the digestive column during the mechanical processes of movement and transport.

By **digestion** we mean the transformation of the food supplied into a form suitable for absorption by the body fluids: the insoluble or barely soluble macromolecular organic nutrients are transformed into water soluble compounds of low molecular weight by the digestive juices. Water, vitamins and most salts are taken up by the body through the gut wall, without being altered.

With proteins the specific structure (see p. 47) is destroyed by digestion and breakdown to amino acids. Otherwise the passage of proteins as such which are foreign to the body, into the blood stream would cause a severe shock reaction (anaphylactic shock).

Digestion is carried out by the digestive enzymes, dissolved in the digestive juices. They are synthesised in special organs (e.g. in the salivary glands in the mucous membranes of the stomach and small intestine and in the pancreas) and pass into the digestive system where they break down macromolecular nutrients, such as protein, fat and carbohydrates into low molecular weight soluble substances.

The food consumed is mixed with salivary amylase (formerly called ptyalin) during mastication in the mouth. Starch is only split there to maltose to a very small extent. The mucins present in saliva render the chyme slippery and it reaches the stomach after being swallowed. There it is acidified by the gastric acid of the gastric juices, which contain about 0.5% hydrochloric acid. This makes it possible for the gastric enzymes to attack the chyme. Pepsin, the main enzyme of the gastric juices, breaks the protein down into smaller fragments, the so-called polypeptides. A second enzyme which splits protein in the stomach is

stomach cathepsin, in addition there is **stomach lipase** with a weak action. The latter can only begin to attack finely emulsified fats such as milk fat and the fat in egg yolk, but not fats in more compact form. The acid environment in the stomach interrupts the effect of the amylase and with the decomposition of starch.

In a normal mixed diet, the food remains in the stomach for about three hours, though a fatty meal remains longer. The exit from the stomach is controlled by the **pylorus**, which only opens briefly for a time. This ensures that the intestine remains equally full for a longer period. At the same time the intestine receives further stomach contents when the acidity of the previous portion has been neutralised and stabilised at about pH 8 by the sodium hydrogen carbonate from the pancreatic juices.

The duodenum, which receives the chyme in portions, contains the entrance of the bile duct and the excretory duct of the pancreas. The pancreas secretion contains **trypsinogen** in inactive form in addition to the sodium hydrogen carbonate already mentioned. The enzyme is activated in the duodenum and transformed into the protein splitting enzyme **trypsin**.

The **trypsin** from the pancreas, together with erepsin and carbohydrates continues the decomposition of the protein which started in the stomach down to the basic building blocks, the amino acids. The **lipase** in the pancreatic and the gastric juices splits the fats to a mixture of fatty acids, mono- and diglycerides and glycerol. This cleavage is greatly speeded up by the previous emulsification of the fats by the monoglycerides which have been formed previously. They have a strong surface action, as also do the bile acids. The decomposition of starch, which started in the mouth is carried on by the amylases of the pancreatic juices. Disaccharides, such as sucrose are also split in the small intestine.

When digestion is complete, the macromolecular components of the food are only present as fragments: while the proteins, fats and carbohydrates have been changed into forms which can pass into the body through the gut wall. This process is called absorption. Absorption of the completely broken down nutrients from the chyme occurs through the wall of the small intestine (jejeunum). Amino acids, monosaccharides and shorter chain fatty acids (up to C_{10}) pass into the blood vessels of the intestinal wall. These vessels join to form the portal vein and flow through the liver, the chemical factory of the body. The fat building blocks absorbed by the intestinal villi of the small intestine, are resynthesised to triglycerides in the cells of the intestine. The resynthesised fat passes through via the lymphatic vessels to the thoracic duct and so reaches the blood bypassing the liver.

Gut peristalsis, the progressive contractions which run along the intestine in wave movements, mixing the contents and pushing them on, occurs by the combined action of circular and longitudinal muscles of the gut. These involuntary movements, helped by the so-called **ballast materials** (indigestible, slippery substances) only take place when the gut is sufficiently full.

At the lower end of the small intestine (ileum) and in the large intestine (colon) almost nothing but water is absorbed and so the chyme becomes thicker. When the chyme has reached the large intestine, rapid bacterial decomposition sets in due to the bacterial flora of the large intestine. (The contents of the small intestine are practically free from bacteria).

The bacteria decompose materials which have passed through the small intestine without being absorbed. Incidentally the digestive enzymes cannot pass through the cell walls because they are macromolecular, but some of their decomposition products are taken up in the large intestine. The bacteria build up their own substance from such materials. Apart from water, the faeces contain indigestible food residues and decomposition products of food fragments, such as the unpleasant smelling indole or skatole derived from amino acids. The faeces collect in the rectum and are then excreted.

CHAPTER 12

Behaviour of Foods during Preparation and Cooking

As has already been mentioned in detail in the introduction, the diet must be of full nutritional value. This means that it should contain all the necessary nutrients in optimal amounts, in the right relationship to each other and useable form. However it is not always easy to fulfill these requirements because generally foods are made up of mainly organic substances subject to more or less rapid decay. This has various causes: in fresh fruit and vegetables the metabolic processes continue, depending on the external circumstances, and cause greater or lesser decomposition of the nutrients; in addition attack by microorganisms or pests can lead to more or less rapid spoilage. Other reactions and changes which affect the nutritive value take place during preparation (cleaning, washing, scouring and chopping), of foods for cooking and during storage.

A special effort must be made to reduce the loss of nutrients during the preparation of food in the kitchen. These losses usually occur in water-soluble nutrients, which are partially or wholly leached out when the food is prepared or pretreated or kept in water in the kitchen. There may also be loss of nutrients sensitive to air, light and heat when food is treated or stored unsuitably.

It is only possible to keep these losses to a minimum when one knows the sensitive nutrients and what measures can be taken to prevent their rapid destruction. The following paragraphs therefore give details of the ways in which preparation can reduce the loss of nutrients sensitive to heat, light and air:

Unsuitable methods of washing and cleaning may lead to considerable losses of vitamins and minerals, especially water soluble vitamin C, and also from the B vitamin group, where vitamin B_1 in particular is affected. Sodium, potassium and calcium, among the minerals, may also be lost. The size of the loss of these nutrients is very variable and depends on the method of processing the food in question. For instance the loss of vitamin C in whole peeled potatoes into water is about 9% in 24 hours, whereas cut up potatoes lose about double as much vitamin C in the same time. Prolonged soaking in water in the kitchen also causes loss of minerals from cut up potatoes and vegetables.

Even cutting foods into small pieces during preparation in the kitchen results in a certain loss of the vitamins sensitive to oxygen, particularly vitamin C. Cutting should only be done immediately before further processing in the kitchen.

Cooking processes. Cleaning and cutting up of food in the kitchen is usually followed by cooking. Heat is necessary to release many of the nutrients in food. The structure of many foods is loosened by heat which improves the digestibility and the utilisation of the nutrients. In addition, some materials which decrease digestibility and utilisation are destroyed by heat, for instance the trypsin inhibitors in Leguminosae. Flavour and acceptability of foods are often increased by cooking; many nutrients, for instance, starch in potatoes, only become digestible and pleasant after heating (boiling, frying). Heat treatment of many foods is essential because of bacteria. The aroma materials which are formed during roasting and baking are important.

The following types of heat treatment of foods must be distinguished because of their different effects on loss of nutrients.

1. Boiling in water (heating in water, 100°C)
2. Steaming (heating in water vapour, 100°C)
3. Cooking (in its own juice, 100°C)
4. Braising (heating of sauted meat pieces in liquid, 100°C)
5. Deep frying in fat (heating in hot fat, 180-200°C)
6. Baking uncovered in air (radiant heat, 100-250°C)
7. Grilling and roasting (radiant heat, 300-350°C)

Before treating any food with heat one should decide on the most suitable form of cooking. Each food has a specific cooking period. During this time various processes occur which are affected by heat and water. Protein coagulates and starch becomes sticky. The food cells may take up water to exude it. They may burst through excess pressure or shrink. The various food contents are to some extent dissolved and pass into the cooking water. Larger amounts of liquid dissolve more nutrients and flavour materials out from food, as the food and water try to attain an equilibrium. It is therefore understandable that the degree of leaching varies in the different methods of cooking. In steaming, cooking in its own juice or braising it can be regulated to a minimum. Steaming and cooking in its own juice should give better results than boiling because relatively small amounts of liquids are used and leaching of the nutrients contained can be reduced. A logical result of this is that the cooking liquids should not be discarded but used in soups and suchlike.

A pressure cooker is frequently used nowadays instead of cooking in an open pan at 100°C. With the pressure cooker, higher temperatures are reached, as 1 atmosphere absolute pressure corresponds to a temperature of 120°C. The cooking time is reduced because of the higher temperature, but at the usual pressure of 0.5–1 atmospheres there is no increased loss of heat sensitive vitamins such as vitamin C compared with cooking in an open saucepan.

Cooking by frying, baking, grilling and roasting are methods which use different temperatures and are used for certain groups of food:

In frying of meat over an open flame or in the pan, the heat is transferred by the fat, and at 180–200°C a crust is formed which seals in the juices. The temperature inside the meat does not go above 100°C. When meat is roasted in the English manner the internal temperature often does not rise above 80°C.

Baking is the cooking of flour mixtures with addition of raising agents (yeast, sour dough, baking powder). The heated gases enclosed in the mixture expand the mass. Heat is conveyed by radiation in the baking oven. The exterior of the baked goods are heated to about 180°C with the formation of a brown crust, whereas the inside is only heated to about 100°C.

Grilling and roasting of meat are usually done without the addition of water. Fish, pies or bread can also be grilled or roasted. Heat is applied by radiation and temperatures up to 350°C may be reached on the surface. At the same time the food becomes brown and appetising. Typical flavour materials are formed which stimulate salivary secretion and the digestive glands. Coffee derives its flavour from roasting.

Greater loss of vitamins often occurs when cooked food is kept warm rather than during cooking. It is necessary however to keep food warm in large scale catering and restaurants. On the other hand losses due to keeping food warm can be kept to a minimum in the home. Reheating does not have as bad an effect on the vitamin content of food as keeping it warm.

With loss of vitamins sensitive to oxygen and the deterioration of fats from autoxidation, the type of material used for the cooking utensils is also important. Minute traces of certain heavy metals can act as catalysts for the oxidation of the vitamins and unsaturated fatty acids. The most active catalyst is copper. Preferably copper utensils, and those made of copper alloys should not be used for the preparation of food.

CHAPTER 13

Preservation of Food

Vegetable tissues continue to respire although they have been separated from the parent body. Animal tissues do not continue to respire but enzymatic changes can take place. Even though some desirable materials, which increase flavour and palatability are formed during storage, for instance aroma materials in fruit and vegetables, sugar from starch and so on, a considerable amount of the constituents in them disappear during ordinary storage. Gradually, far reaching changes take place, especially when the cell is dead, which usually result in a loss of food value. These include deterioration of colour, loss of flavour leading to increasing decay and complete unusability. The causes of these phenomena are the effect of water, air, light and warmth, enzymatic decomposition and inorganic catalysts (e.g. traces of metals), but above all microbes. In short, physical, chemical, biochemical and microbial changes occur, which are all closely related to each other.

Physical processes: Colloidal chemical changes, swelling, drying out, reduction in flavour through loss of volatiles.
Chemical reaction: Oxidation in fats, vitamin C, aromatic materials pigments etc.
Biochemical changes: Due to the activity of native enzymes in foods. Examples are: fat cleavage by lipases, breakdown in protein by proteases, carbohydrate splitting by carbohydrases, enzymatic oxidation by phenoloses, lipoxyenases, peroxidases.
Microbial processes: Fermentation, mouldiness, rottenness due to microorganisms, formation of mycotoxins, e.g. aflatoxin infection of food causing poisoning (Salmonella) and production of toxins in food (by *Clostridium botulinum* and *Staphylococcus aureus*).

The easiest way to prevent these processes is to use the food while it is fresh, but this is made difficult because in certain seasons, especially at harvest time, many food are overabundant (fishing, harvest time for vegetables, fruit, cereals). Attempts have therefore always been made to preserve the surplus from the harvest. Increasing industrialisation created new problems of supply, due to the concentration of large numbers of people in towns. This made it necessary to transport large quantities of food long distances, to keep adequate stores and to preserve food. At the same time it is important to prevent the loss of valuable nutrients or to reduce the loss as far as possible.

All the methods of food preservation, from the early purely empirical ones up to the most recent which are used in the modern food industry, which are based on scientific knowledge, aim to prolong the keepability, palatability and nutritional quality of the natural products of animal and vegetable origin.

However, the modern concept of nutrition requires that the preservation – by whatever method it is carried out – should not only retain the nutritional value of the food but should also, where suitable methods are available, preserve the food in a form which differs as little as possible from the original, biologically unchanged 'fresh' state either in its nutritive value or in its palatability (flavour, aroma, colour). In addition, the retention of its nutritive effect, that is the physiological usability after absorption of the food, is desirable (see section on cooling and freezing, drying processes).

Methods of preservation can be divided into physical and chemical procedures. In many cases they are combined.

1. Physical methods: Cooling processes, freezing processes, sterilisation, pasteurisation, drying, filtration, other newer processes (e.g. irradiation).
2. Chemical methods: Salting and pickling, smoking, curing, sugaring, acidifying, chemical preservation.

13.1 PHYSICAL PROCESSES

Cooling and freezing processes

Cold preservation using cooling and freezing processes is based on the fact that with lower temperatures, the changes which normally occur in foods and which affect palatability and eventually lead to decay are slowed down considerably (Van't HOFF's equation, see p. 249).

Cooling and freezing processes differ from each other in the range of temperatures used. For most commonly preserved fruit and vegetables, temperatures between +4 and −2°C (or even down to the freezing point of the cell plasma) are used in cool or cold storage. In many vegetable and animal foods the enzymatic processes continue during cold storage, although at a slower rate; this automatically limits the usefulness of preservation using these temperatures. Storage cannot be continued for an unlimited period because of the biochemical decomposition which underlies all life processes. Cold storage can be continued longer if at the same time inert, neutral gases are used for atmospheric protection.

Controlled atmosphere storage for fruit, uses nitrogen and carbon dioxide-oxygen mixtures to delay ripening by slowing down the breathing process, e.g. carbon dioxide-nitrogen mixtures are used with eggs. The addition of CO_2 (up to

12%) to the cold storage area has proved useful for meat, for instance on long sea journeys. For storage of meat the cold air should be sterile, and this can be achieved by bacteria filters in the forced air circulation, by continuous irradiation with UV light or less satisfactorily – because of the oxidative effect on the fat in the meat – by using ozone. Vacuum cooling is especially suitable for vegetables. Controlled atmosphere storage for fruit has been utilised for many years.

Freezing is used to preserve food which may deteriorate if held for longer periods:

In preservation by freezing, in contrast to cooling, the temperatures used freeze the greater part of the cell fluids in the animal and plant tissues and so kill the cell. The freezing of the cell contents leads to greater keepability than could be expected from the reduction in temperature in the Van't HOFF's equation. This is because the enzymes no longer have an aqueous or liquid environment for their reactions leading to decomposition, see p. 245.

After the cells have been killed by freezing the enzymes present are still active and the enzymatic processes in the cell do not necessarily cease until a temperature of –40°C is reached. However they are much retarded at customary commercial storage temperatures (around –22°C) so that products made from top quality raw materials approach the quality of natural fresh foods, in particular with regard to properties of taste, colour, consistency and nutrient content.

Commercial frozen food firms are now introducing a bulk storage temperature of –70°C to ensure that no changes occur at all. The food is then brought to –22°C for distribution.

Blanching is carried out first in the case of vegetable foods to prevent any enzyme activity before freezing. The enzymes are inactivated by a brief treatment in hot water or in an atmosphere of steam (see p. 249). This scalding can be used for foods normally eaten raw but is often avoided because we aim to preserve the fresh characteristics of such foods; this applies to all types of fruit (with the exception of some stone fruits which discolour badly when thawed) as well as to tomatoes, cucumbers and onions. On the other hand heat inactivation by blanching before freezing makes it possible to prevent almost all the undesirable enzymatic changes in nutritive value and palatability which occur during storage in those vegetables which are normally cooked. For inactivation of enzymes in fruit see p. 298. Recently microwave procedures have been used (see p. 307) for blanching, but the costs are still relatively high.

Microbes are also affected by cold. In general no bacteria or fungi grow at below –12°C. Vegetative organisms are, however, only partly killed by freezing; the spores of bacteria and fungi and also viruses are very resistant to cold. Therefore vegetable products are by no means sterile when thawed out. Specific food poisoning bacteria cannot grow below +3°C; their growth and toxin formation is therefore limited to the period before freezing and after thawing, (e.g. *Clostridium botulinum*). But toxins which may have been formed before freezing will not be destroyed by freezing. This fact demands both careful hygienic measures in harvesting and preparing foods before freezing (personnel, factory, water) but also quick utilisation of the foods after thawing out. Adequate hygienic processing can be monitored bacteriologically by counting the numbers and types of organisms.

1. **Slow** freezing of packaged and unpackaged foods can take place in still, cold air or by quick freezing in very rapidly moving air of relative high humidity, as for instance in tunnel freezers (HECKERMANN, MURPHY); in the latter process the foods can be frozen in free flow form using eddy layer or fluidized bed processes in trays with perforated bottoms, see p. 304.

2. **Contact** freezing can take place by dipping or flooding with direct contact between the food and cold brine, where the temperature is close to the freezing point of the salt solution (OTTESEN process, e.g. for whole fish). Sugar solutions are also used (e.g. for fruit) or glycerol-alcohol-water mixtures (BLAND process).

3. **Indirect** freezing of food may be carried out in metal cans or canisters or in thin closely fitting packaging materials, sometimes previously evacuated (as in the Cryovac process). These are then dipped in the freezing liquid, for instance in propylene glycol baths at −32°C. In continuous freezers, quick freezing is ensured by the rapid heat transfer. The packaging materials are made of rubber-hydrochloride foils (Pliofilm) or from polyvinyl-polyvinylidene chloride copolymer (Saran), a shrinkable film. For the freezing of packaged foods in quick freeze apparatus (BIRDSEYE system), the foods ready for cooking or ready for eating (fruits sprinkled with sugar or covered with sugar solution) are frozen in rectangular, water vapour resistant packs, up to 7 cm thick (containing 0.25 kg to 3 kg) by arranging them in single layers and lightly pressing them between two hollow metal plates cooled with cold air, brine or an evaporating coolant (e.g. NH_3 [contact process]).

4. **Direct contact** freezing with evaporating coolants, that is by using liquid gases with extremely low boiling points, the so-called cryogenic liquids, such as carbon dioxide, liquid dimethyl ether, liquid nitrogen, methyl chloride, dichloro-difluoromethane (Freon 12, b.p. −30°C). In this process, which is also called 'immersion freezing', the refrigeration results from the JOULE THOMSON effect.

Special attention must be paid to the **rate of freezing** in the freezing processes. Rapid freezing (> 5 cm/h) only allows small crystals to be formed – mainly inside the cell. The fine crystals in the frozen material destroy the cell structures less than the larger ice crystals formed in the space between cell contents and the cell membrane during slower freezing. In many products the large crystals damage the cell walls and lead to a high loss of cell fluid on thawing. Chemical losses (e.g. vitamin C) and biochemical (enzymatic) changes in the cells are less, the faster the temperature is lowered, as in fast freezing. With a high rate of freezing the temperature range during which spoilage by bacteria and moulds can occur is passed through rapidly, so that possible deterioration is avoided.

The concept of a linear freezing rate (W) has been proposed for the measurement of the effectiveness of freezing (and so also of the efficiency of the freezing installations). It is defined as the relationship of the path of the whole freezing front (d_0) to the freezing time (Z_0) and is usually expressed in cm/h:

$$W = \frac{d_0}{Z_0} = \frac{cm}{h}$$

In the present state of knowledge about the effect of speed of freezing on food, the following classifications are used:

Type of freezing	extremely rapid	quick	slow
Average linear speed of freezing in cm/h	over 5	1 to 5	0.2 to 1

The freezing industry today generally uses quick freezing (1 to 5 cm/h) for economic reasons in order to obtain good quality frozen foods. It has the advantage of speed (freezing time) with acceptable refrigeration costs. In co-operative freezing plants, freezing is usually carried out at a rate of about 1 cm/h. Quick freezing is in practice adequate – except for poultry which needs to have the surface first frozen rapidly to retain the colour – to obtain the best results for quality. Co-operative freezing and home freezers manage with a slow freezing rate of 0.2 to 0.5 cm/h for frozen goods which are not necessarily stored for long periods.

It is most important to maintain a sufficiently **low storage temperature** after freezing, which must not exceed −18°C, and appropriate packaging material which is impervious to water vapour and flavour compounds should be used for the food being stored. Frozen foods do not need ventilation, they should be stored in airtight packing because the effect of atmospheric oxygen and the evaporation of ice can lead to serious reduction in quality and weight (drying out, discoloration, loss of aroma, freezer burn). Larger pieces of meat are stored after freezing, in a storage chamber at −18° to −23°C at a humidity of 90% R.H.

The packaging material must be stable (neutral) to hydrophilic and lipophilic materials contained in food, and against condensation inside and outside. (In the UK the word foil is used mainly for metal, film for non-metal). It must be also watertight against the liquid formed when the food thaws. Finally it must show sufficient mechanical strength (even at the lowest temperatures) during storage and transport, during filling, freezing, sealing and closing. Nowadays these requirements can be met by some single layer foils, but more often by foils of cellulose combined with polyethylene, polyethylene, or by polyethylene parchment lined with aluminium. Single layer films which are satisfactory in practice include rubber-hydrochloride foils (Pliofilm), shrink films made of polyvinyl- and polyvinyldiene chloride (Saran), of polyesters (e.g. from terephthalic acid and ethylene glycol) and single films made of polypropylene. Pressed or drawn aluminium foil is popular for packaging and storing convenience foods (see p. 306).

During transport, the 'cold chain', (i.e. a continuous series of measures which prevent frozen goods freezing and unfreezing), must never be broken. The temperature of deep frozen foods must never rise above −18°C from manufacture to sale; only during transport and when ultimately being taken from the store may the temperature rise briefly by 3°C, that is to −15°C (3°C = company practice but not obligatory. −12°C is often used by Local Authority Inspectors and some companies).

Thawing out of frozen goods has a serious effect on the quality of the final product.

Frozen fruit which is normally eaten raw, and vegetables in salt solution are thawed slowly in the closed package. This is best carried out at an even rate in the refrigerator or under the tap (so long as the package is watertight) or in a container of luke warm (never warm or hot) water. Frozen fruit or vegetables which are to be cooked, are best put into a small amount of boiling water or steamed until cooked, so that thawing is immediately followed by cooking. The cooking time of frozen vegetables will be shorter than of fresh vegetables because the blanching and freezing processes have already softened the cellular structure to some extent.

Thawing of large pieces of meat should not be carried out at room temperature or above but at temperatures below +5°C. (3 lb frozen steak takes 24 hrs to defrost in refrigerator but 6–8 hours in ambient temperature and 3–4 hours in an oven at 72°C (163°F). With slower thawing time or higher temperatures the pathogenic bacteria, which may possibly be present, for instance Salmonella or Staphylococcus, can increase rapidly at room temperature.

High frequency heating gives the fastest and most even thaw. This procedure is already used in the U.S.A. (see p. 307).

Freezing is the most natural and the ideal method for keeping fresh many foods of animal and vegetable origin. The nutrients, and especially the active materials, for instance the oxygen sensitive vitamin C, can be preserved to a great extent, in contrast to other methods of preservation. This also applies to the specific properties which help to determine the palatability of the food, such as taste, aroma and colour.

The frozen food industry is expanding, especially in the large scale manufacture of ready to eat convenience foods. These ready prepared dishes (convenience foods) make it easier to supply hospitals, large canteens or schools with meals.

Freezing is especially suitable for fish and fish products, because they deteriorate rapidly and should be treated as soon as possible after catching.

However, not all kinds of fruit and vegetables are suitable for freezing. In some cases canning or drying is more suitable.

Low temperatures are also used in the preparation of fruit juice concentrates, for instance in the KRAUSE-LINDE process. When fruit juices, which have been clarified enzymatically (with pectin enzymes), are frozen slowly, the pure water freezes out first; the fruit juice concentrate is separated by centrifugation or sieving. After this procedure has been repeated a number of times a concentrated juice is obtained. This is usually further concentrated by vacuum and stored at about −18°C after a small amount of fresh juice has been added to strengthen the aroma. The concentrate, which is semi-solid at this temperature, gives a drink which is considered to be almost exactly similar to the natural juice when diluted with water.

Freeze drying is another field in which cold is used, (see p. 305).

Sterilisation and pasteurisation (heat preservation)

Adequate heating at high temperatures (100°C and higher) kills microorganisms and so protects foods from microbial decay, as long as new organisms are excluded. At the temperatures employed for this purpose the native enzymes which cause decay in food (lipases, proteases and oxidases etc) are almost entirely inactivated. Heat conservation is still the most important procedure for preserving foods for a longer period.

Temperatures of 100°C and 130°C are used for **sterilisation**. Industry usually uses temperatures between 115–123°C. This is frequently combined with first removing the air from the cans, either by preheating of open cans ('exhausting') or by extraction of air under vacuum ('evacuation'). The bacterial spores, the most resistant form of bacteria, are frequently not killed off at 100°C, so that the heat treatment must be repeated again a few days later after the viable spores have germinated. This is called fractionated sterilisation or **Tyndallisation** (treatment on 5 successive days, helpful in a laboratory but not factory-wise). The temperature-time relationship for adequate sterilisation is expressed as the so-called F value.

Both in the home and in industry, glass containers and metal cans are used. Industry uses steel plate thinly coated with tin or lacquer to protect the can from corrosion by the contents and to protect the contents from the metal. The steel plate cans, developed during copper shortages, are now of high quality and are widely used. More and more aluminium cans are now being used. The aluminium is made resistant to corrosion by a surface layer of Al_2O_3 produced by the Floxal process (electrolytic oxidation), and then lacquered.

Heating is either carried out at atmospheric pressure in open boiling water baths or in steam at atmospheric pressure, or in autoclaves (closed pressure vessels) with steam or water under pressure. During long heating it is not possible to avoid some undesirable changes to the contents due to chemical interaction. This is reduced in new methods, such as **high temperature short time** heating (HTST) and **rotational sterilisation** in rotary autoclaves.

This process rotates the cans end over end round a common axis during heating, so that the contents are continually in motion. This prevents overheating of the contents near the walls of the can; at the same time a faster heat exchange is obtained and a shorter heating time is necessary for sterilisation. The products are of higher quality and keep better compared with those from older methods.

A shorter time-temperature can also be achieved during sterilisation of foods with a weak acid or neutral reaction (pH 4.5–7.0) by the addition of **nisin** or **subtilin** to aid sterilisation. These materials, polypeptidic in nature, are isolated from the microbes (*Streptococcus lactis* and *Bacillus subtilis*) or the products of their metabolism. They reduce the heat resistance of the spores of Bacillus and Clostridium species which spoil foods.

Sterilisation is particularly suitable for all those foods whose structure and consistency are not changed greatly by higher temperatures, especially meat, vegetables and fruit. They must be tested for sterility by differential incubation over days or weeks. Heat is also used to preserve jams and marmalades by killing the microbes. Here combined methods are used, for instance heating and sugaring. Chemical preservatives can be used.

Pasteurisation at under 100°C is used for foods such as milk, and fish whose protein carbohydrate and vitamin content would be damaged by the higher temperatures used in sterilisation. This procedure only achieves a weakening or a limited reduction in bacterial activity, which prevents spoilage for a limited time. Pasteurisation is used especially for milk, fruit juices, apple juice and fish preserves.

The products obtained by sterilisation and pasteurisation behave in general in the same way as fresh foods prepared in the home.

Drying of foods

The aim of drying food is to withdraw enough water to inhibit the chemical and enzymatic reactions but more especially to prevent microorganisms from developing further. Microorganisms need their nutrients in dissolved (diffused) form in order to grow and therefore foods have to be sufficiently moist for the microorganisms to grow. The microorganisms are by no means always killed by drying, some remain alive, but in an inactive state (spores). When water is taken up, new growth will occur. The withdrawal of water by natural or artificial drying is therefore an effective means of protecting food from microbial decay; however reactions of a particular kind may take place during drying and later in dried food with low water content when it is stored. These take place at a slower rate but limit the keepability of foods which are to be stored for a long time.

Basically this is because structural changes may take place in the whole texture of the food when water is removed during drying. Many of the components of the food, which are chemically very diverse, go through a graduated series of far reaching chemical and physical interrelationships as the water is removed in the drying process.

Water is 'bound' to food in many different ways: **water which drips out, surface-held water, capillary water** (in the smallest interstices between the cells, slow diffusion water) and water which is bound by absorption (= **Langmuir** water). The last named fills the inter- and intramolecular interstices in meat, fruit and vegetables and is of decisive importance in the drying of food. **Water of crystallisation**, which is bound chemically, is only of subsidiary importance in drying. The **sorption** of water, which contains dissolved salts (ions) carbohydrates, proteins, acids and so on, is important in the structure of foods because during the drying process, the **increase in concentration** of these dissolved materials, depending on the direction and speed, may lead to various chemical or enzymatic changes: hydrolysis, browning and other chemical rearrangements, protein denaturation or even enzymatic conversions. Each of these reactions may lead – unless special measures are taken during pretreatment and drying – to a more or less rapid reduction in quality and palatability and finally to decay of dried foods (low in water). It is therefore obvious that the drying processes in the varied groups of foods can never all be carried out by the same technological means and that the technique must suit the food in question.

The characteristic changes which take place in dried foods may have the following causes:
Growth of microbes (bacterial, yeasts, moulds) and their enzymes,
native enzymes in the food (e.g. esterases, oxidases),
non-enzymatic reactions (purely chemical processes),
physico-chemical causes.

All these changes are influenced further by humidity (relative humidity of the air, formation of water of condensation), temperature, oxygen in the air, composition of food, arrangement and distribution of the individual components (tissue structure, particle size).

Influence of microbes: Microbes need sufficient water to absorb nutrients from the foodstuffs which they attack. However, for microbes to develop in dry food, the absolute water content is less important than the relative humidity of the air with which the food is in equilibrium. 'Dry' foods are nearly always hygroscopic, that is their water content is dependent on the relative humidity of the surrounding air.

For most microbes the relative air humidity of air (R.H.) for optimum growth lies between 90 and 100%, but there are considerable differences between individual microorganisms. **Moulds** grow above 70% relative humidity, although some growth is possible between 70 and 75% relative humidity, growth is restricted and can only be detected after several weeks. Only a very few moulds can grow at about 65% R.H. There are some osmotolerant **yeasts** which can grow in concentrates with a high sugar content such as malt extracts, syrups or dried fruit and even reproduce themselves at relative humidities of $<60\%$. In solid foods they can, however, only grow when the equilibrium humidity is above 75% R.H. Most bacteria only grow at above 75% relative humidity. In foods whose water content corresponds to a lower relative humidity, microbes can no longer adapt themselves or absorb nutrients.

In foods, the equilibrium humidity varies according to the **structure** (cell network and capillary structure) and **composition** (proportion of soluble materials, ability to swell, pH value, amount of acid etc.).

The following table shows the threshold value of water content for growth of various microbes.

The **equilibrium humidity** gives the water content of the material when it is in hygroscopic equilibrium with the surrounding air containing water vapour, that is it has adapted its water content to the partial pressure of the water vapour in the surrounding atmosphere.

Threshold water content of some dry foods

(as the equilibrium water content with a relative humidity of 70% at 20°C)

Food	*Water content in %*
dried fruit	18-25
starch	18
dried vegetables (according to type)	14-20
milk powder (fat free)	15
wheat flour	13-15
whole egg powder	10-11
full fat milk powder	about 8

Microbial growth need no longer be feared when foods are adequately dried and suitably stored, so long as the water content remains below the characteristic **equilibrium humidity value** (threshold humidity, critical threshold water content) for the food in question.

Enzymatic reactions: Enzymatic reations, caused by the native enzymes in the food or enzymes from microbes (which have died) occur – much more slowly – when the water content is below the limits for growth of moulds. They reduce the storability of those materials intended for lengthy storage which usually absorb moisture from the surrounding air. The lipolytic processes, which occur when the substrate is in a state of liquid aggregation, are an exception. In this case the substrate is able to diffuse to the enzyme of its own accord and does not need the help of the water phase. See p. 302.

However, there is no linear connection. Each dried food has its own threshold value for water content (which has so far been established empirically), below which enzymes appear to have little or no effect.

In many grain products, the enzymes which attack fat, especially the **esterases** (lipases) and **lipoxygenases** which cause oats to become bitter, start oxidative changes in the free fatty acids with the help of oxygen from the air. The retrogression of lecithin in egg pasta is an enzymatic process, determined by the **phospholipases** which attack to a greater or lesser degree and which increase with the water content of the dry products. In dried fruit and dried vegetable, native **oxidases** can regenerate during storage and produce a hay-like aroma and flavour substances if these enzymes are not completely inactivated by adequate blanching before processing (see pp. 293 and 298). Heat treatment or blanching to inactivate enzymes is not possible or only to a limited extent for grains and their products, because of the damage heat would cause; in addition the enzymes may become reactivated. For this reason the effect of the water content on the course of enzymatic reactions is particularly interesting in this group of foods.

Chemical changes: Purely chemical changes play a decisive part in the processes of decay in dry (low moisture content) foods. **Non-enzymatic browning reactions** (see MAILLARD reaction p. 137) can occur in almost all foods when the necessary conditions are fulfilled; the presence of compounds containing reducing sugars and NH_2 groups (such as amino acids, proteins, cephalins), for instance in milk powder, egg powder, dried potatoes, tomato powder and dried fruit. The moisture content of the dry food has a direct influence in non-enzymatic browning, because the changes can start at relatively **lower** water contents (from about 3%) than those for enzymatic reactions. The speed of this reaction increases initially with increasing water content, reaches a maximum at a certain water content and decreases when the water content rises further. The MAILLARD reaction is dependent on pH, but independent of O_2.

On the other hand, in dry products containing fats and oils, for instance in whole milk powder, wheat flour, milled products, '**oxidative rancidity**' is promoted by a low water content ($\lesssim 2\%$) and frequently further increased by lowering the water content still more. The lack of a protective water layer at the reactive sites is thought to be responsible, for example in egg powder.

The significance of the sorption isotherms: The relationship between the water content of dry products and the reactions possible in them during storage has only recently become known. Recent research has shown that it is not the absolute water content alone, but rather the **type of water binding** in the tissues of foods (which is in direct relationship with the surrounding relative humidity) which is responsible for the possibility of onset and the quantitative course of enzymatic and non-enzymatic (chemical) processes.

The functional relationship between the (equilibrium) **water content** of a food and the water vapour partial pressure, that is the **relative humidity** of the surrounding air, is given by the so-called sorption isotherm. This is usually obtained in the form of an S-shaped curve, when the water content of the dry product is plotted against the relative air humidity in which it stands in equilibrium (see Fig. 18).

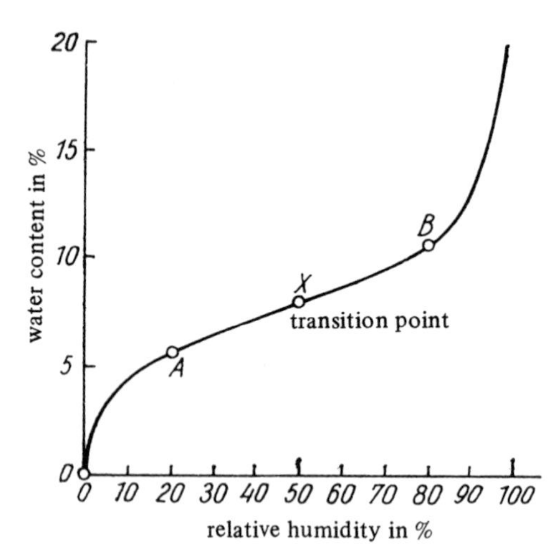

Fig. 18 Sorption isotherm

The first part of the curve (to point A) corresponds with the LANGMUIR sorption isotherm; the water bound in this area is considered to be a monomolecular layer of absorptively firmly bound water, which is very difficult and slow to remove by drying. The second part of the curve (above point B, usually marked by a steep rise with higher relative humidity) is thought to be due to the appearance of **capillary condensation**. In a hygroscopic material rich in capillaries (dry food), lowering of the water vapour pressure is followed by a condensation, so that, with rising relative humidity first the small and then the larger capillaries become filled with (mobile) water. In the straight part of the sorption isotherm, (approximately between A and B) there is assumed to be a multi-layer water deposition. The **sorption isotherm**, which is largely dependent on the degree of distribution of the components, and on the inner and outer surface constitution of a dried foodstuff, gives information about the type of water binding and distribution of the water, whether it is more or less firmly bound, or is easily mobilised and available for certain reactions.

It seems that the relationship between the water content and the relative humidity, and in the characteristic pattern of the sorption isotherm, contributes to the enzymatic processes in decomposition. Therefore foods whose water content lies in the area of the first part of the curve (monomolecular absorption), are to a large extent, protected from enzymatic decomposition.

Enzymatic changes only occur in significant amounts when the water content lies above point A on the sorption isotherm. The lipolytic processes are an exception and occur when the substrate is present in the form of a liquid aggregate (see p. 300).

Non-enzymatic reactions such as the MAILLARD reaction can already commence in the area of purely absorptive bonding and then increase in rate with rising water content. On the other hand autoxidation of fats occurs most rapidly in the water-free state and gradually decreases with rising water content.

Drying Processes. There are serious problems for the modern food drying industry in avoiding the variety of undesirable changes which may occur during the drying process and during the storage and dried products.

A dried food should if possible contain almost all of the nutrients present in a complete form which can be utilised by the body (in accordance with modern nutritional thinking). This can only be done by careful processing and the observance of safety procedures; it is very important to use **packaging materials** for dried foods which are non-permeable to light, air and water. An inert gas, and **cool storage** may possibly be used for longer storage periods.

Instant products are a particular type of dried food. They are extracts in dried powder form, which are easily wettable and can be completely reconstituted with water. Hygroscopic products of this type, such as milk, coffee, fruit juice powder must be protected from water vapour by special packaging or by coating the particles to prevent them from sticking together.

Liquid or pulpy foods, for example fruit juices, milk, are frequently only **partially dried**, producing so-called **concentrates**: modern processing techniques have developed a number of different types of equipment which protect the food through the use of brief application of heat, with or without vacuum, relatively low temperatures during concentration, an adjustable degree of concentration. These are used to produce a variety of end products. The most common are: surface film evaporator as a rising or climbing film evaporator, falling film or falling stream evaporator, rotary evaporator, centrifugal evaporator (e.g. centritherm evaporator) and special plate evaporators, as used for the production of concentrates of citrus and stone fruit and grape juices, for dried milk, for pectin concentrates and aroma concentrates from fruit juices.

Removal of water to obtain dried products may be carried out as a vaporisation or evaporation drying process, and the moisture is converted into a gas by the use of low temperature heat and removed as a steam or gas mixture. If drying is carried out at low vapour pressures and low temperatures, it is called **vaporisation drying**. If the temperatures and vapour pressure are close to the boiling point of water, the process is called **evaporation drying**. Evaporation drying under vacuum is generally called vacuum drying.

The technical processes used in evaporation drying vary according to the type of food to be dried: shelf or chamber driers, tunnel driers, moving belt driers, plate or shelf driers, trickle driers, drum driers, spray driers and flow driers. Vaporisation drying uses single or double roller driers (thin layer apparatus), palette or scoop driers, screw driers and tunnel driers. Vacuum drying can be carried out in drying chambers, in vacuum roller or scoop driers.

Heat is introduced and applied to the materials to be dried in various ways; by convection drying (using hot air), by conduction of heat (contact drying), by radiation (radiation drying) and by dielectrical means (high frequency drying). When continuous convection drying is used, the hot air can be introduced either in parallel flow, counter current flow or cross current flow. These methods are frequently combined in order to protect foods which are sensitive to heat.

Wherever possible vacuum drying is now used. This protects the heat sensitive proteins and carbohydrates (see Browning reaction) and also heat sensitive vitamins such as B_1, folic acid, pantotechnic acid, and especially those vitamins liable to oxidation, A, C, D and E as well as carotene. In vacuum drying, heat cannot be applied through convection by heated gases, but only by radiation, conduction or by previous heating of the material. Vacuum drying is especially important in the manufacture of foods rich in vitamins and for fruit, meat, fish and egg products.

Examples of the use of modern drying processes:

Roller (contact) and **spray drying**: Liquid or pulpy foods are commonly converted into powder by contact drying using warmed rollers (roller driers). The material, applied to the revolving rollers (which are heated from inside) as a thin layer, is continuously dried during one revolution and scraped off by a scraper.

Spray or **atomisation** drying is particularly suitable for some kinds of food. The liquid material, converted into a spray or a mist by a nozzle or disc spray, has the water removed by a hot air or gas stream in a drying tower. The drying of the particles in the mist occurs as they float, not on hot drying surfaces. This method is used particularly for materials where damage to the protein must be avoided: eggs, milk, gelatine, blood. Material which contain little protein, such as extracts of vegetable pectin, coffee or foods containing sugar can also be spray dried. Although the hot air, heated by heat exchangers, is introduced at 120-180°C, the actual temperature in the material to be dried is only 50-55°C, because quick cooling results from the vaporisation of water due to the spraying. Denaturation changes in the protein or undesirable reactions between proteins and carbohydrates (see Browning reaction, p. 139) are thus reduced to a minimum.

The BIRS process, using cool air, is a variation of ordinary spray-drying. The material is dried in large, tall towers (up to 72 m high and 15-36 m diameter) at temperatures < 30°C. An arrangement at the top of the tower distributes the liquid as a flowing material, and cool, dry air (maximum humidity 0.12-0.8 g water) rises up in counter current at a speed of 0.1-0.5 m/sec. The powder which collects at the bottom of the tower needs to be dried further according to the water content, for milk, for example, taking another 20-30 minutes. However, expectations for this process have not been justified and it can only be used to a limited extent.

Foam mat drying (foam drying, whirling drying): The principle of this method is to build stable foams from the material to be dried and to dry them with warm air as thin layers. The idea of drying viscous or thin liquid foods in this way is not new and is also used, for example, in spray drying. The difference is that the material is brought into fine suspension by certain additives so the time taken for drying is very much reduced.

The food or food concentrate to be dried is mixed with a physiologically safe stabilizer, such as soya protein, glycerol monostearate, sucrose fatty acid esters. Compressed gas is then led into the mixture and a stable foam is produced. The foam product is dried to a water content of 1-2% by filtered hot air at 60–100°C, as a thin layer on perforated conveyor belts or on tin sheets with holes in them. The temperature of the dried material does not exceed 60°C. The products, which are spongy and very porous, can then be ground, pressed or milled. Tomato puree, citrus juice concentrates, grapes, apples, pineapple, fruit juices, mashed potato, mashed pumpkin, tea, coffee, meat extracts and sometimes milk are dried by this method. Dried fruit powders have to have additional aroma added to the concentrate and to be encapsulated (e.g. in sugar). These products also need suitable packaging (see p. 295).

Freeze drying: New industrial methods of preservation are being developed, in addition to the classical methods of drying and freezing, which combine the advantages of freezing and drying: **freeze drying** and **freeze (vacuum) drying**. The latter is a combination of sublimation and vacuum drying.

Freeze drying can also be called sublimation drying, lyophilisation or deep freeze drying; the French use the word 'lyophilisation', whereas in English the terms 'freeze drying' or 'drying by sublimation' are used. These terms must not be confused with the technical term 'dehydro-freezing'.

In contrast to the usual drying processes, the water in the material to be dried is frozen and passes directly from the solid to the gaseous state (sublimation; example, air drying of frozen washing). The same process also takes place in commercial freeze drying, but it is speeded up by producing a vacuum of 0.2–2 Torr with a steam injector which produces the necessary fall in concentration (in relation to the water vapour pressure of 4.5 Torr over ice at 0°C). In sublimation, the material to be dried has a considerable amount of latent heat (heat of sublimation) to be removed, amounting to 680 kcal/kg of sublimed ice. This heat must be returned to the material, otherwise there would only be a cooling of the ice and lowering of the necessary pressure gradient between the ice in the frozen food and the apparatus for removal of the water vapour (steam injector).

The 'warmer' the ice in the material to be dried, the faster the drying. The quality of material dried will only be maintained if the ice does not melt during the process, the melting temperature depends on the material. It is characteristic that the dried material retains its outer form in freeze drying, because each time only the outermost molecules of the ice layer are 'sublimed'. Therefore the layer of the ice becomes continually smaller and finally vanishes altogether whereas the drying, ice-free part grows bigger, leaving a porous, dry product, free of water. Because there is no liquid zone between the drying material and the ice nucleus, it is possible to avoid shrinkage. In many cases, for instance with some varieties of vegetable such as cauliflower, the materials do not have to be previously frozen. Freezing and drying can be combined, so that some of the water is vaporised in the entrance lock of the vacuum drying tunnel. This automatically leads to rapid cooling of water in the food (vacuum cooling). In the subsequent part of the tunnel, the frozen material is freeze dried. Ice condensers or steam blast injectors are used to produce the necessary vacuum in large scale industrial freeze drying. Temperatures used are between 30 and $> 100°C$, and the amount of heat introduced in any phase of drying is never enough even to cause partial melting of the ice; the ice temperatures in the drying material is measured on the barometric principle, without measuring probes, and controlled automatically.

Freeze drying is a procedure in which flavour, aroma, colour and odour materials and other sensitive substances are to a large extent preserved. The low temperatures used prevents any increase of aerobic or anaerobic microorganisms during the process. The relatively high vacuum (lack of O_2) means that materials susceptible to oxidation, as for instance vitamin C and aroma materials, are protected, whereas in ordinary methods of drying, high losses of such active materials and other sensitive nutrients must be accepted.

In principal all foods can be freeze dried. Many vegetable and animal foods are already dried to a larger or smaller extent. These include meat, milk, eggs, fish, fruit juices, green vegetables, fungi, onions, potatoes, ready prepared camping meals, soups, coffee and so on. Freeze drying is an ideal method for producing biological materials which are sensitive to temperature such as blood plasma, penicillin, enzyme preparations, blood serum, and innoculation materials.

The advantages of freeze drying are that the texture and the outer form of powder products are retained. They are very soluble and reconstitute quickly (rehydration) when water is added because of their porous nature.

There are problems in retaining the quality of freeze dried products, because they are particularly sensitive to oxygen in the air, to water vapour and to foreign odours, so that they can easily suffer chemical, physico-chemical and enzymatic changes, and lose structure which is sensitive to mechanical damage. For this reason the packaging industry has to provide suitable packaging. Cans of thin aluminium or tin plate are used, or cups of aluminium foil, multi-layer films, or polyethylene lined steel cans. It is necessary to provide special protection from the oxygen in the air and its critical concentration in the dry substance is given as 0.1 mg O_2/kg. It is also advisable to pack under vacuum or inert gas (N_2). These measures are particularly important in the case of substances particularly sensitive to aroma changes such as freeze dried coffee or other instant products.

Sterility by filtration

Cold filtration (cold process) which filters out the microbes, is sometimes used to prevent the activity of microorganisms. When the fruit juices have been clarified by de-sliming and fining, the clear juice is passed through especially fine filter plates (SEITZ filters), which hold back the microbes, so that the materials will keep so long as reinfection is prevented. Another advantage to this process is that the materials which give the natural bouquet are retained and can develop during subsequent ripening. In practice, filter aggregates are used in which several filter discs made of pressed sterile asbestos and cellulose are inserted one behind the other. The filtered juices are best stored under carbon dioxide pressure (BOHI-SEITZ process).

Irradiation procedures

There are two sources of irradiation used in irradiation processes: **high energy electrons** and **electromagnetic radiation**. Both are high energy rich sources (X-rays and γ-rays), while UV rays and infra-red radiation are less energy rich.

Energy rich radiation. There is a promising future for the use of the following radiation in food preservation.

1. High energy electron radiation produced by van de GRAAF accelerators and linear accelerators (output from 250 to several thousand watts);
2. Gamma rays, from radioactive isotopes such as ^{60}Co or ^{137}Cs. X-rays are in practice too expensive.

The energy given off or taken up in irradiation of the material is called the irradiation dose. The rad unit of irradiation corresponds to the release of energy of 100 erg when the radiation passes through 1 g of material. 1 krad (kilorad) = 1000 rad; 1 Mrad (megarad) = 1000 krad; 1 Mrad = 2.4 kal/kg. When 1 kg of water or aqueous material is irradiated with 1 Mrad there is a rise in temperature of 2.4°C. The maximum dose for the sterilisation of foods is 5 Mrad, so that the rise in temperature is not more than 12°C. The term 'cold sterilisation' is therefore used. The electron generators have a relatively high dosage output so that, (compared with gamma rays) a relatively short irradiation time of the food is necessary.

Cold sterilisation can only be used to a limited extent to destroy microorganisms, because the high irradiation dose necessary (3-5 Mrad) produces undesirable changes in taste, appearance and consistency. Also these dosages do not inactivate the endogenous enzymes which can cause decay in food. Irradiation combined with a brief heat treatment (blanching) does not lead to changes in the flavour of food and is therefore a suitable method for inactivation of both microbes and native enzymes.

Pasteurisation irradiation does not achieve complete destruction of microorganisms though it greatly reduces their numbers and increases the length of time for which meat, fish, poultry and some kinds of fruits can be stored. The same dosage is suitable as a hygienic measure to guard against Salmonella.

When foods are carefully irradiated there is practically no reduction in nutritive value, no vitamin loss, nor is there a reduction in the biological value of the proteins. The vitamin loss in sterilisation irradiation corresponds to that in heat sterilisation in certain cases.

Energy rich irradiation treatment is permitted to a limited extent in the USA, USSR, Great Britain, Norway, Israel, Canada, Japan, for the destruction of insects in grain; to prevent the chitting of potatoes and onions. It is also used for smoked and fresh meat and for spices to reduce the microbial content. It is a certain guard against Salmonella in heat sensitive foods: dried egg, frozen meat, poultry and game. Industrial irradiation of foods is only permitted for special germ free diets for patients under germ free conditions in the United Kingdom.

UV irradiation. UV radiation has been used for a number of years, in contrast to gamma .radiation. A wavelength of 265 nm is most effective for most microbes, and only *Escherichia coli* requires 254 nm, so that the stabilised lamps for destruction of bacteria are usually mercury vapour, low pressure lamps, with a wavelength range from 240-300 nm. In practice UV irradiation is only suitable for the sterilisation of surfaces and transparent materials such as water and air because it does not have enough penetration power. One of the most important uses is sterilisation of storage areas, where care must be taken to avoid damage to the skin and especially the eyes of personnel. UV light is also used for disinfection of instruments, vessels and water used for washing, for example in fish processing factories and dairies, and also for surface irradiation of meat. For the latter use the additional use of UV rays in cold storage areas is promising. Experiments have shown that after double storage time and at a temperature of 5°C and a relative humidity of 75–85%, irradiated meat has only half the surface bacterial count of non-irradiated meat.

Control of moulds in seeds (seed sterilisation) is also carried out with low pressure mercury lamps, with the seed carried over shaking seives set at an angle in a series of different levels. The ozone formed by the UV irradiation probably contributes to the effectiveness.

Ozone itself is also used for sterilisation and removal of odours. The cheapest and safest method to produce it is with UV light (at a wave length of 185–195 nm). It is important to have optimum humidity when using ozone to sterilise air, whereas humidity has no effect in the removal of foreign smells.

Infrared irradiation. Infrared wave lengths form part of the electromagnetic spectrum. Infrared radiation transfers electrical energy to the material treated. It is partially absorbed and so heats the absorbing layers of the food. (Infrared waves are therefore not 'heat rays', as they are frequently called.)

Drying and heating with infrared waves is characterised by the fact that no intermediate carriers, such as metal or a liquid, are needed for the heat: the heating only at the point of absorption.

Of some practical importance are some temperature irradiators, such as surface irradiators, lamp irradiators, tube irradiators, which emit waves on several wave lengths which during irradiation are absorbed by the material according to its capacity to absorb particular wavelengths in the spectrum.

High frequency heating (Microwave)

In conventional cooking processes used for foods for centuries, the heat necessary for the breaking down of the food, depending on the physical conditions and circumstances prevailing, is applied to the materials to be cooked from the outside, via another medium (the metal of the cooking pot, water). The cooking time therefore depends on the temperature gradient between the source of energy and the food, and in particular on the heat conductivity from the outside to inside of the material to be cooked.

In high frequency heating of foods, the heat does not have to penetrate from the outside to the inside, it is produced everywhere in the material simultaneously. The electrical energy necessary for this is not changed primarily into heat applied directly to the food; it is transformed into vibration energy of very high frequency (microwaves up to about 5000 MHz) by means of a special tube, the so-called magnetron.

These microwaves are able to penetrate organic materials in definable thickness layers to cause rapid vibration of the molecules of these materials. The molecules produce heat of friction at these vibration frequencies and the food is evenly heated in a very short time.

Only glass, porcelain and certain artificial plastic materials can be used as containers for the food; as metals reflect microwaves, metal objects or containers must not be placed in the cooking area.

High frequency generators with a frequency of between 1000 and 5000 MHz have been used experimentally for pasteurising milk, beer and wrapped bread and are used for roasting coffee, for treating cereals, and particularly for thawing frozen ready prepared meals and for cooking and warming up foods of all kinds.

Treatment of flour with high frequency energy kills insects, spores and moulds. Other food processes suitable for the use of high frequency energy are the blanching of fruit and vegetables before drying or deep freezing and in freeze drying, because the heat generated inside the material to be dried is particularly helpful.

The localisation of heat on the surface of the food in conventional cooking methods does not occur in high frequency heating, so that there is no browning or formation of a crust, for example in bread baking or in roasting of meat. High frequency cooking is particularly suitable for instance in a diet kitchen when the formation of a brown crust is undesirable, or when the cooking is to be done without any conditions (fat, water).

Although this process is still comparatively expensive, it is considered likely that microwave energy, as a very specific source of heat in treating food, will soon be used in many areas, for instance in large scale catering. It is also very suitable for the production of fruit powders.

Storage under gas pressure

Liquids, such as fruit juices and milk, can be preserved under pressure with CO_2 gas. In the BOHI-SEITZ process, fruit juices are stored under carbon dioxide pressure of 7–8 atmospheres pressure at 15°C or 4 atmospheres pressure at 0°C, which suppresses microbial activity and prevents fermentation. The HOFIUS process attempts to inhibit the enzymes in milk and cream with an oxygen pressure of 8–10 atmospheres pressure.

13.2 PRESERVATION BY PROCESSING

Chemical preservation by salting, pickling, acidifying, smoking, sugaring, using sulphur and preservation in alcohol, have long been known and used and are combined with **preparation** of food from animal or vegetable sources. Frequently desirable basic chemical and physical changes occur during the preservation procedures: changes in structure, for example; removal of water from tissues during salting, pickling and smoking; protein and carbohydrate changes in natural and artificial acidification; changes in aroma with the formation of new desirable odour and flavour materials. These processes also delay or reduce the decay due to microorganisms (microbial decomposition) and the influence of air (oxygen), trace metals, water (chemical-physical or chemical decomposition).

However, the preservative effect of one of these processes alone is often not adequate, so that **combined** methods such as that of following pickling, by smoking, or heating (sterilising) must be used after the addition of sugar (marmalade, fruit syrups, jams). In many cases the use of chemical additives is the final effective method for safe storage, for example (benzoic acid, sorbic acid, sulphur dioxide (see p. 320). During transport and in unfavourable storage conditions (as in the tropics), cold or frozen storage may have to be used in addition.

Salting and pickling

Salting tends to prevent undesirable microbial activity in food. Salt, however, is only effective at high concentrations (15–25% solution) and usually does not kill all the microbes, but only weakens them; those which cause putrefaction, cocci and wild yeasts are clearly inhibited. The growth of other microorganisms is often increased by small amounts of NaCl. These include those involved in the so-called biological acidification of food (salting olives, gherkins and sauerkraut). Stronger salting, which removes water from the cells (osmotic effect) frequently causes changes in food, which may be detectable in odour and taste, as with meat and fish.

Dry pickling of meat involves rubbing with a mixture of salt and saltpetre, and excluding air and storing for a few weeks in oak casks, whereas in **wet pickling** the meat is placed in a solution of salt and saltpetre.

The saltpetre also preserves the colour of the meat, once it has been partially reduced enzymatically to nitrite. In the United Kingdom, USA and other countries, the pickling solution may be made up from a solution of sodium chloride, nitrate and nitrite. Because of the potential carcinogenicity of nitrosamines formed by the reaction of secondary amines with nitrites and the dangers of excessive nitrate the levels of both nitrate and nitrite are strictly controlled (ppm levels). Nowadays mild salting is usually preferred. In order to increase keepability and to improve the flavour, salting may be followed by smoking.

The nitric oxide (NO) which is formed from the nitrite in acid solution (pH value of meat 5.5-6.5), forms red nitric oxide myoglobin (nitroso-myoglobin) with muscle myoglobulin and this is transformed into stable nitric oxide myochromogen, which is also red, when the meat is stored for longer periods and the temperature is appropriate.

Smoking

Smoking is used mainly to preserve meat, sausages and fish. In smoking, the water binding effect of a large amount of cooking salt in the pickled material is combined with drying in warm smoke which also has a bactericidal effect. This property of smoke derives from its content of phenols, cresols, acetic acid and formic acid, methyl alcohol, formaldehyde and so on. The desirable typical smoky aroma and taste develop at the same time. Wood from deciduous trees, in the form of sawdust or shavings, is used to produce smoke but wood of the elder, oak, beech and hickory are the most valued. Coniferous or pine wood shavings give a turpentine flavour; peat, lignite and coal make the meat unpalatable. Large scale operations generally use smoke with as little soot as possible; nowadays liquid smokes, made from a mixture of particularly aromatic wood shavings are used increasingly.

According to the intensity and temperature of the smoke one distinguishes between **cold smoking** and **hot smoking**. In cold smoking, the smoke slowly and deeply penetrates the meat at a temperature of about 20°C, giving a good preservation action. However, there is a considerable weight loss. In hot smoking, carried out between 50 and 100°C, the keeping quality of the meat is slightly improved, but little weight is lost. At higher temperatures the fat melts and penetrates into the meat. With smoking, there is a drying of the tissue on the outside of the meat with consequent flavour and texture changes.

Wet smoking is a modification of cold smoking, in which the humidity is kept as high as possible. This process should prevent the formation of a hard outer layer which does not allow the smoke to penetrate. Country, farm smoking is the name given to smoking in very sooty smoke. In quick or artificial smoking, permitted in some countries including the United Kingdom but not in others, the pickled meat is treated with smoke liquors. These include liquid smoke, or smoke seasoning, which contains the individual components of smoke as well as wood vinegar or beechwood tar compounds and ethereal oils and salts. The so-called smoking salts, sold in some countries, consist mainly of salt, saltpetre and wood tar.

Acidification

Edible acids are also used to suppress microorganisms and preserve foods. The concentrations of acid used are not sufficient to completely inactivate microbial activity.

Natural fermentation acidification is frequently used to preserve vegetables. The acids formed when carbohydrates are fermented, such as acetic acid, malic acid, citric acid and especially lactic acid, reduce the pH value and so preserve the product. Fermented vegetables such as sauerkraut and salted beans are examples of this method. The cell structure of vegetables is also loosened enzymatically and the digestibility of the product increased. Salting is an important part of the process of acidification and fermentation, because salt keeps other undesirable salt sensitive microorganisms away. Other biological acidifications, due to bacteria, are also used in the production of milk products, such as soured milk, yoghurt, kefir and kumyss.

Biological acidification is not suitable for meat, because toxic decomposition products can be formed with undesirable changes in colour and flavour. Meat, however, can be preserved by adding acid, as for instance by marinading.

These acidified products are very different from the original material. Colloidal chemical changes occur as well as carbohydrate and protein decomposition in both enzymatic and artifical acidification. They are valued by some people because of their reputed beneficial effects and palatability.

Preservation in alcohol

Ethanol (ethyl alcohol) produced by alcoholic fermentation of berries and stone fruits prevents the growth of undesirable microorganisms and has a similar preservative action as acetic acid and lactic acid produced in fermentation. A minimum alcoholic content of 14% (by volume) is necessary for a preservative action. The keepability of wine, egg liqueur and other alcoholic drinks depends on their alcoholic content. Fruit may be placed in spirits, for example fruits preserved in rum.

Fermentation is halted by adding additional alcohol (making dumb) to dessert wines: sherry and port are examples of fortified wines.

Preservations with sugar

The water binding effect of the addition of large amounts of sugar (cane sugar, invert sugar, grape sugar, glucose or starch syrup) reduces the water content of foods below the limit which most bacteria and fungi can grow. At the same time the osmotic effect of a concentrated sugar solution on their enzyme systems inhibits the microbes. This double withdrawal of water (dehydration) achieves protection from microbial decomposition. This process is used especially for preserving fruit and in the manufacture of candied fruit (orange and lemon), fruit syrups, marmalades, jams, jellies and other products. Addition of acid increases the preservative action of sugar. Different amounts of sugar are necessary according to the water content of the products to be preserved: jams and marmalades need 50–55%, pulps about 40–45%, syrups need 60–65% additional sugar. Sugar is also used to preserve milk as in sweetened condensed milk.

13.3 CHEMICAL ADDITIVES IN FOOD PRESERVATION

The use of a few chemical preservatives can be traced back to the early time in human history.

Examples are the use of salt, vinegar, sugar or honey for preservation, the use of sulphur in the cask when making wine, saltpetre for pickling, or smoke to increase the keepability of meat and fish.

Intentional and non-intentional additives

For various reasons the number of individual additives has increased to such an extent that recently from the point of view of health, it seems desirable to have clearer definitions of additives. Apart from the actual chemical preservatives which arrest microbial decay in the widest sense, all materials which prevent or retard chemical, physical and biological changes in foods must be included under the headings of preservatives. Scientific literature and legislation uses the terms **'additives'** and 'non-intentional additives' without having adequate distinctions between them. These terms overlap to a large extent, but are not identical. Additives are those materials which have been intentionally added to the food at some stage of processing, but they need not be foreign materials to the food. Examples are vitamins, amino acids and minerals, which increase the nutritive value of groups of foods; moreover salt, spices and water should be included here. Foreign materials, on the other hand, are materials which do not occur in the food in nature, but are intentionally or unintentionally added or get taken up into the food and become part of it. Examples of foreign materials which are not additives are the residues of many materials used in agriculture, such as pesticides, materials used to protect stored material, contaminants of all kinds. Materials which are formed in the food by some process of treatment (heat, cold, irradiation) are also foreign materials rather than additives.

In the Anglo-Saxon literature, additives and foreign materials are usually both included under additives, but they distinguish between intentional additives and those (non-intentional) incidental additives which have reached the food unintentionally.

The food laws of the German Federal Republic define the term 'foreign materials' as those with no nutritional value, added to the food.

The food laws of other countries contain other definitions for additives and foreign materials. There are at present no uniform regulations.

The advisability of using additives to protect the quality of our food in suitable cases is undisputed and generally recognised. The huge increase in world population makes it essential to increase yields, to find new sources of food and to protect the available food from decay.

Many factors combine to provide justification for the use of intentional additives as an additional measure to preserve the quality of our food. Included here are the crowding together of large masses of population in industrial centres who must be supplied with fresh and preserved food together with a large reserve stock of food which must also retain its quality. Added to this is the long time frequently needed for transport of food from the production area to the area where it is consumed. Finally there is the change to foods more easily digested which is connected with the general reduction in physical activity of the aged.

Pesticides are a special group of incidental additives which protect the harvest from many types of spoilage. If they were no longer used it would mean death from hunger for many people. Other special incidental additives are chemical additives such as sulphur dioxide which prevent the decomposition of semi preserved products. Some gelling agents such as emulsifiers, baking acids, moistening agents and colouring matters have an advantageous effect on the structure and appearance of our food and so add directly or indirectly to the nutritive effect.

Technical additives are special types of additives. This group includes, subject to the food laws of various countries, those materials which are used in the manufacture and processing of food, but are not intended for consumption. These materials may only be present in food in amounts which cannot be technically avoided, or in amounts which do not exceed a fixed permitted quantity. These include solvents, extractants, precipitating agents, catalysts and enzymes, hydrolysing agents, neutralising agents, buffers, cleansing and disinfecting materials as well as absorbants which clarify or bleach, for example, bleaching earths. When these are carefully used there is little danger of them building up a residue in the food, or being transferred into the food. This means that, apart from exceptions, they present no special hygienic or toxicological problem.

About 120 additives are presently in use in many countries, which does not include those which have unintentionally reached the food in very small amounts.

Various countries have made regulations about permitted additives both intentional and unintentional, and their general basic principles are internationally recognised.

The following general principles apply for permitted incidental additives: The use of additives is only justified if there is demonstrably a real need from the consumers point of view or if a technical necessity has been shown. Also no more material may be used than the absolute minimum.

From a health point of view, the safety of additives can be ensured by setting up a '**positive list**', in which are listed all materials which have been proved, as far as is known, to be harmless and are therefore permitted.

The laws in some countries operate this procedure; in place of the 'principle of misuse' the principle of 'forbidden' is applied. This means that lists of materials which endanger health and are therefore forbidden are not used. Other countries, however, do not follow this procedure.

For safety with continuous consumption an adequate safety factor is necessary as well as tests on the possible synergistic effect of taking several additives at the same time and the possible effects of decomposition products or toxic materials in the food have all to be considered. This means that to avoid possible danger to health, toxicological, biochemical and nutritional tests are carried out continuously according to internationally agreed principles (WHO/FAO).

It can be seen that **staple foods** which need to be of high quality such as dietetic foods, baby foods and those which the consumer expects to be **fresh foods**, must be substantially free of incidental additives. Special criteria should apply to any exceptions.

Special mention should be made of the fact that additives must only be used for perfect foods, and then only if the addition does not give them the appearance of being better than it is and so deceive the consumer about the true value of the food. The presence of additives should also be made known to the user by suitable declaration on the label.

Classification

There has been no lack of attempts to arrange the huge fields of additives into groups. Either the method of action or the purpose of the additive has been used as the principle for classification. The latter method applies to the following classification:

(a) materials added **intentionally**

(b) materials which reach the food **unintentionally**

Chemical preservatives added intentionally are not only those active against microbial changes (chemical additives in the narrow sense), but also those which prevent chemical, physical and biochemical changes in the food (chemical preservatives in the wider sense). Preservatives added **intentionally** are therefore those materials which delay or are designed to delay or prevent undesirable changes in foods, in as far as they themselves are components of the food. This definition considerably extends the concept of foreign materials which are deliberately added to the foods as 'materials to preserve quality'. All materials which prevent undesirable changes, including those which reduce the nutritional value (loss of vitamins, changes in fats and proteins), and those of an economic nature, (e.g. the decreased value of stale bread, softening of fruit, limpness of vegetables) leading to unsuitability for consumption are included here.

Intentionally added materials (a) can be subdivided into those with chemical action; those with physical action; those with physiological action and other additives.

Intentional additives with **chemical** action can be conveniently classified as follows:

1. antimicrobial agents (**preservatives** in the narrow sense)

2. antioxidants
3. complexing agents
4. agents which improve colour (materials to initiate colour changes such as nitrite, nitrate, copper salts and materials which prevent undesirable discolouration)
5. bleaching agents (oxidising and reducing agents)
6. enzymes (e.g. papain)

Chemical additives with mainly **physical** action include:

1. colouring agents
2. thickening and gelling agents
3. surface active agents (emulsifiers, stabilisers, solvents, materials to prevent bread staling)
4. foaming and stabilising agents
5. agents which prevent foaming
6. baking powders
7. materials which retain moisture
8. coating materials (coverings to prevent drying out or attack by microbes, glazing materials)
9. firming agents for vegetable tissue (certain calcium salts)
10. material to prevent separation in drinks
11. separating agents

Additives with **physiological** action include:

1. materials to improve the nutritive value (vitamins and provitamins, amino acids, mineral salts)
2. materials to improve the flavour and aroma (sweetening materials, salty and acidic materials, bitter tasting materials, natural and artificial flavour materials, ethereal oils, flavour enhancing materials such as sodium glutamate, components of smoke etc.)
3. materials with special dietetic action, stimulants such as caffeine.

Other additives include the reference materials, starch, sesame oil, hydroxymethylfurfural and the denaturing or bittering materials.

The second main group (b), the additives which reach the food **unintentionally** is also varied and can be divided as follows:

1. materials used in cattle raising and poultry production
2. materials used to treat soil,
3. materials used to protect plants (pesticides),
4. materials used to protect stored food (antimicrobial agents, gassing agents, and materials which affect ripening)

Technological additives are used in industrial processes. They include solvents, extraction agents, precipitating agents, absorption agents in the treatment of drinking water, decolourising or clarifying agents, catalysts, enzymes, cleansing and disinfecting agents, acids for hydrolysis, caustic alkalies, neutralising agents and buffering agents.

Materials which are produced during technological processing can also be included here. They are materials which can be formed through physical processing (e.g. by heat, cold or irradiation), or chemical processing.

A final group includes those materials which contaminate foods in general, grease on parts of the processing machinery, residues of packaging materials, dirt, radioactive fall out and chemical materials.

It is beyond the scope of this book to discuss all these in detail. Therefore only chemical additives, which are particularly important will be mentioned here.

Chemical preservatives

The permitted levels of preservatives vary from one country to another. The European Economic Community has published Directives and The World Health Organisation/Food and Agriculture Organisation of the United Nations have also published recommendations which it is hoped will lead eventually to standardisation.

The United Kingdom is developing a system of 'Permitted Lists'. The permitted preservatives at present number 34 and include:

Nitrates and **nitrites** in meat products;

Sulphur dioxide and **benzoic acid** in fruit and vegetable products;

Antioxidants, 14 in number, are permitted in foods, and six of these are specifically controlled. Their main use is either in oils or fats or in foods containing oils and fats. Restrictions are imposed on the propyl, octyl and dodecyl gallates; butylated hydroxyanisole (BHA), butylated hydroxytoluene (BHT) and ethoxyquin. The other eight antioxidants which include ascorbic acid, ascorbates and the tocopherols are permitted in any food, subject to the general provisions of the Food and Drugs Act 1955, with later amendments. **Propionic acid** and **sorbic acid** are both permitted up to a given level in cakes, but only propionic acid in bread.

Some preservatives are permitted at specific low levels. Additional evidence for safety and about toxicology is necessary before suitable levels can finally be determined. Nitrates and nitrites belong to this group. Other preservatives such as 'Liquid Smokes' have now been withdrawn from use as preservatives.

In the United Kingdom preservatives are used in bacon, ham and cured meats; beer; dried fruit; fruit juices and fruit pulps; jams; potatoes, both raw, peeled and dehydrated; sausages and sausage meats; soft drinks; dried vegetables, other than brussel sprouts, cabbage and potatoes; bread and cakes. The antioxidants are permitted in edible oils and fats, in butter for manufacturing purposes and for apples and pears.

In the United States, the Food and Drug Administration controls the use of preservatives and other additives. The lists of permitted preservatives and their levels are slightly different from the United Kingdom.

The United States have the special classification GRAS (generally recognised as safe) which can be used for food additives for which long usage indicates that they are safe but for which safety and toxicological testing is lacking. The U.S. authorities, during the past few years, have removed a number of compounds from the GRAS classification and have asked for proof of safety. The aromatic acid, para-hydroxybenzoic acid and its esters may be used as preservatives in the States. The methyl and propyl esters are GRAS and are called methyl and propyl parabens. Sorbic acid and its salts are effective antifungal agents used in cheese products, fruit juices, wines, baked products and other foods. Sorbic acid is also classified as GRAS.

Preservative mechanism. Investigation of the **kinetics** of the **antimicrobial action** of preservatives has shown that the mechanism is a **reaction of the first order**; when the concentration of the preservative is high enough, the microorganisms are progressively killed off, so that the substrate gradually becomes sterile if no new infection occurs. However, if the killing of the microorganisms is stopped too soon, either by dilution or evaporation of the preservative, by alteration of the pH value of the substrate or for other reasons, renewed growth of those microbes which had not been killed can occur. The **rapidity** with which death occurs is given by the **death constant**. It is not always the same, but rather depends on the **type of preservative**, on its **concentration**, on the **type of organism** and on the **temperature**. It is independent of the **number of microorganism cells present**. The death process, which can be described mathematically, presupposes uniform cell material and therefore a single culture; which in practice is never the case.

If the amount of preservative is **inadequate**, the microorganisms present will multiply. This increase, however, will be slow compared with a test sample with no preservative, and slower the more preservative is added. After a shorter or longer period of time, if the preservative is insufficient, the food will decay. To **summarise, a substrate which is mixed with preservative will tend either to killing off of all microbes (sterility of the substrate) or towards decay.**

The usual distinction between **bactericidal** and **bacteriostatic** effects is not confirmed by modern work. With adequate concentrations of preservative, both effects only differ from each other in the size of the death constant, i.e. the greater or lesser speed with which the microorganisms are killed.

There are still particular difficulties in identifying all the factors in the antimicrobial action of a preservative, and in no case have they been adequately and comprehensively explained. As well as physical and physico-chemical mechanisms, such as absorption, diffusion, resorption, denaturation and destruction of the cell membranes, there are also **chemical** reactions. An example of the latter is the exchange of the terminal groups of enzyme proteins with prothetic groups on enzymes, as well as inactivation (e.g. of sulphydryl groups of the enzymes) which leads to blocking of vital cell metabolism of carbohydrate, fat or protein in the microorganism.

Special attention must be paid to **combinations of different food preservatives**; the bactericidal effect of some combinations considerably exceeds the effect of one component alone and so make it possible to economise on preservative. The combination may also contain materials which themselves have no antimicrobial action, such as salts, especially cooking salt, which increases cell membrane penetrability by swelling, or acids, which improve the activity of preservatives dependent on pH by improved permeability.

The properties of the food itself as a substrate, especially it **pH value**, influence the process of preservation and may be very different from food to food.

The (non-dissociating) **esters of p-hydroxybenzoic acid are independent of pH** and barely dissociate in neutral and weak acid reaction substrates (foods) where they are effective as **antimicrobial** agents.

However, most permitted preservatives are dependent on pH for their effectiveness. They are themselves weak acids and are added in the form of salts. They only have an antimicrobial effect in more or less acid foods in an environment where they are present as acids in a non- or slightly dissociated state.

As far as is known at present, only the **non-dissociated** lipid soluble compounds, not the hydrophilic acid ions, pass through the semi-permeable cell membranes of the microorganisms and are able to distribute themselves in protoplasm. This would explain the fact, long known in practice, that sodium benzoate, or the salts of other acids whose use as preservatives is permitted, only have a preservative effect in a strongly acid environment, but are inadequate in weak acid or neutral foods.

Table 25 gives the **dissociation** values of some preservatives at **various pH values**. The figures apply to the pH values after the preservative has been added to the substrate, and it is immaterial whether the free acid or its salt was added. Table 25 shows that, in accordance with the law of Mass Action, the dissociation of an acid is determined by the pH value and that, with decreasing strength (i.e. with decreasing value of the dissociation constant) of the acid preservative, even in a neutral or weak acid reaction, increasing numbers of undissociated molecules are present, which have an antimicrobial effect. The weaker the acid preservative (i.e. the smaller its dissociation constant), the closer the pH value of the whole system can come to the neutral point, without effecting the preservative action. **Non-dissociating** preservatives are therefore relatively **independent of pH** in their action, that is, they can be used in the neutral state.

Table 25 The proportion of undissociated (antimicrobially active) acid for various preservatives dependent on the pH value of the substrate.

Preservative	Dissociation constant K	Undissociated portion in % at: pH 2	pH 3	pH 4	pH 5	pH 6	pH 7
sulphurous acid	$1.7 \cdot 10^{-2}$	37	5.5	0.55	0.04	0.001	0.0000
bromacetic acid	$2.05 \cdot 10^{-3}$	83	32.8	4.65	0.48	0.049	0.0049
salicylic acid	$1.06 \cdot 10^{-3}$	90	49	8.6	0.94	0.094	0.0094
formic acid	$1.77 \cdot 10^{-4}$	98.3	85.0	36.1	5.35	0.56	0.056
p-chlorobenzoic acid	$9.3 \cdot 10^{-5}$	99	91	52	9.7	1.06	0.107
benzoic acid	$6.46 \cdot 10^{-5}$	99.3	93.9	60.7	13.4	1.52	0.15
p-hydroxybenzoic acid	$3.3 \cdot 10^{-5}$	99.7	96.8	75.2	23.2	2.94	0.30
acetic acid	$1.8 \cdot 10^{-5}$	99.8	98.2	84.7	35.7	5.26	0.55
sorbic acid	$1.73 \cdot 10^{-5}$	99.8	98.3	85.2	36.6	5.46	0.57
propionic acid	$1.4 \cdot 10^{-5}$	100	99	88	42	6.7	0.71
hydrogen sulphite	$1.02 \cdot 10^{-7}$	100	100	99.9	99.0	90.7	49.5
boric acid	$8.3 \cdot 10^{-10}$	100	100	100	100	100	100

When the dissociation constant (K) of a preservative is known, as well as the hydrogen ion concentration $[H^+]$ or the pH value of the substrate (food), the amount (ρ) of **non-dissociated** acid can be calculated from the following formula:

$$\rho = \frac{[H^+]}{[H^+] + K}$$

Acids which readily dissociate, only develop a preservative action in foods with a strong acid reaction. This includes **formic acid, benzoic acid, sorbic acid** and also **salicylic acid**, formerly much used, now forbidden. The proportion of sulphurous acid dissociated increases with increasing pH value but HSO_3' and SO_3'' ions still have a certain preservative action.

Hydrogen sulphite ions have a small dissociation constant, and they are still partially present to some extent at high pH, so that sulphurous acid can develop some antimicrobial action in substrates which are not markedly acid. Boric acid is completely undissociated even at neutrality. This may be the reason why it is still used in some countries in spite of reservations about its safety, for example in the fish industry to preserve crabs and caviar, and also in liquid egg products.

A number of other properties of the substrate have an indirect influence on the antimicrobial action of preservatives, because they improve or worsen the essential conditions for microbial life. These include the **oxygen partial pressure** and **redox potential** (pH of the substrate), the **water content** and **osmotic pressure**, the **temperature** and the **relative humidity** of the air as well as the absorption of the food substrate.

The effect of the antimicrobial action of the preservative may also be the reason why foods can act as **ion exchangers**. This may also be advantageous for example, when **calcium acetate** or **calcium propionate** are used for preservation. The acids which alone are effective, may be set free by ion exchange. There may also be chemical reactions between the substrate and preservative which can affect the preservative effect unfavourably. An example of this is the reaction between aldehydes such as acetaldehyde or carbohydrates (aldoses) with sulphurous acid to form hydroxysulphonic acids, in which the **sulphur is bound**.

Preservatives may affect the colour of foods by deepening or lightening (bleaching) the colour, or by causing new colours to appear. **Phenolic antioxidants** (see p. 118) especially can bring about **blue, violet** or **brown** discoloration when **iron salts** are present. The addition of **sulphurous acid** or **hydrogen peroxide** usually causes **bleaching**. SO_2 sometimes does the same so that a better colour is obtained for instance, in dried fruit. Frequently the bleaching or retention of colour is particularly desired in addition to preservation.

There is also the possibility that the **taste or smell** of the preservative will be transferred to the material to be preserved, for example with **benzoic acid, p-hydoxybenzoic acid esters, diphenyl, sulphurous acid, formic acid** and esters of **gallic acid**. Gases used for gas storage can be absorbed by the foods and trapped there, causing changes in flavour. Unpleasant changes in flavour may also arise from secondary reactions between preservatives. and food. When salicylic acid has been used a phenolic flavour may easily arise during cooking whereas boric acid and borates can intensify unpleasant flavours in crabs and preserved eggs. Vitamins may also be subject to many changes. Oxidising preservatives, such as hydrogen peroxide destroy, more or less rapidly, vitamin A, carotene, vitamin C as well as other vitamins sensitive to oxidation. The destructive action of sulphurous acid and its salts on vitamin B_1 and on dehydroascorbic acid (see vitamin C, p. 227) is typical. In general, however, these effects on foods are negligible, and only when uncontrolled or excessive amounts of preservatives are used. The food laws are enacted to guard against such occurrences.

Important Preservatives. In addition to the general description, some details will be given here concerning the use of these preservatives.

Sorbic acid and its salts.

$$CH_3{-}CH{=}CH{-}CH{=}CH{-}COOH:$$

Sorbic acid, a 2,4-hexadiene acid, is used preferentially in the preservation of foods, because of its effectiveness in inhibiting the growth of yeasts and moulds. It is found as a lactone in the fruit of the mountain rowan tree and is a very stable compound although it has a conjugated double bond. In spite of its marked antimicrobial action it does not affect the odour, flavour or structure of foods. The antimicrobial action of sorbic acid is mainly due to **undissociated** molecules, which is why this preservative is used preferentially in an acid environment (see Table 25, p. 320).

A certain minimum concentration of sorbic acid is necessary for adequate action, analogous to the antimicrobial effectiveness of other straight chain, aliphatic, unsaturated monocarboxylic acids. These minimum concentrations usually lie between 0.1 and 0.25% to 1.2%. However, lower level doses of sorbic acid are useless if preservation is aimed at because moulds, in particular, can break down small amounts of sorbic acid to CO_2 and water during metabolism and certain less sensitive bacteria are even able to hydrate the double bond in the sorbic acid molecule.

Sodium chloride solutions (8% NaCl) increase the action of sorbic acid by three to five times and sugar also has a synergistic effect.

Sorbic acid is considered to be safe from a physiological point of view because in human metabolism it is broken down like a normal edible fat. For this reason sorbic acid and its sodium, potassium and calcium compounds (also in **combination** of other preservatives) is widely used particularly where protection is needed or required against moulds and yeasts.

Benzoic acid and its salts:

$$C_6H_5\text{—COOH(Na)}$$

Benzoic acid in amounts between 0.1 and 0.4% and in special cases higher amounts, are permitted as preservatives in many foods and groups of foods.

The **antimicrobial action** depends on pH, only the free **non-dissociated** acid is effective. By reducing the pH value from 7 to 4 the activity becomes 400 times as high (see Table 25).

Whereas protein reduces the antimicrobial effectiveness of benzoic acid, phosphates and chlorides have a synergistical, or supporting effect. As its action depends on pH, benzoic acid is an especially suitable preservative for acid foods, such as semi-manufactured goods from fruit (basic materials for fruit juices, fruit purees and fruit pulps), and those manufactured from fruits (marmalades, jams, jellies and fruit drinks). Benzoic acid occurs naturally in cranberries and acts as a natural preservative. Benzoic acid can also be used for products made from vegetables (onion, pickled conserves, sauces for seasoning). In foods which are high in protein, benzoic acid can also be used: for fish products, crab and lobster products, for egg products with a long shelf life, whole liquid egg, liquid egg yolk, table mustard, and marzipan or products similar to marzipan.

In practice, mixtures of benzoic acid with formic acid or citric acid (where permitted) are used to increase effectiveness, particularly with semi-preserved foods.

p-Hydroxybenzoic acid ester and its sodium salts:

$$\text{(Na)HO—}C_6H_4\text{—COOC}_2H_5(C_3H_7)$$

Free p-hydroxybenzoic acid has practically no preservative action, although it has an antimicrobial effect. The esters of p-hydroxybenzoic acid which have long been used as preservatives are much more effective than the free acid and, as they depend very little on pH, are therefore also very suitable for preserving neutral foods. Combining the various esters increases the antimicrobial effectiveness. The quantities necessary for preservation are relatively small (about 0.05–0.1%). The esters used have practically no effect on smell or taste and have good preservative action. The activity of the sodium compounds is only slightly less than that of the free esters. The preservative action of the esters increases rapidly from the methyl esters through the n-butyl ester to the benzyl ester. Details about their use in various foods and groups of foods can be found in the various food regulations of different countries.

Formic acid and its derivatives:

The antimicrobial action of **formic acid**, HCOOH, and its salts affects bacteria, moulds and yeasts. The optimum effectiveness of formic acid as a preservative is in the acid region; around pH 3 about 85%, and at pH 6 only 0.56% of the antimicrobially active, undissociated acid is present (see Table 25, p. 320).

Sulphurous acid and its derivatives:

Sulphurous acids and its derivatives have long been known and widely used as preservatives and antioxidants. (The sodium salt, sodium metabisulphite, is sold as Campden Tablets in the United Kingdom). Sulphur dioxide is a useful preservative, inhibiting the Maillard or 'Browning Reaction' as well as many enzymatic processes. For these reasons it has been widely used in many foods such as dried fruit, meat products such as fresh sausages and in wines. The W.H.O./F.A.O. has laid down guidelines for the Acceptable Daily Intake of sulphur dioxide and its compounds; hence its use is now being much more carefully controlled.

In addition to sulphur dioxide (SO_2), or sulphurous acid (H_2SO_3), sodium sulphite ($Na_2SO_3 \cdot 7H_2O$), sodium hydrogen sulphite or sodium bisulphite ($NaHSO_3$), sodium pyrosulphite or sodium metabisulphite ($Na_2S_2O_5$), potassium pyrosulphite or potassium metabisulphite ($K_2S_2O_5$), calcium hydrogen sulphite or calcium bisulphite [$Ca(HSO_3)_2$] are used.

Sulphurous acid (as an aqueous solution of SO_2) is largely undissociated, so that the majority of it is present as unchanged SO_2 in equilibrium.

The microbicidal action of the sulphurous acids and their derivatives is strongly dependent on pH and environment. It commences at concentrations of about 0.05% and becomes optimal at 0.5% at pH 4.

The reversible action of sulphurous acid with reducing carbohydrates and the formation of **'bound sulphurous acid'** (glucosulphurous acid etc) is particularly important in commercial practice. The equilibrium is strongly dependent on pH.

At pH between 3.0 and 5.8 the addition products between reducing carbohydrates and sulphurous acid are most stable. In the stomach sulphurous acid is set free from them. The glucose sulphite complex, in contrast to other adducts, which occur in practice (such as the formaldehyde and acetaldehyde compounds) is broken down especially rapidly and is therefore of particular interest to toxicologists. In contrast to the adducts of sulphurous acid with formaldehyde and acetaldehyde, the **glucose-sulphurous acid** adduct has a definite effect in inhibiting bacteria and moulds.

Sulphurous acid inhibits the action of various vegetable enzymes (phenolases, dehydrogenases); it also changes the S–S bonds in cystine into sulphhydryl groups. Sulphur dioxide destroys vitamin B_1. It does not accumulate in the body.

In addition to the antimicrobial action, sulphurous acid in many cases also has a distinct effect in preserving colour and also exhibits an antioxidant effect in the preparation of dried fruits.

People with a low acid or subacid stomach have considerable sensitivity to sulphurous acid, a fact which, even today, makes the use of sulphurous acid questionable, although at present (e.g. in the manufacture of wine and grape juice) it cannot be replaced by any other food additive. Sulphurous acid is also used to treat peeled apples, peaches, nuts, mushrooms and fungi and potatoes in canning and bottling, in preparing dried fruit (apples, pears, apricots, peaches, raisins) and dried vegetables (celery, cauliflower, asparagus) as well as in the commercial production of sucrose, and starch products (potato starch, sago). It plays a considerable role, essential to the technology of production in the potato processing industry, for example in frozen, ready prepared dishes, mashed potato, in fried potato products (crisps, chips) and finally in dehydrated potatoes. Other uses are in the preparation of dried mixed peel (for cake baking) and in preparing sugar preparations from starch (syrup). Finally, there is the large field of use in preserving fruit pulps, fruit pastes, concentrated fruit juices, juices from citrus fruit, fruit juices (pectin concentrates). Sulphurous acid is also still essential for rendering germ free containers such as bottles, barrels and vats (Campden tablets).

Propionic acid and its salts.

$$CH_3{-}CH_2{-}COOH:$$

Its preservative action varies greatly from one type of microorganism to another. Only the undissociated molecule is active against microbes. The quantities necessary lie between 0.1 and 6%. Ca-propionate prevents the formation of mould and ropiness in bread, but recently a mixture of the sodium salts of propionic acid and sorbic acid has proved advantageous in preventing mouldiness in bread.

Pyrocarbonic acid diethyl ester.

$$C_2H_5O{-}CO{-}O{-}CO{-}OC_2H_5:$$

This very reactive molecule is difficult to dissolve in water, but readily dissolves in organic solvents and fatty oils. In water the ester hydrolyses to alcohol and carbon dioxide, in foods it reacts to form compounds with phenols (polyphenols), amines, fruit acids (carboxylic acids) and other materials present in food. As well as the esters diethylcarbonate is also formed, which may be important for analytical proof of added pyrocarbonic acid diethyl esters. The pyrocarbonic acid diethyl ester has been suggested for the preservation of beverages such as wine, beer, cider, fruit drinks and fruit juice drinks. The effective dosage is between 200 and 800 mg/l. The antimicrobial action then is active against lactic acid bacteria, yeasts and film forming yeasts and to a slight extent also moulds. As already mentioned, the microbial action of pyrocarbonic acid diethyl ester starts to decrease after a few hours, because of rapid hydrolysis with water, so that in certain cases those spores still present may germinate or secondary infection may occur. At higher concentrations there may be changes in colour and reduction in flavour of the food due to the reaction possibilities described above.

Diphenyl (biphenyl); o-phenylphenol and its sodium salt.

OH(Na)

Diphenyl (biphenyl) is a white, steam volatile substance, insoluble in water and with a pleasant smell. Nowadays it is used all over the world to preserve citrus fruit. The packing material (wrappers, crates or cardboard dividers) are impregnated with diphenyl. The quantities used lie around 0.9 g diphenyl to 1 kg fruit. The high vapour pressure of the compound and its ready solubility in lipids means that diphenyl is transferred mainly to the skins of the citrus fruit and to a slight extent also reaches the flesh and the juice of the fruit. The antimicrobial action of diphenyl is already active at a concentration of 0.08 mg diphenyl/l of air and inhibits the development of green and blue moulds especially. When there is lengthy transport and storage, citrus fruit are always liable to attack by these moulds and high losses ensue.

o-Phenylphenol is widely used in the same way as diphenyl for the preservation of citrus fruits, especially against attack by moulds. The wrappers are either impregnated or the fruit are dipped in a solution of 0.5–2% of sodium-o-phenylphenolate, or the skins are impregnated with the preservative using waxes, wax emulsions or resins. Like diphenyl, o-phenylphenol and its soduim salts penetrate the skin of the fruit, and to a small extent the flesh also, sometimes making their presence known as an off flavour. Thiabendazole, a 2-(4-thiazolyl)-benzimidazole: m.p. 304–305°C, sublimation from 250°C; largely stable in aqueous suspension and in an acid environment (pH 2); practically insoluble in water, only slightly soluble in alcohols, esters, chlorinated solvents; soluble in dimethylformamide, dimethylsulphoxide. This fungicide, absorbed by leaves and roots, is used as a spray powder in the cultivation of bananas and citrus fruits.

Hexamethylenetetramine.

Hexamethylentetramine has no antimicrobial effect itself; its microbial action depends on the splitting off of formaldehyde, which is dependent on pH and occurs much more strongly in an acid environment (foods, gastric juices) than near neutral pH. Hexamethylenetetramine (Hexa) is only active against certain microbes and specific bacteria. The inhibition of growth of budding yeasts and moulds needs 20–100 fold increase in concentration, which is effective against bacteria.

Hexamethylenetetramine has been widely used in many countries to preserve fish products, in spite of reservations about its effect on health. It is very suitable for ensuring keepability of cold marinaded fish or shell fish, for salt fish in oil, for caviar and other fish roe, for anchovies and crab products. The amounts used are between 250–800 mg/kg. Hexa is frequently combined with other preservatives in concentrations of 0.005–0.01%. It has no effect on flavour. Its use is now restricted in many countries.

Nitrite, nitrous acid and its sodium and potassium salts

Nitrite, which can be added direct to a curing solution or pickling solution, can also be produced by bacterial reduction of nitrate. Nitrite reacts with myoglobulin to give the characteristic pink colour of cured meat and bacon, nitroso-myoglobin. Nitrite also contributes to the flavour but its most important role is in inhibiting the growth of development of *Clostridium botulinum* and the deathly botulinus toxin. Strict limits are placed on the usage of nitrites and nitrates could react with secondary amines in food, under acid conditions, to form nitrosamines which are active as carcinogens in animals.

Literature for Further Study

1.1 NUTRITION, NUTRIENT DEMAND AND NUTRIENT CONTENT

Composition of Foods, R. A. McCance and E. M. Widdowson, 4th ed., ed. D. A. T. Southgate and A. Paul, HMSO, London, 1978

Recommended Daily Intakes of Nutrients for Food Groups in the United Kingdom, DHSS, HMSO, London, 1979

Reports of the Nutrition Board, *Recommended Daily Allowances,* National Research Council, 9th ed., Washington, 1980

Human Nutrition and Dietetics, L. S. P. Davidson, R. Passmore, J. F. Brock and A. S. Truswell, 6th ed., Churchill-Livingstone, Edinburgh, 1975

Introduction to Modern Biochemistry, P. Karlson, 4th ed., Academic Press, New York and London, 1970

Biochemistry, A. H. Lehninger, 2nd ed., Worth Publishers Inc., New York, 1975

Dictionary of Nutrition and Food Technology, A. E. Bender, 4th ed., Newnes-Butterworths, London, 1975

Nutrition and its Disorders, D. S. McLaren, 2nd, Churchill-Livingstone, Edinburgh, 1976

Food Processing and Nutrition, A. E. Bender, Academic Press, New York and London, 1978

The Analysis of Nutrients in Food, D. R. Osborne and P. Voogt, Academic Press, New York and London, 1978

Nutritional Evaluation of Food Processing, ed. R. S. Harris and E. Karmas, 2nd ed., AVI Publishers, Westport, 1975

1.2

Trace Elements in Human and Animal Nutrition, E. J. Underwood, 4th ed., Academic Press, New York and London, 1977

Nutrients in Processed Foods: Fats and Carbohydrates, ed. P. L. White, D. C. Fletcher and M. Ellis, Amer. Med. Assoc., Publishing Sciences Group, Acton, 1975

Nutrients in Processed Foods: Proteins, ed. P. L. White and D. C. Fletcher, Amer. Med. Assoc., Publishing Sciences Group, Acton, 1974

Nutrients in Processed Foods: Vitamins and Minerals, ed. P. L. White and D. C. Fletcher, Amer. Med. Assoc., Publishing Sciences Group, Acton, 1974

2.1 FOOD SCIENCE AND TECHNOLOGY: FOOD PRESERVATION

Advances in Food Research, continuing series, Vol. 1, 1948 to Vol. 24, 1978, Academic Press, New York and London

Principles of Food Science, Part I, *Food Chemistry,* Marcel Dekker, New York, 1976; Part II, *Physical Principles of Food Preservation,* Marcel Dekker, New York, 1975

Food Science, N. N. Potter, 3rd ed., AVI Publishers, Westport, 1978

Principles of Food Chemistry, J. M. de Man, AVI Publishers, Westport, 1976

Principles of Food Science, G. Borgstrom, 2 vols., Academic Press, New York and London, 1968 and 1969

Freezing Preservation of Foods, ed. D. K. Tressler, W. B. van Arsdel and M. J. Copley, 4 vols., AVI Publishers, Westport, 1966, 1968

Encyclopaedia of Food Technology, ed. A. H. Johnson and M. S. Peterson, AVI Publishers, Westport, 1974

Symposium on Foods: Chemistry and Physiology of Flavors, ed. H. W. Schultz, E. A. Day and L. M. Libbey, AVI Publishers, Westport, 1967

Flavor Chemistry, ed. I. Hornstein, Amer. Chem. Soc. Advances in Chemistry Series 56, Amer. Chem. Soc. Washington, 1966

Gustation and Olfaction, ed. G. Ohloff and A. F. Thomas, Academic Press, New York and London, 1971

Food Chemistry, L. W. Awrand and A. E. Woods, AVI Publishers, Westport, 1973

The Flavor Profile, L. B. Sjostrom, A. D. Little Inc., Cambridge, Mass., 1972

Flavor Research – Principles and Techniques, ed. R. Teranishi, I. Hornstein, P. Issenberg and E. L. Wick, Marcel Dekker, New York, 1971

Fruit and Vegetable Juice Production, D. K. Tressler and M. A. Joslyn, AVI Publishers, Westport, 1954

Commercial Vegetable Processing, ed. J. G. Woodroof and B. H. Luh, AVI Publishers, Westport, 1975

Food Texture, S. A. Matz, AVI Publishers, Westport, 1962

Industrial Rheology, ed. P. Sherman, Academic Press, New York and London, 1970

Viscosity and Flow Measurement, J. R. van Wazer, J. W. Lyons, K. Y. Kim and R. E. Colwell, John Wiley & Sons, New York, 1966

Food Microscopy, ed. J. G. Vaughan, Academic Press, New York and London, 1979

Encylopaedia of Food Science, M. S. Peterson and A. H. Johnson, AVI Publishers, 1978

Cheese and Fermented Milk Foods, F. Kosikowski, 2nd ed., Edwards Bros, Ann Arbor. 1977

Elements of Food Technology, ed. N. W. Desrosier, AVI Publishers, Westport, 1977

The Technology of Food Preservation, N. W. Desrosier and J. N. Desrosier, AVI Publishers, Westport, 1977

Thermobacteriology in Food Processing, C. R. Stumbo, 2nd ed., Academic Press, New York and London, 1973

Food Colloids, H. D. Graham, AVI Publishers, Westport, 1977

Tomato Production, Processing and Quality Evaluation, W. A. Gould, AVI Publishers, Westport, 1974

Improvement of Protein Nutrition, Committee of Amino Acids, Foods and Nutrition Board, National Research Council, National Academy of Sciences, Washington, 1974

Food Microbiology: Public Health and Spoilage Aspects, M. P. de Fiquieredo and D. F. Splittstaesser, AVI Publishers, Westport, 1976

Quality Control for the Food Industry, Vol. 1, *Fundamentals,* Vol. 2, *Applications,* A. Kramer and B. A. Twiggs, AVI Publishers, Westport, 1970 and 1973

Food Proteins, J. R. Whitaker and S. R. Tannenbaum, AVI Publishers, 1977

Principles of Dairy Chemistry, R. Jenness and S. Patton, 2nd ed., John Wiley & Sons, New York, 1976

Food Theory and Applications, ed. P. C. Paul and H. H. Palmer, John Wiley & Sons, New York, 1972

Methods in Food Analysis, M. A. Joslyn, 2nd ed., Academic Press, New York and London, 1970

Chemical Analysis of Foods and Food Products, M. B. Jacobs, 3rd ed., Van Nostrand Co., New York, 1962

Instrumental Methods of Food Analysis, A. J. McLeod, Elek Press, London, 1973

Chemical Analysis of Food, D. Pearson, 7th ed., Churchill-Livingstone, London, 1976

Food Industries Manual, ed. A. H. Wollen, 20th ed., Leonard Hill, London, 1969

2.2 FOOD ENGINEERING

Food Engineering; Principles and Selected Applications, M. Loncin and R. C. Merson, Academic Press, New York and London, 1979

Freeze Drying and Advanced Food Technology, ed. S. A. Goldblith, L. Rey and W. W. Rothmayr, Academic Press, New York and London, 1975

Food Engineering Operations, J. G. Brennan, J. R. Butters, N. D. Cowell and A. E. V. Lilly, 2nd ed., Applied Science Pub., London, 1976

Encyclopaedia of Food Engineering, C. H. Hall, A. W. Farral and A. L. Rippen, AVI Publishers, Westport, 1971

Food Process Engineering, D. R. Heldman, 2nd ed., AVI Publishers, Westport, 1977

Fundamentals of Food Engineering, S. E. Charm, 3rd ed., AVI Publishers, Westport, 1978

Dairy Technology and Engineering, W. J. Harper and C. H. Hall, AVI Publishers, Westport, 1976

Food Dehydration, Vol. 1, *Drying Methods and Phenomena,* Vol. 2, *Practices and Applications,* W. D. van Arsdel and A. I. Morgan, AVI Publishers, Westport, 1973

Microwave Heating, D. A. Copson, AVI Publishers, Westport, 1975

Unit Operations in Food Processing, R. L. Earle, Pergamon Press, Oxford, 1966

Handbook of Package Engineering, J. F. Hanlon, McGraw-Hill, New York, 1971

3 CONSTITUENTS OF FOOD; GENERAL

Introduction to the Biochemistry of Foods, J. B. S. Braverman, 2nd edn., Z. Berk, Elsevier, Amsterdam, 1976

Comprehensive Biochemistry, ed. M. Florkin and E. H. Stotz, Vols. 5-21, Elsevier, Amsterdam, 1963-1971

Quality Control in the Food Industry, ed. S. M. Herschdoerfer, 3 vols., Academic Press, New York and London, 1967-1970

Fruit and Vegetables, R. B. Duckworth, Pergamon Press, Oxford, 1966

Technology of Cereals, N. L, Kent, 2nd ed., Pergamon Press, Oxford, 1975

Meat Science, R. A. Lawrie, 3rd ed., Pergamon Press, Oxford, 1977

Recent Advances in Food Science, ed. J. Hawthorn and J. M. Leitch, 4 vols., Butterworths, London, 1962-1968

Fundamentals of Dairy Chemistry, ed. B. H. Webb, A. H. Johnson and J. A. Alford, AVI Publishers, Westport, 1974

Plant Pigments, Flavors and Textures; The Chemistry and Biochemistry of Selected Compounds, N. A. M. Eskin, Academic Press, New York and London, 1979

Modern Dairy Products, L. M. Lampert, 3rd ed., Chem. Pub. Co., New York, 1975

3.1 PROTEINS

The Science of Meat and Meat Products, ed. J. F. Price and B. S. Schweigert, W. H. Freeman & Co, San Francisco, 1971

Advances in Protein Chemistry, ed. J. T. Edsall and others, continuing series, Vol. 1, 1944 to Vol. 38, 1978, Academic Press, New York and London

Proteins in Human Nutrition, ed. J. W. G. Porter and B. A. Rolls, Academic Press, New York and London, 1973

Industrial Uses of Cereals, ed. Y. Pomeranz, Amer. Assn. Cereal Chemists, St. Paul, Minn., 1973

Soybeans: Chemistry and Technology, Vol. 1, *Proteins,* ed. A. K. Smith and S. J. Circle, AVI Publishers, Westport, 1978

Symposium of Foods: Proteins and Reactions, ed. H. W. Schultz and A. F. Anglemeier, AVI Publishers, Westport, 1964

New Protein Foods, ed. A. M. Altschul, Vol. 1, *Technology* Part A, 1974, Vol. 2, *Technology* Part B, 1976, Academic Press, New York and London

The Science and Technology of Gelatin, ed. A. G. Ward and A. Courts, Academic Press, New York and London, 1976

3.2 FATS AND ASSOCIATED SUBSTANCES (LIPIDS)

Margarine, A. J. C. Andersen and P. N. Williams, 2nd ed., Pergamon Press, Oxford, 1965

Bailey's Industrial Oil and Fat Products, ed. D. Swern, 4th ed., Interscience, New York, 1978

Chemical Constitution of Natural Fats, T. P. Hilditch and P. N. Williams, 2nd ed., Chapman & Hall, London, 1964

Progress in the Chemistry of Fats and Other Lipids, ed. R. T. Holman, continuing series, Vols. 1-15, Pergamon Press, Oxford, 1951-1978

Advances in Lipid Research, ed. R. Paoletti and D. Kritchevsky, continuing series, Vols. 1-17, Academic Press, New York and London, 1963-1979

Topics in Lipid Chemistry, ed. F. D. Gunstone, several volumes, Logos Press, London, 1970 onwards

An Introduction to the Chemistry and Biochemistry of Fatty Acids and their Glycerides, F. D. Gunstone, Chapman & Hall, London, 1967

Symposium on Foods: Lipids and their Oxidation, ed. H. W. Schultz, E. A. Day and R. O. Sinnhuber, AVI Publishers, Westport, 1962

Dairy Lipids and Lipid Metabolism, ed M. F. Brink and D. Kritchevsky, AVI Publishers, Westport, 1968

Analysis of Fats and Oils, K. A. Williams, 4th ed., Churchill, London, 1966

Functional Properties of Fats in Foods, J. Solms, 2nd ed., Forster Verlag, Zurich, 1973

The Structure and Utilisation of Oil Seeds, J. G. Vaughan, Chapman & Hall, London, 1970

3.3 CARBOHYDRATES

Advances in Carbohydrate Chemistry and Biochemistry, ed. M. L. Wolfram and others, continuing series Vol. 1, 1945 to Vol. 34, 1979, Academic Press, New York and London

Pectic Substances, Z. I. Kertesz, Interscience, New York, 1951

Symposium: *Sweeteners,* ed. G. E. Inglett, AVI Publishers, Westport, 1974

Sugar Chemistry, R. S. Shallenberger and G. G. Birch, AVI Publishers, Westport, 1975

Symposium on Foods: Carbohydrates and their Roles, ed. H. W. Schultz, R. F. Cain and R. W. Wrolstad, AVI Publishers, Westport, 1969

Starch: Chemistry and Technology, ed. R. L. Whistler and E. F. Paschall, 2 Vols., Academic Press, New York and London, 1967

Sweeteners and Sugars, ed. G. G. Birch, L. F. Green and C. B. Coulson, Applied Science Pub., London, 1971

Determination of Food Carbohydrates, D. A. T. Southgate, Applied Science Publ., London, 1976

Examination and Analysis of Starch and Starch Products, ed. J. A. Radley, Applied Science Pub., London, 1976

Sugars in Nutrition, H. L. Sipple and K. W. McNutt, Academic Press, New York and London, 1974

Sugar, Science and Technology, ed. G. G. Birch and K. J. Parker, Applied Science Pub., London, 1979

3.4 VITAMINS, ENZYMES, MINERALS AND WATER

The Enzymes, ed. P. D. Boyer, H. Lardy and K. Myrback, 3rd ed., 12 volumes, Academic Press, New York and London, 1970–75

Enzymes in Food Processing, G. Reed, 2nd ed., Academic Press, New York and London, 1975

Symposium on Foods; Food Enzymes, ed. H. W. Schultz, AVI Publishers, Eestport, 1960

Phosphates in Food Processing, ed. J. M. deMan and P. Melnychyn, AVI Publishers, Westport, 1971

The Biochemistry of Fruits and their Products, ed. A. C. Hulme, 2 vols., Academic Press, New York and London, 1970 and 1971

Principles of Enzymology for the Food Sciences, J. R. Whitaker, Marcel Dekker, New York, 1972

Water Activity and Food, J. A. Troller and J. H. B. Christian, Academic Press, New York and London, 1978

Vitamins, ed. M. Stein, Churchill-Livingstone, Edinburgh, 1971

Food Related Enzymes, ed. J. R. Whitaker, Advances in Chemistry Series 136, Amer. Chem. Soc., Washington, 1974

Advances in Enzymology, ed. by A. Meister and others, continuing series, Vol. 1, 1941 to Vol. 49, 1979, Interscience, New York.

Modern Food Microbiology, J. M. Jay, 2nd ed., Van Nostrand Co., New York, 1978

Water Relations of Food, ed. R. B. Duckworth, Academic Press, New York and London, 1975

Food Microbiology, W. C. Frazier, 2nd ed., McGraw-Hill, New York, 1967

3.5 ADDITIVES

Handbook of Food Additives, ed. T. E. Furia, Chemical Rubber Co., Cleveland, 1968

The Use of Chemicals in Food Production, Processing, Storage and Distribution, National Academy of Sciences, Washington, 1973
Toxicants occurring naturally in Foods, 2nd ed., National Academy of Sciences, Washington, 1973
Why Additives? The Safety of Foods, ed. The British Nutrition Foundation, Forbes Publications, London, 1977
Toxic Constituents of Animal Foodstuffs, ed. I. S. Liener, 2nd ed., Academic Press, New York and London, 1974
Chemicals in Food and Environment, Medical Department, The British Council, London, 1976
Microbial Inhibitors in Food, ed. N. Molen and A. Erichsen, Almqvist and Wiksell, Stockholm, 1964
GRAS – Concept and Application, R. L. Hall, *Food Technology,* **29**, 48, (1975) and later papers in the same journal
Code of Federal Regulations 21: *Food and Drugs* Parts 1 to 99 and 100 to 199, Revised annually, U.S. Govt. Printing Office, Washington
Processed Foods and the Consumer: Additives, Labeling, Standards and Nutrition, V. S. Packard Jr., University of Minnesota Press, Minneapolis, 1976
Bell and O'Keefe's Sale of Food and Drugs, Sir W. J. Bell and J. A. O'Keefe, 14th ed., revised J. A. O'Keefe, Butterworths, London, 1968
Concise Guide to Food Legislation (U.K.), D. Pearson, Nat. Coll. Food Technol., Weybridge, Surrey, 1977

SELECTION OF GERMAN BIBLIOGRAPHY

For the benefit of readers of this English edition, the Series Editor has compiled the following list of selected German Language books from the Bibliography published in the 3rd German edition, 1976.

1 NUTRITION, NUTRIENT DEMAND AND NUTRIENT CONTENT

Biochemisches Taschenbuch, H. M. Rauen, Berlin, 1964
Die Zusammensetzung der Lebensmittel (Naehrwerttabellen), S. W. Souci, W. Fachmann and H. Kraut Stuttgart, 1969
Lebensmitteltabellen fuer Naehrwertberechnung, S. W. Souci and H. Bosch, Stuttgart, 1967
Toxikologie der Naehrungsmittel, E. Lindner, Stuttgart, 1974

2 FOOD SCIENCE AND TECHNOLOGY: FOOD PRESERVATION

Das Kuehlen und Gefrieren von Lebensmitteln im Haushalt und in Gemeinschaftsanlagen, J. Gutschmidt, Frankfurt/M, 1964
Haltbarkeit und Sorptionsverhalten wasserarmer Lebensmittel, R. Heiss, Heidelberg, 1968

Tiefgefrorene Lebensmittel, K. H. Herrmann, Berlin – Hamburg, 1970
Untersuchungsmethoden in der Konservenindustrie, H. J. Lange, Berlin – Hamburg, 1972
Die Grundlagen der Verfahrenstechnik in der Lebensmittelindustrie, M. Loncin, Aarau and Frankfurt/M, 1969
Industrielle Mikrobiologie, H-J. Rehm, Heidelberg, 1967
Technologie des Zuckers, F. Schneider, Hannover, 1968
Leitfaden moderner Methoden in der Lebensmittelanalytik, I Optische Methoden, II Chromatographische Methoden, H. G. Maier, Darmstadt, 1966 and 1974
Sorbinsaeure Bde 1-4, E. Lueck, Hamburg, 1969-1973
Lehrbuch der Lebensmittelchemie, J. Schormueller, Berlin, 1974
Grundlagen der Lebensmittelmikrobiologie, G. Mueller, Leipzig, 1974

3 CONSTITUENTS OF FOOD: GENERAL

Praktisches Handbuch der Lebensmittel, A. Glas, Munich, 1965
Handbuch der Lebensmittelchemie, I Bestandteile der Lebensmittel, IIA Analytik der Lebensmittel, Berlin – Heidelberg, 1965 and 1967

3.1 PROTEINS

Chemie der Eiweisskoerper, E. Waldschmidt-Leitz and O. Kirchmeier, Stuttgart, 1968

3.2 FATS AND ASSOCIATED SUBSTANCES (LIPIDS)

Neuzeitliche Technologie der Fette und Fettprodukte, Lieferung I-IV, H. P. Kaufmann, Muenster, 1956-1962
Die Gewinnung von Fetten und fetten Oelen, R. Luede, Dresden – Leipzig, 1954
Die Raffination von Fetten und fetten Oelen, R. Luede, Dresden – Leipzig, 1957
Oele und Fette, Teil I Anayse der Nahrungsfette, Teil II Gewinnung und Verarbeitung von Nahrungsfetten, W. Wachs, Berlin – Hamburg, 1961 and 1963

3.3 CARBOHYDRATES

Zuckerchemie, G. Henseke, Berlin, 1966
Die Technologie der Kohlenhydrate – in – Chemische Technologie, Organische Technologie II. Bd. IV, Mrs. K. Winnacker and L. Kuechler, Munich, 1960 and 1961
Chemie der Zucker und Polysaccharide, F. Michael, Leipzig, 1956

3.4 VITAMINS, ENZYMES, MINERALS AND WATER

Vitamine, Baessler/Lang, Darmstadt, 1975
Biochemie der Vitamine, Th. Bersin, Frankfurt/M, 1968
Vitaminbestimmungen, R. Strohecker and H. M. Henning, Weinheim, 1963
Wasser, Mineralstoffe, Spurenelemente, K. Lang, Darmstadt, 1974

Index

Index